APS

Advances in Pharmacological Sciences

Effects of Nicotine on Biological Systems II

Edited by
P.B.S. Clarke
M. Quik
F. Adlkofer
K. Thurau

Birkhäuser Verlag
Basel · Boston · Berlin

Editors:

Professor Paul Brian Sydenham Clarke
Department of Pharmacology
and Therapeutics
McGill University
3655 Drummond Street
Montreal
Québec
Canada H3G 1Y6

Professor Maryka Quik
Department of Pharmacology
and Therapeutics
McGill University
3655 Drummond Street
Montreal
Québec
Canada H3G 1Y6

Professor Dr. Franz Adlkofer
Sekretär
Stiftung für Verhalten und Umwelt
Pettenkoferstrasse 12
D-80336 München
Germany

Professor Dr. Dr.h.c. Klaus Thurau
Vorstand
Physiologisches Institut
Universität München
Pettenkoferstrasse 12
D-80336 München
Germany

A CIP catalogue record for this book is available from the Library of Congress, Washington D.C., USA

Deutsche Bibliothek Cataloging-in-Publication Data
Effects of nicotine on biological systems II / International Symposium on Nicotine, Satellite Symposium of the XIIth International Congress of Pharmacology, Montreal, Canada, Juli 21 – 24, 1994. Ed. by P. B. S. Clarke ... – Basel ; Boston ; Berlin : Birkhäuser.
NE: Clarke, Paul B. S. [Hrsg.], International Sympopsium on Nicotine <1994, Sainte Adèle, Québec; International Congress of Pharmacology <12, 1994, Montréal>
[Proceedings]. – 1995
(Advances in pharmacological sciences)
ISBN 3-7643-5083-0 (Basel ...)
ISBN 0-8176-5083-0 (Boston)

Camera-ready copy prepared by the editors and authors
Printed on acid-free paper produced from chlorine-free pulp
Printed in Germany

ISBN 3-7643-5083-0
ISBN 0-8176-5083-0

9 8 7 6 5 4 3 2 1

Contents

INTRODUCTORY REMARKS

This book presents the proceedings of the conference "The effects of nicotine on biological systems II", held from July 21-24 1994 in Sainte-Adèle, near Montreal. The meeting is the third in a series of international symposia held as satellite symposia to the congresses of the International Union of Pharmacology (IUPhar) and which are devoted to the topic of nicotine and its pharmacological and toxicological actions. The first of these was held in Brisbane in 1987, the second in Hamburg in 1990.

The field of nicotine research is enjoying a resurgence of interest, and the meeting attracted 292 scientists, making it the largest such gathering yet. Scientists came to Quebec from 18 different countries on 5 continents. The scientific heart of the conference was the 138 posters, displayed for the duration of the meeting, and the 47 invited lectures. The scientific sessions included a series of overviews incorporating such questions as: what we do not know and what we think we know - often on the basis of too little evidence - about nicotine's actions: nicotine's acute effects on physiological systems; the pharmacological actions of nicotine metabolites and the definition of nicotinic involvement in smoking related pathology.

The structure and function of nicotinic receptors from a molecular viewpoint was an important item on the agenda. Key issues included the question of which nicotinic receptor subunits coassemble to form receptors and the functional role of the different combinations. Approaches under investigation include in this respect the development of subunit-specific antibodies and ligands (including epibatidine) and the use of nicotinic-receptor-subunit-transfected cell lines and of primary neurons in culture. One receptor subunit currently under intensive scrutiny is the α-bungarotoxin sensitive nicotinic receptor containing $\alpha 7$.

A further general theme was nicotinic receptor regulation and tolerance in the CNS. Application of nanomolar concentrations of nicotine, in concentrations much below those required to induce significant neuropharmacological actions, can promote a desensitization which is largely mediated at the receptor level. More than one receptor subtype may be involved in the nicotinic responses that are measured, but not all tolerance to nicotine appears to be mediated at the receptor level.

Evidence was provided that nicotine is a neuroteratogen, and that nicotinic receptor subtypes change during development of autonomic ganglia. The trophic actions of nicotine and related compounds provided another novel topic. Effects on cell growth mediated by α-bungarotoxin sensitive receptors were described, supporting the notion that this receptor may exert a trophic function. The ability of nicotine to activate c-fos, proenkephalin and thromboxane and leukotriene synthesis may provide clues as to potential mechanisms by which this agent results in its diverse pharmacologic actions.

More than a dozen speakers were invited to tackle controversies related to the role of nicotine in smoking. Some debate surrounded the question of whether nicotine is addictive and proponents stressed the compulsive nature of nicotine consumption. Others described the psychologically beneficial effects of nicotine. Although it seems clear that rats self-administer nicotine acutely because the drug activates the mesolimbic dopamine system, continuous administration of 'smoking doses' has recently been found to render this system highly refractory to acute nicotine challenge. Individual differences in tobacco use may be related to genetically-determined responsiveness to nicotine, for which there is ample evidence in mice.

One session addressed the emerging issue of the potentially therapeutic benefits of nicotine and related compounds. The well-established negative association between Parkinson's disease and smoking was reviewed, and the role and therapeutic potential of nicotine, in Alzheimer's dementia and schizophrenia discussed. Beneficial clinical effects were described in patients suffering from ulcerative colitis and Tourette's syndrome.

The meeting and the publication of its proceedings could not have been a success without the support of generous donors. In addition to the principal sponsor (VERUM Foundation, Germany), the following companies and organizations provided financial support: Abbott Laboratories, Astra Arcus AB, BAT Ltd., Council for Tobacco Research, McGill University, Faculty of Medicine, Fisons Pharmaceuticals, International Society for Neurochemistry, Japan Tobacco Inc., Kabi Pharmacia AB, Medical Research Council of Canada, Merck Frosst Canada Inc., Pfizer Inc., Philip Morris Europe, R.J. Reynolds Tobacco Co., Royal Bank of Canada, and Zyma GmbH.

The scientific programme was deliberately broad in scope and in the planning stages we received valuable help from the International Advisory Committee comprising: D.J.K. Balfour (UK), K. Bättig (Switzerland), N.L. Benowitz (USA), D.K. Berg (USA), D. Bertrand (Switzerland), J.-P. Changeux (France), F. Clementi (Italy), A.C. Collins (USA), P. Froggatt (UK), K. Fuxe (Sweden), G. Lunt (UK), J. Patrick (USA), M.J. Rand (Australia), L.W. Role (USA), M.A.H. Russell (UK), M. Steriade (Canada), and I.P. Stolerman (UK). We thank them all for their helpful comments and suggestions.

Special thanks are due to the conference coordinator, Anne Roussell, who from the earliest stage organized virtually all the logistics for the conference. The assembly of this book was aided by the tenacious efforts of Angela Hund (VERUM Foundation).

Paul B.S. Clarke
Franz Adlkofer

Maryka Quik
Klaus Thurau

that seemed to validate this general approach, results from different studies have been contradictory. This approach is also complicated by difficulties in quantifying receptor density and by uncertainties about what triggers receptor up- or down-regulation (see below), not to mention possible direct actions of AChEIs on nAChRs (e.g. 6). A more direct demonstration of nicotinic cholinergic transmission would involve selective activation of the putative cholinergic input combined with local pharmacological interventions at the putative site of neurotransmission. For several reasons, this is very hard to do in the brain, most notably because to selectively stimulate identified cholinergic pathways is well nigh impossible. Thus, evidence of nicotinic cholinergic transmission in mammalian brain is remarkably slight, being limited to only a few brain nuclei (7).

What are the consequences of CNS nAChR blockade ?

If nicotinic cholinergic transmission is an important feature of brain function, it should be possible to detect behavioural or other changes when nAChRs are blocked. The nicotinic antagonist used in the great majority of behavioural experiments is mecamylamine. Typically, mecamylamine blocks nicotine's CNS actions completely when it is systemically administered in a dose of 1 mg/kg or below. Mecamylamine appears to act insurmountably in the CNS (8), and thus it should not be necessary to administer a higher dose in order to block CNS nicotinic cholinergic transmission. Indeed, it is not clear whether a selective *nicotinic* block can be achieved at doses much above 1 mg/kg, as high concentrations of mecamylamine block NMDA-type glutamate receptors (9). Thus, cognitive and other deficits that are obtained only at high doses of mecamylamine (e.g. 5 or 10 mg/kg sc) do not provide strong evidence for the existence of nicotinic cholinergic transmission in the brain. Although rather low doses of mecamylamine have been shown to impair mental functioning in human subjects, it is not yet clear, I believe, whether these effects are due to a *central* action of the antagonist or whether they derive indirectly from ganglion block.

Are brain nAChRs precisely located across the neuronal surface ?

If acetylcholine (ACh) is rapidly hydrolyzed after release in the brain, one might expect nAChRs to be preferentially located at synapses. However, the limited data available are equivocal on this point (3,4). Moreover, several neural pathways have been identified in land), J.-P. Changeux (France), F. Clementi (Italy), A.C. Collins (USA), P. Froggatt (UK), K. Fuxe (Sweden), G. Lunt (UK), J. Patrick (USA), M.J. Rand (Australia), L.W. Role (USA), M.A.H. Russell (UK), M. Steriade (Canada), and I.P. Stolerman (UK). We thank them all for their helpful comments and suggestions.

Special thanks are due to the conference coordinator, Anne Roussell, who from the earliest stage organized virtually all the logistics for the conference. The assembly of this book was aided by the tenacious efforts of Angela Hund (VERUM Foundation).

Paul B.S. Clarke
Franz Adlkofer

Maryka Quik
Klaus Thurau

Overview

Effects of Nicotine on
Biological Systems II
Advances in Pharmacological Sciences

CURRENT CONTROVERSIES IN NICOTINE RESEARCH

Paul B.S. Clarke

Dept. of Pharmacology and Therapeutics, McGill University, 3655 Drummond Street Room 1325, Montréal, Québec, Canada H3G 1Y6

Summary: Nicotine research has proliferated alarmingly in recent years and the profusion of data makes it hard to maintain a critical eye on developments outside one's immediate speciality. In this brief review, I discuss a number of widely-held views which I believe deserve to be questioned.

Is nicotinic cholinergic transmission an established feature of the mammalian brain ?

Cholinergic pathways in the brain have been elucidated in several mammalian species, most thoroughly in the rat. In this species, transcripts encoding various nicotinic cholinoceptor (nAChR) subunits have been mapped by in situ hybridization histochemistry (e.g. 1). At the protein level, certain nAChRs have also been mapped anatomically by radioligand autoradiography (e.g. 2) sometimes combined with lesions, and by immunohistochemistry (e.g. 3,4). Electron microscopic localization has also been reported using certain probes that recognize nAChRs (e.g. 4). At a functional level, electrophysiological studies have identified many brain nuclei in which neurons are sensitive to the application of nicotinic agonists, and nicotinic agonists have been shown to increase transmitter release in a number of brain areas, via direct actions on nerve terminals (5).

Despite the plethora of anatomical and functional information available on cholinergic pathways and nAChRs, there have been very few attempts to demonstrate sites of nicotinic cholinergic transmission in the mammalian brain. There are several ways to address this question, some yielding results more easily than others, but none easy to interpret. A rather global approach is to examine the effects of chronic in vivo treatment with acetylcholinesterase inhibitors (AChEIs) on radioligand binding to nAChRs. However, despite early findings that seemed to validate this general approach, results from different studies have been contradictory. This approach is also complicated by difficulties in quantifying receptor density and by uncertainties about what triggers receptor up- or down-regulation (see below), not to mention possible direct actions of AChEIs on nAChRs (e.g. 6). A more direct demonstration of nicotinic cholinergic transmission would involve selective activation of the putative cholinergic input combined with local pharmacological interventions at the putative site of neurotransmission. For several reasons, this is very hard to do in the brain, most notably because to selectively stimulate identified cholinergic pathways is well nigh impossible. Thus, evidence of nicotinic cholinergic transmission in mammalian brain is remarkably slight, being limited to only a few brain nuclei (7).

What are the consequences of CNS nAChR blockade ?

If nicotinic cholinergic transmission is an important feature of brain function, it should be possible to detect behavioural or other changes when nAChRs are blocked. The nicotinic antagonist used in the great majority of behavioural experiments is mecamylamine. Typically, mecamylamine blocks nicotine's CNS actions completely when it is systemically administered in a dose of 1 mg/kg or below. Mecamylamine appears to act insurmountably in the CNS (8), and thus it should not be necessary to administer a higher dose in order to block CNS nicotinic cholinergic transmission. Indeed, it is not clear whether a selective *nicotinic* block can be achieved at doses much above 1 mg/kg, as high concentrations of mecamylamine block NMDA-type glutamate receptors (9). Thus, cognitive and other deficits that are obtained only at high doses of mecamylamine (e.g. 5 or 10 mg/kg sc) do not provide strong evidence for the existence of nicotinic cholinergic transmission in the brain. Although rather low doses of mecamylamine have been shown to impair mental functioning in human subjects, it is not yet clear, I believe, whether these effects are due to a *central* action of the antagonist or whether they derive indirectly from ganglion block.

Are brain nAChRs precisely located across the neuronal surface ?

If acetylcholine (ACh) is rapidly hydrolyzed after release in the brain, one might expect nAChRs to be preferentially located at synapses. However, the limited data available are equivocal on this point (3,4). Moreover, several neural pathways have been identified in mammalian brain where nicotinic receptors are present both at the level of cell bodies and terminals (see ref. 7). This arrangement suggests that nAChRs, once made, are not precisely transported to particular loci on the cell membrane.

Are α4/ß2-containing nAChRs particularly prevalent in brain ?

Two major problems are encountered in trying to assess the relative prevalence of nAChR subtypes in the brain. First, the number of such subtypes is unknown and potentially large, and selective probes are only available for a few of them; tissue levels of mRNAs encoding different nAChR subunits offer a poor substitute for protein measurements. The second problem lies in the use of radioligands such as ^{3}H-nicotine, ^{3}H-ACh and ^{3}H-cytisine to label nAChRs containing α4 and ß2 subunits (10). The binding of these radioligands appears to rely on a shift in the equilibrium towards a high-affinity state of the receptor. This shift may not be absolute, implying that the B_{max} obtained in binding experiments may only approximate the true density of receptors, a suggestion supported by a report that the measured B_{max} of high-affinity ^{3}H-agonist binding can be increased by the mere addition of drugs to the in vitro binding assay (11).

Do α4/ß2-containing nAChRs mediate actions of smoking doses of nicotine ?

The concentration of nicotine in the brain of human tobacco smokers is likely to be around 1 μM, possibly peaking to 10 μM after a puff is taken. Correlative evidence of several kinds is consistent with the possibility that α4/ß2-containing nAChRs mediate some of these effects. For example, the autoradiographic distribution of ^{3}H-nicotine (2), which labels these receptors (10), is similar to the anatomical pattern of neuronal activation shown by 2-deoxy-glucose uptake after systemic injection of nicotine (12). However, given that we have few markers for other nicotinic receptor subtypes, the considerable overlap that exists in the anatomical distribution of nAChR subunits precludes a clear conclusion. The elevations of ^{3}H-nicotine binding density found in the post mortem brains of smokers (13) indicates that these receptors are targets for nicotine, but does not necessarily imply that they are *activated* during smoking.

Is chronic nicotine administration necessary for the chronic effects of nicotine ?

Some of the behavioural effects of nicotine in animals are lastingly affected by brief, even single pre-exposure to the drug (14,15). In contrast, the up-regulation of high-affinity ^{3}H-agonist binding that occurs in the brain of rodents chronically treated with nicotine has been assumed to require chronic treatment, since it takes several days to appear (16). The experimental designs used to date have not generally discriminated between effects of nicotine pre-exposure that require multiple nicotine pretreatments and those that simply require time to develop after an initial exposure. This issue is given new impetus by the report of an acute effect of nicotine that develops over several days after drug administration (17).

What triggers nicotinic "receptor up-regulation" ?

Chronic treatment with several nicotinic agonists reliably increases the density of high-affinity ^{3}H-agonist and ^{125}I-alpha-bungarotoxin binding sites in rodent brain. It has been suggested that this "paradoxical" up-regulation occurs through a time-averaged antagonistic effect of nicotine. Were some sort of *functional* blockade the stimulus for up-regulation, nicotinic antagonists should also be effective. However, published findings are mixed, and it is not clear whether appropriate doses of mecamylamine were given. Recently, we have shown that chlorisondamine neither mimics nor blocks ^{3}H-nicotine binding up-regulation, despite producing chronic central nicotinic blockade (18). This suggests that functional blockade of these receptors may *not* be the trigger for up-regulation. In addition, our results suggested that up-regulation of ^{125}I-alpha-bungarotoxin binding by nicotine may require receptor *activation*.

Are actions of nicotine in the brain important in tobacco smoking ?

Most of the evidence that *central* actions of nicotine may be important regulators of smoking was provided twenty years ago, by the observation that the centrally-active antagonist mecamylamine increased smoking behaviour whereas the quaternary antagonist pentolinium did not (19). Only recently were the reinforcing effects of nicotine shown to be of central origin in animals self-administering the drug (20). How closely does this paradigm model tobacco smoking in humans ? One should not forget that tobacco smoking is richly context-dependent in humans. Furthermore, it should be borne in mind that both the anatomical location and the receptor activation and desensitization kinetics of nAChRs may well vary between species. Certain authors have stressed the role that peripheral cues may play in maintaining smoking behaviour (see Rose, this volume), and in this context, it may be as well to recall that nicotine can stimulate neuronal firing in the brain even before it reaches this organ (21).

Why does nicotine replacement therapy not help everyone ?

Nicotine replacement therapy, whether in the form of gum or transdermal patch, only helps a small minority of smokers to quit for periods of a year or more (22). In absolute numbers, this still represents an important medical advance, but one is left wondering why relapse is the rule rather than the exception. The key, I believe, is to see tobacco smoking as an over-learnt behaviour; merely replacing the nicotine does not remove the smoker's history of repeatedly being reinforced for smoking. For this reason, it may be important to consider the development of more selective antagonists (23).

References

1. Wada E, Wada K, Boulter J, Deneris E, Heinemann S, Patrick J, Swanson LW. Distribution of alpha 2, alpha 3, alpha 4, and beta 2 neuronal nicotinic receptor subunit mRNAs in the central nervous system: a hybridization histochemical study in the rat. J Comp Neurol 1989; 284:314-335.
2. Clarke PBS, Schwartz RD, Paul SM, Pert CB, Pert A. Nicotinic binding in rat brain: autoradiographic comparison of ^{3}H-acetylcholine, ^{3}H-nicotine, and ^{125}I-alpha-bungarotoxin. J Neurosci 1985; 5:1307-1315.
3. Schroder H, Zilles K, Maelicke A, Hajos F. Immunohisto- and cytochemical localization of cortical nicotinic cholinoceptors in rat and man. Brain Res 1989; 502:287-295.
4. Hill JA Jr, Zoli M, Bourgeois J-P, Changeux J-P. Immunocytochemical localization of a neuronal nicotinic receptor: The ß2-subunit. J Neurosci 1993; 13:1551-1568.
5. Wonnacott S, Irons J, Rapier C, Thorne B, Lunt GG. Presynaptic modulation of transmitter release by nicotinic receptors. Prog Brain Res 1989; 79:157-163.
6. Clarke PBS, Reuben M, El-Bizri H. Blockade of nicotinic responses by physostigmine, tacrine and other cholinesterase inhibitors in rat striatum. Br J Pharmacol 1994; 111:695-702.
7. Clarke PBS. Nicotinic receptors in mammalian brain: localization and relation to cholinergic function. Prog Brain Res 1993; 98:77-83.
8. El-Bizri H, Clarke PBS. Blockade of nicotinic receptor-mediated release of dopamine from striatal synaptosomes by chlorisondamine and other nicotinic antagonists administered *in vitro*. Br J Pharmacol 1994; 111:406-413.
9. Clarke PBS, Chaudieu I, El-Bizri H, Boksa P, Quik M, Esplin BA, Capek R. The pharmacology of the nicotinic antagonist, chlorisondamine, investigated in rat brain and autonomic ganglion. Br J Pharmacol 1994; 111:397-405.
10. Flores CM, Rogers SW, Pabreza LA, Wolfe BB, Kellar KJ. A subtype of nicotinic cholinergic receptor in rat brain is composed of $\alpha 4$ and ß2 subunits and is up-regulated by chronic nicotine treatment. Mol Pharmacol 1992; 41:31-37.

11. Takayama H, Majewska MD, London ED. Interactions of noncompetitive inhibitors with nicotinic receptors in the rat brain. J Pharmacol Exp Ther 1989; 251:1083-1089.
12. London ED, Connolly RJ, Szikszay M, Wamsley JK, Dam M. Effects of nicotine on local cerebral glucose utilization in the rat. J Neurosci 1988; 8:3920-3928.
13. Benwell ME, Balfour DJ, Anderson JM. Evidence that tobacco smoking increases the density of ( ) [^{3}H]nicotine binding sites in human brain. J Neurochem 1988; 50:1243-1247.
14. Stolerman IP, Fink R, Jarvik ME. Acute and chronic tolerance to nicotine measured by activity in rats. Psychopharmacologia 1973; 30:329-342.
15. Clarke PBS, Kumar R. The effects of nicotine on locomotor activity in non-tolerant and tolerant rats. Br J Pharmacol 1983; 78:329-337.
16. Marks MJ, Stitzel JA, Collins AC. Time course study of the effects of chronic nicotine infusion on drug response and brain receptors. J Pharmacol Exp Ther 1985; 235:619-628.
17. Mitchell SN, Smith KM, Joseph MH, Gray JA. Increases in tyrosine hydroxylase messenger RNA in the locus coeruleus after a single dose of nicotine are followed by time-dependent increases in enzyme activity and noradrenaline release. Neuroscience 1993; 56:989-997.
18. El-Bizri H, Clarke PBS. Regulation of nicotinic receptors in rat brain following quasi-irreversible nicotinic blockade by chlorisondamine and chronic treatment with nicotine. Br J Pharmacol 1994: in press.
19. Stolerman IP, Goldfarb T, Fink R, Jarvik ME. Influencing cigarette smoking with nicotine antagonists. Psychopharmacologia 1973; 28:247-259.
20. Corrigall WA, Franklin KBJ, Coen KM, Clarke PBS. The mesolimbic dopaminergic system is implicated in the reinforcing effects of nicotine. Psychopharmacology (Berl) 1992; 107:285-289.
21. Engberg G, Hajos M. Nicotine-induced activation of locus coeruleus neurons--An analysis of peripheral versus central induction. Naunyn Schmiedebergs Arch Pharmacol 1994; 349:443-446.
22. Tang JL, Law M, Wald N. How effective is nicotine replacement therapy in helping people to stop smoking. BMJ 1994; 308:21-26.
23. Clarke PBS. Nicotinic receptor blockade therapy and smoking cessation. Br J Addict 1991; 86:501-505.

Effects of Nicotine on
Biological Systems II
Advances in Pharmacological Sciences

ACUTE BIOLOGICAL EFFECTS OF NICOTINE AND ITS METABOLITES

Neal L. Benowitz

Division of Clinical Pharmacology and Experimental Therapeutics
University of California, San Francisco, San Francisco General Hospital Medical Center
San Francisco, California 94110

Summary: The actions of nicotine in humans, with emphasis on recent research findings, are reviewed. Issues that are discussed include the importance of rate of dosing and development of tolerance in determining nicotine effects, quantitative studies of nicotine metabolism, and cardiovascular, endocrine and metabolic effects of nicotine.

Introduction

This introductory paper will briefly review the human pharmacology of nicotine and its metabolites, with emphasis on newer findings since the last international nicotine meeting. For a more comprehensive review of older research, readers are referred to earlier reviews (1,2).

Neurochemical Effects of Nicotine

Nicotine acts on nicotinic cholinergic receptors in the brain and other parts of the nervous system, in large part by releasing or facilitating the release of various neurotransmitters. These include dopamine, norepinephrine, acetylcholine, serotonin, vasopressin and beta endorphin. Release of this variety of neurotransmitters may explain the diversity of nicotine effects, some seemingly opposing (such as arousal and relaxation) reported by smokers (Fig. 1).

With repetitive exposure to nicotine, there is neural adaptation and tolerance develops, resulting in diminished neurotransmitter release (3). Such tolerance underlies the development of physical dependence to nicotine, which contributes to nicotine addiction. Additionally, some tobacco users may dose themselves with nicotine so as to maintain a desensitized state of nicotinic receptors, which could contribute to the reinforcing effects of tobacco.

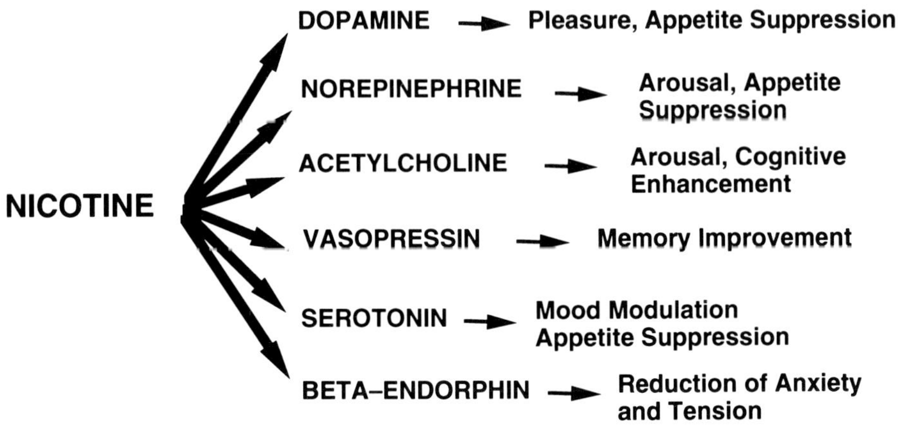

Figure 1. Neurochemical effects of nicotine.

Pharmacokinetics

The pharmacokinetics and metabolism of nicotine have been described in great detail elsewhere (4). Of particular interest is the importance of the rate of nicotine dosing and relevant kinetics to pharmacological action. Cigarette smoking results in rapid delivery through the lungs, rapid entry of nicotine into the arterial blood stream, and delivery within 10–15 seconds to the brain. Peak arterial concentrations of nicotine are as high as 100 ng/ml (6×10^{-7} M) (5). Exposure to transient high levels of nicotine with intermittent troughs favors the development of more intense pharmacologic effects with less time for the development of tolerance, as well as allowing time between doses for some degree of receptor resensitization.

In contrast, smokeless tobacco (snuff and chewing tobacco) and nicotine pharmaceutical products such as transdermal nicotine and nicotine gum, deliver nicotine to the circulation gradually. Arterial and venous blood levels remain in near equilibrium, with levels of 10–30 ng/ml ($1\text{–}2 \times 10^{-7}$ M). Gradual absorption permits considerable time for the development of tolerance, and the intensity of pharmacologic effects from nicotine delivered in this way appears to be less.

Metabolism

The metabolic fate of nicotine has been nearly fully characterized (6-8) (Fig. 2). Newly

characterized metabolites of nicotine include nicotine and cotinine N–glucuronides and trans–3'–hydroxycotinine–O–glucuronide (9,10). Nornicotine has been confirmed to be a metabolite of nicotine in humans, although it is also derived in part from the tobacco itself (11). Quantitative studies of nicotine metabolism indicate that on average 72% of nicotine is metabolized via C–oxidation to cotinine, with a range of 55 to 92% (12). Of interest is a recent characterization of individuals who have deficient C–oxidation of nicotine and produce very little cotinine (13).

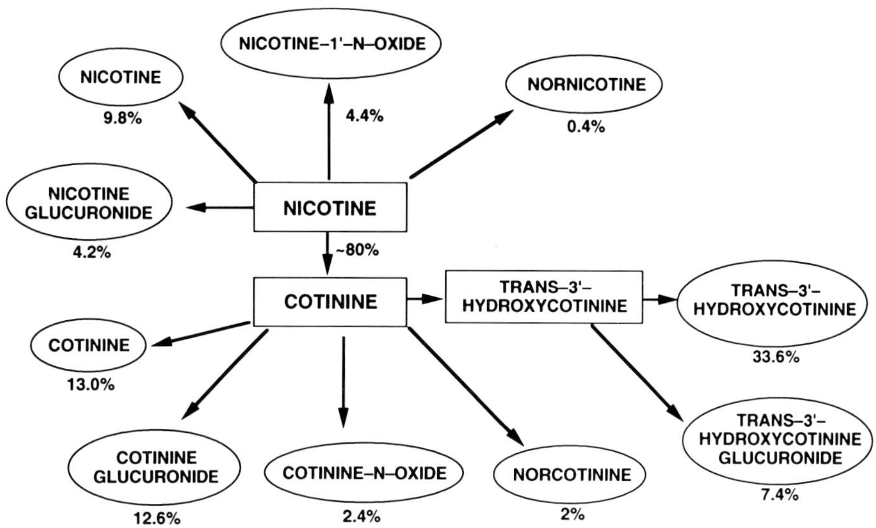

Figure 2. Quantitative scheme of nicotine metabolism, based on average excretion of metabolites as percent of systemic dose during transdermal nicotine application. Circled compounds indicate excretion in the urine, and associated numbers indicate percent of systemic dose of nicotine (6).

Cardiovascular Effects

The main cardiovascular effects of nicotine are mediated through the activation of the sympathetic nervous system, with neural release of norepinephrine and adrenal release of epinephrine (Fig. 3). Vasopressin release may also contribute to hemodynamic effects, which has been shown for cutaneous vasoconstriction. The hemodynamic effects of nicotine — increased heart rate, blood pressure, myocardial contractility and cardiac output as well as vasoconstriction of particular vascular beds — have been well characterized. Recently, there is evidence that nicotine constricts coronary arteries as well (14). While cigarette smoking clearly results in platelet activation and enhanced blood coagulation, it is unclear if nicotine per se contributes to this effect. Using urinary excretion of thromboxane A2 metabolites, and other markers of platelet activation, it appears that neither the use of smokeless tobacco nor transdermal nicotine, both resulting in venous plasma nicotine concentrations similar to those of cigarette smoking, results in platelet activation (15,16). Whether rapid delivery of nicotine, such as with cigarette smoking, or some other chemical in cigarette smoke, is responsible for platelet activation remains to be determined.

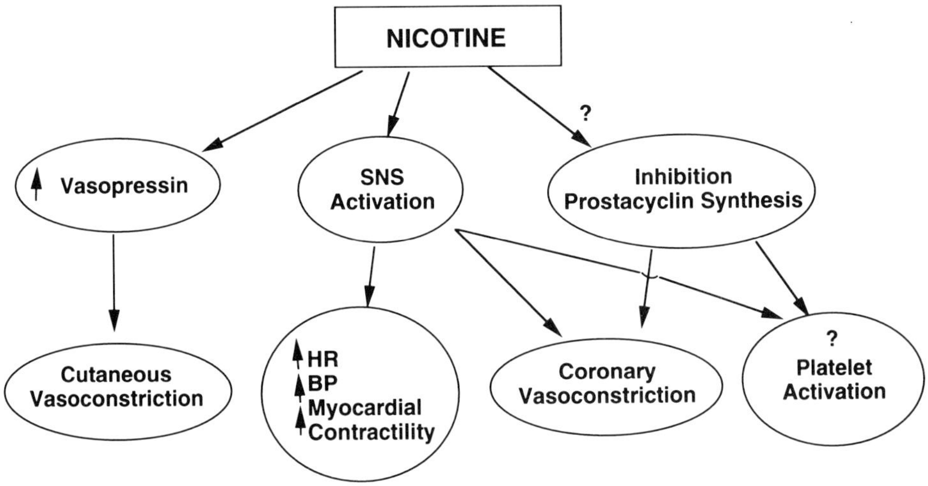

Figure 3. Cardiovascular effects of nicotine.

Endocrine Effects

The systemic release by nicotine of a number of hormones has been described (Fig. 4). Many of the studies describing these effects have been performed with acute challenges with subjects rapidly smoking two or three cigarettes, which would produce high arterial nicotine levels and in some cases resulted in symptoms of acute nicotine intoxication. The endocrine effects of regular smoking, where there may also be the development of tolerance, have been less well studied. There is evidence, based on excretion studies, that catecholamine release is persistently increased with regular daily smoking, but data on other hormones are conflicting.

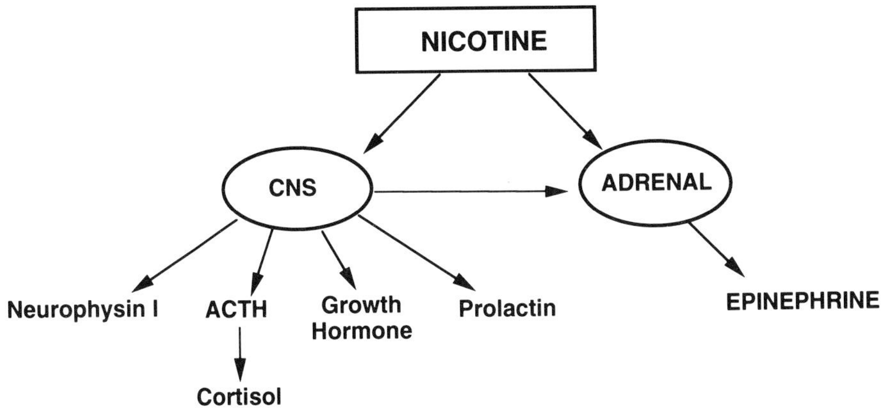

Figure 4. Systemic endocrine effects of nicotine.

Metabolic Effects

Metabolic effects of cigarette smoking of current interest are lower body weights and more atherogenic lipoprotein profiles in smokers compared with nonsmokers. Current concepts of the metabolic effects of nicotine are summarized in Fig. 5. Smokers weigh an average 3–4 kg less than nonsmokers. A major determinant of lower body weight in smokers appears to be increased energy expenditure, as evidenced by a higher metabolic rate (17). Studies of cigarette smoking, nasal nicotine and intravenous nicotine have shown that nicotine is responsible for the increase in metabolic rate (17,18). That caloric consumption does not increase to counteract the increased energy expenditure related to an increased metabolic rate suggests that nicotine also directly suppresses appetite, consistent with the central nervous system release of norepinephrine and serotonin. The mechanism of nicotine effects appears to involve activation of the sympathetic nervous system. Some effects of nicotine on metabolic rate may be explained by futile cycling of triglycerides (i.e., cycling from lipolysis to re–esterification). However, quantitatively futile cycles do not appear to account for the excess energy expenditure produced by nicotine, so other mechanisms must also be involved (19).

Cigarette smoking is associated with higher low density lipoprotein and lower high density lipoprotein cholesterol levels, a profile associated with a higher risk of coronary artery disease. One mechanism may be sympathetic nervous system–mediated lipolysis with release of free fatty acids. Acute cigarette smoking as well as intravenous nicotine are associated with an increase in free fatty acid concentrations. Regular cigarette smoking increases free fatty acid release from adipose tissue by 77% (19). In metabolic turnover studies, it has been demonstrated that most of the free fatty acid that is released is recycled via hepatic uptake (19). Hepatic metabolism of free fatty acids is known to result in increased production of low density lipoproteins. Thus, nicotine, in accelerating lipolysis, could contribute to an atherogenic lipid profile.

Pharmacology of Nicotine Metabolites

Nicotine metabolism results in the generation of a number of metabolites, as described previously. Some of these metabolites have pharmacologic activities that could contribute to the effects of nicotine in people. Of much current interest is the potential activity of cotinine. Cotinine is present in plasma at concentrations approximately 15 times those of nicotine, and cotinine levels persist at high levels for 24 hours a day. Cotinine has little or no activity at nicotinic cholinergic receptors (20). However, cotinine does have effects on the release of neurotransmitters in the brain, and affects a number of enzymes including those involved in the synthesis of estrogen and testosterone (21-24). Cotinine decreases vascular resistance and decreases blood pressure in animals, although this occurs only at high concentrations (25). Blood pressure in a cohort of bus drivers has been found to be inversely

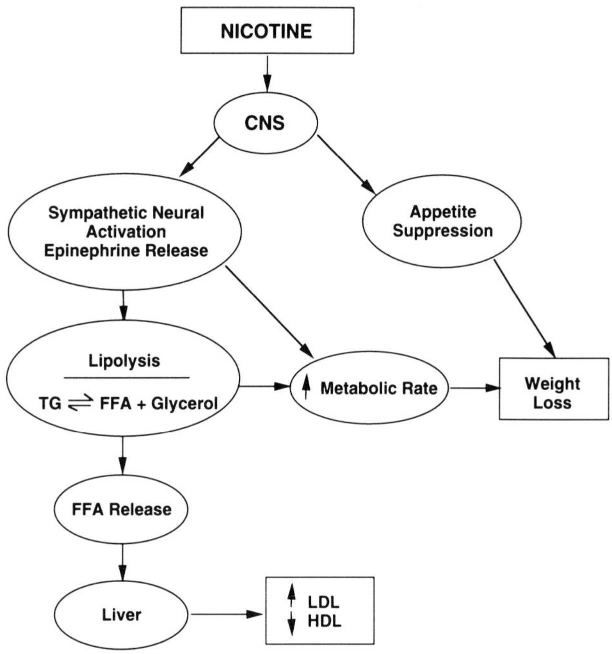

Figure 5. Metabolic effects of nicotine. Abbreviations: CNS — Central nervous system; TG — triglycerides; FFA — free fatty acids; LDL — low density lipoproteins; HDL — high density lipoproteins.

correlated to serum cotinine, raising the possibility that cotinine could contribute to the cardiovascular effects of cigarette smoking (26). Recent human studies suggest that cotinine modifies the symptoms of tobacco withdrawal, although it does so in a way that is dissimilar to the effects of nicotine (27). Cotinine appears to produce some stimulant–like effects, and reduces overall withdrawal score. The relevance of cotinine to the pharmacologic effects of chronic nicotine exposure is as yet unresolved, but clearly deserves future research.

***Acknowledgements**: Research supported in part by grants DA02277 and DA01696 from the National Institute on Drug Abuse. I wish to thank Ms. Kaye Welch for assistance in preparation of this manuscript and the figures.*

References

1. Benowitz NL. Pharmacologic aspects of cigarette smoking and nicotine addiction. N Engl J Med 1988;319:1318-1330.
2. Department of Health and Human Services, Public Health Service: The Health Consequences of Smoking: Nicotine Addiction. A Report of the Surgeon General. DHHS (CDC) Publication No. 88–8406. Washington DC, US Government Printing Office, 1988.
3. Balfour DJK. The neural mechanisms underlying the rewarding properties of nicotine. J Smoking–Related Dis 1994;5(Suppl. 1), in press.
4. Benowitz NL, Porchet H, Jacob P III. Pharmacokinetics, metabolism, and pharmacodynamics of nicotine. In: Wonnacott S, Russell MAH, Stolerman IP, Wonnacott S, Russell MAH, Stolerman IP. Nicotine Psychopharmacology: Molecular, Cellular and Behavioral Aspects. Oxford: Oxford University Press, 1990: 112-157.
5. Henningfield JE, Stapleton JM, Benowitz NL, Grayson RF, London ED. Higher levels of nicotine in arterial than in venous blood after cigarette smoking. Drug Alc Depend 1993; 33:23-29.
6. Benowitz NL, Jacob P III, Fong I, Gupta S. Nicotine metabolic profile in man: Comparison of cigarette smoking and transdermal nicotine. J Pharmacol Exp Ther 1994;268:296-303.
7. Curvall M, Kazemi–Vala E, Englund G. Conjugation pathways in nicotine metabolism. In: Adlkofer F, Thurau K, Adlkofer F, Thurau K. Effects of Nicotine on Biological Systems. Basel: Birkhauser-Verlag, 1991: 69-75.
8. Byrd GD, Chang K, Greene JM, deBethizy JD. Evidence for urinary excretion of glucuronide conjugates of nicotine, cotinine, and *trans*–3'–hydroxycotinine in smokers. Drug Metab Dispos 1992;20:192-197.
9. Caldwell WS, Greene JM, Byrd GD, Chang KM, Uhrig MS, deBethizy JD. Characterization of the glucuronide conjugate of cotinine: A previously unidentified major metabolite of nicotine in smokers' urine. Chem Res Toxicol 1992;5:280-285.
10. Schepers G, Demetriou D, Rustemeier K, Voncken P, Diehl B. Nicotine phase 2 metabolites in human urine — structure of metabolically formed *trans*-3'-hydroxycotinine glucuronide. Med Sci Res 1992;20:863-865.
11. Jacob P III, Benowitz NL. Oxidative metabolism of nicotine in vivo. In: Effects of Nicotine on Biological Systems, ed. by F. Adlkofer and K. Thuraus, pp. 35-44, Birkhauser–Verlag, Basel, 1991.
12. Benowitz NL, Jacob P III. Metabolism of nicotine to cotinine studied by a dual stable isotope method. Clin Pharmacol Ther 1994, in press.
13. Benowitz NL, Sachs D, Jacob P III. Deficient C-oxidation of nicotine (Abst). Clin Pharmacol Ther 1994;55:181.

14. Quillen JE, Rossen JD, Oskarsson HJ, Minor RL Jr., Lopez JAG, Winniford MD. Acute effect of cigarette smoking on the coronary circulation: Constriction of epicardial and resistance vessels. J Am Coll Cardiol 1993;22:642-647.
15. Wennmalm A, Benthin G, Granström EF, Persson L, Peterson A, Winell S. Relation between tobacco use and urinary excretion of thromboxane A2 and prostacyclin metabolites in young men. Circulation 1991;83:1698-1704.
16. Benowitz NL, Fitzgerald GA, Wilson M, Zhang Q. Nicotine effects on eicosanoid formation and hemostatic function: Comparison of transdermal nicotine and cigarette smoking. J Am Coll Cardiol 1993;22:1159-1167.
17. Perkins KA. Metabolic effects of cigarette smoking. J Appl Physiol 1992;72:401-409.
18. Arcavi L, Jacob P III, Hellerstein M, Benowitz NL. Divergent tolerance to metabolic and cardiovascular effects of nicotine in light and heavy cigarette smokers. Clin Pharmacol Ther 1994, in press.
19. Hellerstein MK, Benowitz NL, Neese RA et al. Effects of cigarette smoking and its cessation on lipid metabolism and energy expenditure in heavy smokers. J Clin Invest 1994;93: 265-272.
20. Abood LG, Grassi S, Costanza M. Binding of optically pure (–)-[^{3}H]nicotine to rat brain membranes. FEBS Letters 1983;157:147-149.
21. Fuxe K, Everitt BJ, Hokfelt T. On the action of nicotine and cotinine on central 5-hydroxytryptamine neurons. Pharmacol Biochem Behav 1979;10:671-677.
22. Barbieri RL, Gochberg J, Ryan KJ. Nicotine, cotinine, and anabasine inhibit aromatase in human trophoblast in vitro. J Clin Invest 1986;77:1727-1733.
23. Yeh J, Barbieri RL, Friedman AJ. Nicotine and cotinine inhibit rat testis androgen biosynthesis in vitro. J Steroid Biochem 1989;33:627-630.
24. Patterson TR, Stringham JD, Meikle AW. Nicotine and cotinine inhibit steroidogenesis in mouse Leydig cells. Life Sciences 1990;46:265-272.
25. Dominiak P, Fuchs G, von Toth S, Grobecker H. Effects of nicotine and its major metabolites on blood pressure in anaesthetized rats. Klin Wochenschr 1985;63:90-92.
26. Benowitz NL, Sharp DS. Inverse relation between serum cotinine concentration and blood pressure in cigarette smokers. Circulation 1989;80:1309-1312.
27. Keenan RM, Hatsukami DK, Pentel PR, Thompson T, Grillo MA. Pharmacodynamic effects of cotinine in abstinent cigarette smokers. Clin Pharmacol Ther 1994, in press.

Effects of Nicotine on
Biological Systems II
Advances in Pharmacological Sciences

INVOLVEMENT OF NICOTINE AND ITS METABOLITES IN THE PATHOLOGY OF SMOKING-RELATED DISEASES: FACTS AND HYPOTHESES

F. X. Adlkofer

VERUM, Stiftung für Verhalten und Umwelt, Goethestraße 20, D-80336 München

Summary: The pathogenesis of the major smoking-related diseases, cancer of the lung and coronary heart disease, are poorly understood. There are, however, well founded hypotheses how smoking could contribute to the development of these diseases. On the other hand, the constituents of cigarette smoke that might be the causative agents remain to be identified. At present, there is no evidence at all that nicotine - the substance most smokers smoke for - has any influence on the development of cancer in otherwise healthy smokers, and only spurious evidence that it may be involved in the development of coronary heart disease.

Introduction

The present state of knowledge about the pathogenesis of the major smoking-related diseases, cancer of the lung and coronary heart disease, is still scanty despite the numerous articles which have been published in this field. However, the progress obtained during the last decade has at least enabled us to formulate hypotheses so that we can discuss the role of nicotine in the development of these diseases. As far as this is possible, I have attempted in this review to distinguish fact from fiction.

Nicotine and Cancerogenesis

Since smoking is a risk factor for cancer at several sites, the question arises as to whether nicotine is involved in cancer development. The dose of nicotine absorbed through smoking, through the use of smokeless tobacco products and through nicotine-containing pharmaceutical products can be determined by measuring nicotine or its metabolites in body fluids. Nicotine *per se* is not mutagenic or genotoxic. Therefore, no DNA or protein adducts are formed and no chromosomal aberrations, gene mutations, oncogen activation and tumor suppressor gene inactivations occur. The principle stable urinary metabolites of nicotine, cotinine and nicotine-1-N-oxide, are not carcinogenic or tumor promotors. The significance of the nicotine-1',5'-iminium ion, a putative intermediate in nicotine metabolism to cotinine, which does perhaps not even exist in man (6), is still not known.

Under *in vitro* acidic aqueous conditions, nicotine reacts slowly with nitrite to produce low yields of N-nitrosonornicotine (NNN) and 4-(N-methylnitrosamino)-1-(3-pyridyl)-1-butanone (NNK), two potent carcinogens present in cured tobacco and tobacco smoke (32), and until recently in low concentrations in nicotine patches used in smoking cessation programs (Adlkofer *et al.*, unreported data). It has been speculated that endogenous nitrosation of nicotine and its metabolites occurs in man (10). Our own studies do not support this hypothesis (33).

N-Nitrosamines such as NNK require metabolic activation via α-hydroxylation to exert their genotoxic potential (10). Nicotine is known to lower the *in vitro* mutagenicity and genotoxicity of N-nitrosamines such as NNK (19), probably by inhibiting α-hydroxylation which occurs both *in vitro* (27) and *in vivo* (25). Furthermore, nicotine also diminishes *in vivo* hemoglobin adduct formation by NNK (15). Since cigarette smoke contains 10,000-fold higher levels of nicotine than NNK (32), it may thus protect against the biological action of NNK.

Contrary to the reported inhibitory effect of nicotine on NNK activation, it has been demonstrated that *in vitro* exposure of neuroendocrine cell lines, the precursors of small cell lung cancer (SCLC), to elevated concentrations of CO_2 causes mitogenesis. At 10% CO_2 concentrations, nicotine augments mitogenesis while identical exposure of cell lines in a 5% CO_2 atmosphere has no effect (28). Intraalveolar conditions in healthy humans (18-20% O_2, 3-4% CO_2) are considerably different than in the above experimental models. Whether nicotine - through its mitogenic effect - can contribute to the development of SCLC in subjects with chronic obstructive lung disease (COLD), due to their increased intraalveolar CO_2 levels, remains open to question (28). Patients with COLD do not normally smoke and no data are available which demonstrate an increased SCLC-rate in subjects with this disease (38).

Nicotine and Atherosclerotic Disease

Atherosclerosis, the process underlying coronary heart disease, peripheral vascular disease, and stroke, begins in childhood and progresses through several stages to result in clinically manifest disease in middle age and later life. Elevated blood pressure, elevated plasma LDL cholesterol levels and lowered HDL cholesterol levels and smoking are associated with more extensive and more severe atherosclerosis, and a greater risk of clinical disease.

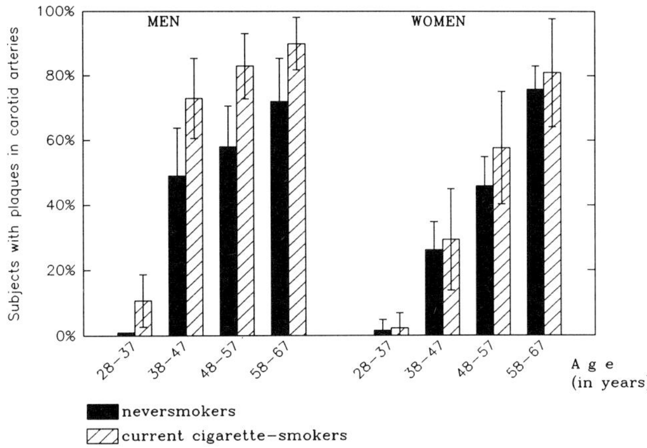

Figure 1. Smoking-related lesions of the carotid arteries in men and women of different age groups after standardization for plasma cholesterol in a representative population in Southern Germany as detected by ß-sonography (Heller et al., unpublished data). Given: percentage per age group and 95%-confidence intervals

The factors that link smoking, or components of cigarette smoke, to mechanisms involved in the pathogenesis of atherosclerotic diseases have not yet been identified. All reported attempts to reproduce human-like atherosclerotic effects by cigarette smoke inhalation or nicotine administration in animal studies have failed or yielded inconsistent results, thus hampering efforts to determine the mechanisms involved (2; 21; 26).

As shown in Figure 1, atherosclerotic lesions of the carotid arteries increase steeply with age, in women later in life than in men, due to the cessation of hormonal protection in the post-menopausal years. After standardizing for hypercholesteremia, which is not uncommon in Southern Germany, smoking resulted in a considerable increase in the frequency of carotid plaques. In sharp contrast, no association between smoking and coronary heart disease can be found in populations with low plasma cholesterol levels and consequently low risk of coronary heart disease (30). This is the reason to suggest that smoking significantly augments atherosclerosis only in the presence of hypercholesteremia, but might not be an independent atherosclerotic risk factor (21).

Even if heavy smokers have on average more extensive and advanced atherosclerosis than nonsmokers (30), the difference does not seem sufficient to account for a two-fold or greater risk to develop clinically manifest coronary heart disease (21). Furthermore, epidemiological studies show a rapid regression of cardiovascular risk within the first year following smoking cessation (30), which cannot be explained by a regression of vascular lesions caused by smoking. Therefore, it has been suggested that smoking may increase the risk of coronary heart disease primarily by influencing the terminal thrombotic episode (21).

Several systemic alterations promoting atherogenic and vasoocclusive processes may be caused by smoking. In the following they are discussed in more detail in order to clarify whether or how nicotine may be involved in the progress of these diseases.

1. *Sympathoadrenal activation*: Smoking a cigarette acutely increases blood pressure and heart rate. Acute myocardial ischemia occurs when the demand for oxygen is compromised by insufficient coronary blood flow. In the case of coronary stenosis due to local vasoconstriction or due to atherosclerotic plaques, the increased O_2-consumption cannot be met, especially when part of the O_2-carrying hemoglobin is blocked by CO as in smokers, and as a consequence angina or myocardial infarction develop. Sudden death occurs when acute myocardial ischemia causes lethal arrhythmia (2). Of course, it cannot be excluded that the transient but frequently repeated blood pressure elevations, may also contribute to endothelial injury. This however seems to be doubtful in view of the often physiologically changing blood pressure episodes during the day, because of physical and emotional stress which occur quite independent of smoking.

It is nicotine in cigarette smoke which plays the decisive role in the stimulation of the cardiovascular system by releasing catecholamines. In the rare cases of pheochromocytoma, blood pressure may even rise extremely due to this mechanism (2). In evaluating the various effects of nicotine, the following findings have to be considered. a) Nicotine increases blood pressure, heart rate, myocardial contractility and cardiac output (2). b) These hemodynamic effects of nicotine result in an increase of coronary blood flow in healthy people, while the coronary perfusion in subjects with coronary heart disease may be diminished, probably due to vasoconstriction or coronary spasm because of α-adrenergic stimulation (2). c) Nicotine use may render arrhythmias that occur in patients with coronary heart disease more dangerous. Several anecdotal case reports claim atrial fibrillation or even sudden death after nicotine administration through chewing gums or patches, mostly in combination with smoking (2). d) On the other hand and quite unexpectedly, smokers tend to have lower blood pressure than nonsmokers, even after controlling for body weight (2). There is some indication that nicotine or cotinine, via their effect on prostaglandin metabolism, may be the causative agents (12).

e) In a randomized multicenter study on 156 patients with coronary heart disease, no increase in angina frequency, overall cardiac symptom status, nocturnal cardiovascular events, or EEG changes were found after transdermal nicotine application. Similarly, patients with the highest random nicotine plasma levels did not appear to have an increased incidence of adverse effects (36). At present, no final conclusion can be drawn whether the sympathoadrenal activation through nicotine materially contributes to the pathogenesis of atherosclerosis and coronary heart disease.

2. *Platelet activation:* Assessment of whether the platelets and also the leukocytes of smokers show increased aggregation and adhesiveness are fraught with methodological problems. Both normal and increased adhesiveness have been observed in *ex vivo* studies. Similarly, various *ex vivo* studies have shown increased, normal and decreased aggregability. Even studies of the acute effects of cigarette smoking have yielded inconclusive results with both increased and unaltered platelet adhesiveness, and increased and decreased platelet aggregability (1). Cigarette smoking may increase the urinary excretion of thromboxane A_2 - and prostacyclin metabolites, measures of platelet activation *in vivo* (18), indicating that smoking may stimulate thromboxane A_2 - and prostacyclin synthesis. Thromboxane A_2, produced by activated platelets, macrophages and neutrophils, as well as prostacyclin, which is synthesized mainly by endothelial cells, but also by smooth muscle cells and fibroblasts, are believed to play an important role in the control of blood pressure, vascular smooth muscle tone and thrombus formation (14). Thromboxane A_2 is an activator of platelet aggregation and a vasoconstrictor, and prostacyclin an inhibitor of platelet aggregation and a vasodilator. While an increase in thromboxane A_2 formation through smoking has been demonstrated (3, 35), an effect of smoking on prostacyclin synthesis has yet not been confirmed (35). There is evidence for an altered platelet function in smokers as demonstrated by a shorter half-life of platelets, but it is not known yet whether this is a direct effect of smoking on platelets or a reaction of platelets on vascular dysfunction due to smoking (8). Activated platelets and leukocytes are thought to contribute not only to the development of atherosclerotic plaques but importantly to the acute process of thrombus formation.

With respect to nicotine as a possible cause of platelet and leucocyte hyperactivity the following findings are of interest: a) Infusion of norepinephrine which is also released into the circulation after nicotine uptake seems to reduce rather than enhance platelet aggregation (29);

b) Nicotine causes desensitization of platelets to the aggregating action of epinephrine which is also released by nicotine (17, 23); c) α-Adrenoreceptor agonists, such as epinephrine and norepinephrine are stimulators of endogenous prostacyclin synthesis (13); d) Nicotine may not interfere with the release of prostacyclin at all (5) or even stimulate basal prostacyclin synthesis in human vascular endothelium (4); e) In contrast to smoking, nicotine and/or cotinine may selectively inhibit thromboxane A_2 formation (9, 31, 37); f) Vitamin C, but not vitamin E, prevents cigarette smoke-induced leukocyte aggregation and adhesion to the endothelium *in vivo* (20) indicating that a radical-dependent mechanism, and not nicotine, may be involved in the process. Thus, evidence is accumulating that nicotine may not be responsible for the platelet and leucocyte activation found in smokers.

3. *Lipid metabolism*: Several studies have shown that smokers have an about 10% lower HDL-cholesterol concentration than nonsmokers, and a tendency for higher LDL-cholesterol levels (2, 30). The mechanisms involved are not known. There is evidence that cigarette smoke exposes LDL via peroxidative modification to enhanced metabolism by macrophages (7). The role of nicotine in this alteration is still hypothetical. Nicotine increases the free fatty acid concentration in plasma and may promote the synthesis of triglycerides and LDL by the liver, which in turn may result in decreased HDL production (2). In a 2 week study in which healthy nonsmokers were administered nicotine chewing gum producing plasma cotinine levels comparable to those in moderate smokers, no change in serum lipid profiles was reported (24). Similar results have been obtained in at least two other studies (16; Assmann, unpublished data). The limitations of studies comparing lipid profiles in smokers with those of users of nicotine chewing gum are obvious. Nicotine uptake is lower and slower than through smoking. Thus a final conclusion on the influence of nicotine on lipid metabolism cannot yet be drawn.

4. *Hemorrheologic alterations*: Many hematological parameters such as hematocrit, white cell count and platelet count, are affected by cigarette smoking (1, 21). Derangement in the coagulation system of smokers may be due to increased levels of fibrinogen in plasma, which has been found to be a risk factor for ischemic heart disease comparable in strength to an elevated serum cholesterol level (22). With the exception of a CO effect on hematocrit, the constituents of cigarette smoke that might create these atherogenic and thrombogenic conditions remain to be identified. There is only poor evidence that they are caused by nicotine,

although it cannot be excluded at present that the stimulation of the sympathoadrenergic system may at least contribute (1).

From these findings it is not possible to conclude that nicotine is implicated in the pathogenesis of atherosclerosis and cardiovascular disease. In the United Kingdom, Wald *et al.* have shown the highest serum levels of cotinine in pipe smokers, who are known to have the lowest risk of chronic heart disease in all smoking groups when compared to nonsmokers (34). In Sweden, Huhtasaari has compared the age-adjusted odds ratio for myocardial infarction in nonsmokers, smokers and snuff dippers. While the risk was significantly increased in smokers, no difference was found between nonsmokers and snuff dippers, although the cotinine plasma concentration was comparable in smokers and snuff dippers (11). These epidemiological data seem to mitigate the possible cardiotoxic effect of nicotine even further. Therefore, the present state of knowledge favours the assumption that the contribution of nicotine to the development of atherosclerosis in smokers must be small, if it occurs at all. In the case of already manifest atherosclerotic disease it is likely that nicotine is not or is possibly only a minor factor in acute coronary events, since the sympathoadrenergic system acquires to a great extent tolerance towards nicotine.

References

1. Bain BJ. The haematological effects of smoking. J Smoking-Related Dis 1992; 3:99-108.
2. Benowitz NL. Nicotine and coronary heart disease. Trends Cardiovasc Med 1991; 1: 315-321.
3. Barrow SE, Ward PS, Sleightholm MA, Ritter JM, Dollery CT. Cigarette smoking: profiles of thromboxane- and postacyclin-derived products in human urine. Biochim Biophys Acta 1989; 993:121-127.
4. Boutherin-Falson O, Blaes N. Nicotine increases basal prostacyclin production and DNA synthesis of human endothelial cells in primary cultures. Nouv Res Fr Hematol 1990; 32:253-258.
5. Bull HA, Pittilo RM, Woolf N, Machin SJ. The effect of nicotine on human endothelial cell release of prostaglandins and ultrastructure. Br J Exp Path 1988; 69:413-421.
6. Caldwell WS, Byrd GD, Dobson GP, Dull GM. Nicotine iminium ions are not detected in smokers' urine. In: International Symposium on Nicotine: The Effects of Nicotine on Biological Systems II. The Abstracts. Clarke PBS, Quik M, Adlkofer FX, Thurau K. Basel: Birkhäuser, 1994: P130.
7. Church DF, Pryor WA. Free-radical chemistry of cigarette smoke and its toxicological implications. Environ Health Perspect 1985; 64:111-126.
8. FitzGerald GA, Oates JA, Nowak J. Cigarette smoking and hemostatic function. Am Heart J 1988; 115:267-271.
9. Goerig M, Ullrich V, Schettler G, Foltis C, Habenicht A. A new role for nicotine: selective inhibition of thromboxane formation by direct interaction with thromboxane synthase in human promyelocytic leukaemia cells differentiating into macrophages. Clin Investig 1992; 70:239-243.
10. Hoffmann D, Hecht SS. Nicotine-derived N-nitrosamines and tobacco-related cancer: current status and future directions. Cancer Res 1985; 45:935-944.

11. Huhtasaari F. Smokeless tobacco *vs* cigarettes as risk factors for myocardial infarction in middle-aged men. International Congress on Smoking Cessation, Glasgow, Great Britain, S.E.C.C., March 5-8, 1994: Final Programme and Abstracts 4.4.
12. Hui SCG, Ogle CW. Chronic nicotine treatment enhances the depressor responses to arachidonic acid in the rat. Pharmacology 1991; 42:257-261.
13. Jeremy JY, Mikhailidis DP, Dandona P. Cigarette smoke extracts but not nicotine inhibit prostacyclin (PGI_2) synthesis in human, rabbit and rat vascular tissue. Prost Leuk Med 1985; 19:261-270.
14. Jeremy JY, Mikhailidis DP. Smoking and vascular prostanoids: relevance to the pathogenesis of atheroma and thrombosis. J Smoking-Related Dis 1990; 1:59-69.
15. Kutzer C, Richter E, Atawodi SE. In: The Effect of Nicotine on Biological Systems II. Clarke PBS, Quik M, Adlkofer FX, Thurau K. Basel: Birkhäuser, 1994: this volume.
16. Lagrue G, Grimaldi B, Martin C, Demaria C, Jacotot B. Gomme nicotine et profil lipidique. Pharmacologie Expérimentale 1989; 37:937-939.
17. Lassila R, Laustiola KE. Physical exercise provokes platelet desensitization in men who smoke cigarettes - involvement of sympathoadrenergic mechanisms - a study of monzygotic twin pairs discordant for smoking. Thromb Res 1988; 51:145-155.
18. Lassila R, Laustiola KE. Cigarette smoking and platelet-vessel wall interactions. Prostaglandins. Leukot Essent Fatty Acid 1992; 46:81-86.
19. Lee C, Fulp C, Bombick E, Doolittle D. Nicotine and cotinine inhibit the mutagenicity of N-nitrosamines present in tobacco smoke. Environ Mol Mutagen 1994; 23(Suppl). 323.
20. Lehr H-A, Frei B, Arfors K-E. Vitamin C prevents cigarette smoke-induced leukocyte aggregation and adhesion to endothelium *in vivo*. Proc Natl Acad Sci USA 1994; 91: 7688-7692
21. McGill HC. The cardiovascular pathology of smoking. Am Heart J 1988; 115:250-257.
22. Meade TW, Brozovic M, Chakrabarti RR, Haines AP, Imeson JD, Mellows S, Miller GJ, North WRS, Stirling Y, Thompson SG. Haemostatic function and ishaemic heart disease: principal results of the Northwick Park heart study. Lancet 1986: 2: 533-537.
23. Motulsky HJ, Shattil SJ, Ferry N, Rozansky D, Insel PA. Desensitization of epinephrine-initiated platelet aggregation does not alter binding to the α_2-adrenergic receptor or receptor coupling to adenylate cyclase. Mol Pharmacol 1986; 29:1-6.
24. Quensel M, Agardh C-D, Nilsson-Ehle P. Nicotine does not affect plasma lipoprotein concentrations in healthy men. Scand J Clin Lab Invest 1989; 49:149-153.
25. Richter E, Tricker AR. Nicotine inhibits the metabolic activation of the tobacco-specific nitrosamine 4-(methylnitrosamino)-1-(3-pyridyl)-1-butanone in rats. Carcinogenesis 1994; 15:1061-1064.
26. Rogers WR, Carey KD, McMahan CA, Montiel MM, Mott GE, Wigodsky HS, McGill HC Jr. Cigarette smoking, dietary hyperlipidemia, and experimental atherosclerosis in the baboon. Exp Mol Pathol 1988; 48:135-151.
27. Schuller HM, Castonguay A, Orloff M, Rossignol G. Modulation of the uptake and metabolism of 4-(methylnitrosamino)-1-(3-pyridyl)-1-butanone by nicotine in hamster lung. Cancer Res 1991; 51:2009-2014.
28. Schuller HM. Mechanisms of nicotine stimulated cell proliferation in normal and neoplastic neuroendocrine lung cells. In: The Effect of Nicotine on Biological Systems II. Clarke PBS, Quick M, Adlkofer FX, Thurau K. Basel: Birkhäuser, 1994: this volume.
29. Siess W, Lorenz R, Roth P, Weber PC. Plasma catecholamines, platelet aggregation and associated thromboxane formation after physical exercise, smoking and norepinephrine infusion. Circulation 1982; 66:44-48.
30. Surgeon Generals Report. The Health Consequences of Smoking. Washington DC: US Government Printing Office, 1983: 13-156.
31. Toivanen J, Ylikorkala O, Viinikka L. Effects of smoking and nicotine on human prostacyclin and thromboxane production *in vivo* and *in vitro*. Toxicol Appl Pharmacol 1986; 82:301-306.
32. Tricker AR, Ditrich C, Preussmann R. N-Nitroso compounds in cigarette tobacco and their occurrence in mainstream tobacco smoke. Carcinogenesis 1991; 12:257-261.
33. Tricker AR, Scherer G, Conze C, Adlkofer F, Pachinger A, Klus H. Evaluation of 4-(N-methylnitrosamino)-4-(3-pyridyl)butyric acid as a potential monitor of endogenous nitrosation of nicotine and its metabolites. Carcinogenesis 1993; 14:1409-1414.

34. Wald NJ, Idle M, Boreham J. Serum cotinine levels in pipe smokers: evidence against nicotine as cause of coronary heart disease. Lancet 1981; 2:775-777.
35. Wennmalm Å, Benthin G, Granström EF, Persson L, Petersson, A-S, Winell S. Relation between tobacco use and urinary excretion of thromboxane A_2 and prostacyclin metabolites in young men. Circulation 1991; 83:1698-1704.
36. Working Group for the Study of Transdermal Nicotine in Patients with Coronary Artery Disease. Nicotine replacement therapy for patients with coronary artery disease. Arch Intern Med 1994; 154:989-995.
37. Ylikorkala O, Viinikka L, Lehtovirta P. Effect of nicotine on fetal prostacyclin and thromboxane in humans. Obstet Gynecol 1985; 66:102-105.
38. Weiss W. COPD and lung cancer. In: Chronic Obstructive Pulmonary Disease. Cherniak NS. Philadelphia: WB Saunders Company, 1991: 344-347.

Structure and function of nicotinic receptors

Effects of Nicotine on
Biological Systems II
Advances in Pharmacological Sciences

A CYCLOPHILIN-DEPENDENT MECHANISM FOR α7 NEURONAL NICOTINIC ACETYLCHOLINE RECEPTOR MATURATION

Santosh A. Helekar, Lorna Colquhoun, Hong Dang, Danong Chen, Finn Goldner and Jim Patrick

Division of Neuroscience, Baylor College of Medicine, Houston, Texas 77030, U.S.A.

Summary: The neuronal nicotinic acetylcholine receptor (nAChR) subunit α7 forms a principal neuronal α-bungarotoxin binding site that is thought to have a neuromodulatory and developmental role. Here we present evidence that the expression of α7 homo-oligomeric receptors in Xenopus oocytes is dependent on the activity of the peptidyl prolyl isomerase (PPIase) cyclophilin. We show further that the expression of the hetero-oligomeric muscle nAChR is devoid of such a requirement, and that the muscle receptor non-alpha subunits relieve the α7 receptor from this requirement. We conclude that cyclophilin might be uniquely required as a PPIase or a molecular chaperone in the folding and assembly of α7 homo-oligomers.

Introduction

Neuronal nicotinic acetylcholine receptors (nAChRs) are expressed in diverse cortical and subcortical regions of the brain as evidenced by both ligand binding and in situ hybridrization studies (1, 2). α-Bungarotoxin (α-BTX) binding sites that are known to be abundantly and widely distributed in the brain constitute one class of nAChRs (3). There is now considerable evidence that an important component of this receptor is the nAChR subunit α7 (4-6). This subunit forms a homo-oligomeric ligand-gated ion channel when expressed in *Xenopus* oocytes (4, 5). This channel is activated by nicotinic agonists, and shows a pharmacological profile that is characteristic of a nAChR (4-7). The agonist-induced current through this channel is rapidly desensitizing, inward rectifying at positive membrane potentials, and is mainly carried by monovalent cations. There is however, a very significant divalent cation permeability that might suggest a role for this channel in neural plasticity (5, 8). Although the precise subcellular locale of this receptor is unclear it has been speculated that it might be localized to the synaptic terminal, and therefore might influence some aspect of the transmitter release mechanism. We have raised polyclonal antibodies against this subunit, and are currently determining its cellular and subcellular distribution in the brain using immunohistochemistry and immuno-electron microscopy.

Because the $\alpha 7$ receptor that is expressed in *Xenopus* oocytes is a homo-oligomer, and a homo-oligomeric receptor is the simplest model of a ligand-gated ion channel, this receptor would be well suited to study many aspects of receptor biology. Three questions are most pertinent in this regard. 1) What is the structure of the native receptor? 2) How is it assembled? 3) What is the precise mechanism underlying its function? We have initiated a study to attempt to answer these questions by transiently expressing the $\alpha 7$ subunit in *Xenopus* oocytes and in the COS cell line. Although there is expression of an $\alpha 7$ subunit containing receptor in these cells the level of expression appears to be very low compared to the expression of the muscle nAChR under similar conditions (9). Since the synthesis of $\alpha 7$ polypeptide is not reduced in these cells the most likely explanation for the reduced receptor expression appears to reside in the folding, assembly, targeting or degradation of mature receptors. This paper describes the findings of a study designed to elucidate one possible mechanism underlying the assembly of $\alpha 7$ homo-oligomeric receptors. We have used the *Xenopus* oocyte expression system for this purpose because the inefficiently expressed $\alpha 7$ receptor-mediated currents can be more easily and stably recorded in these cells.

It is known that de novo folding or refolding of several proteins can be catalysed by a peptidyl prolyl isomerase (PPIase) called cyclophilin (CyP) (10). The PPIase activity of CyP catalyses the rate limiting cis-trans isomerization of prolyl peptide bonds in these proteins. The A-subtype of CyP (CyPA) is a ubiquitous protein that was first shown to possess PPIase activity in vitro (11). It is now also known to assume the role of a molecular chaperone during protein folding (12). Additionally, as a binary complex with CsA CyP can inhibit calcineurin and change the state of phosphorylation of proteins (13). All ligand-gated ion channels including the $\alpha 7$ nAChR show a striking abundance of highly conserved proline residues, suggesting that these residues and/or the unique peptide bond conformation that they impart might be contributing in some significant way to the structure of these channels. These prolyl peptide conformations might also particularly impede the process of subunit folding and receptor assembly. Consequently, this process might require the activity of a PPIase such as CyP. We tested this hypothesis, and provided evidence that CyP is indeed required for the expression of mature α7 homo-oligomeric receptors.

Blockade of Cyclophilins with Cyclosporin A Causes a Reduction in α7 Receptor Expression

We observed that functional α7 homo-oligomeric receptor expression following injection of α7 cDNA subcloned in eucaryotic expression vectors was significantly reduced in *Xenopus* oocytes treated with a selective blocker of CyPs, cyclosporin A (CsA) (Fig. 1A and B). This effect is neither due to a blockade of the function of mature surface receptors nor due to a blockade of endogenous Ca^{2+}-activated Cl^- currents that are known to be activated secondarily by Ca^{2+} influx through α7 receptors (5, 8, 14). Therefore, the CsA-induced reduction in maximal α7 receptor-mediated currents is most likely due to the reduced formation of functional α7 receptors or the formation of α7 receptors with reduced function. If the first alternative is right then the total number of surface α7 receptors should be reduced in CsA-treated compared to untreated oocytes. We tested this possibility by performing α-BTX and α7 anti-peptide antibody binding assays. We observed that both specific toxin and antibody binding was reduced by ~74 % and ~60 %, respectively in CsA-treated compared to untreated oocytes (14). These data indicate that CsA might be blocking the expression or enhancing the degradation of surface α7 receptors.

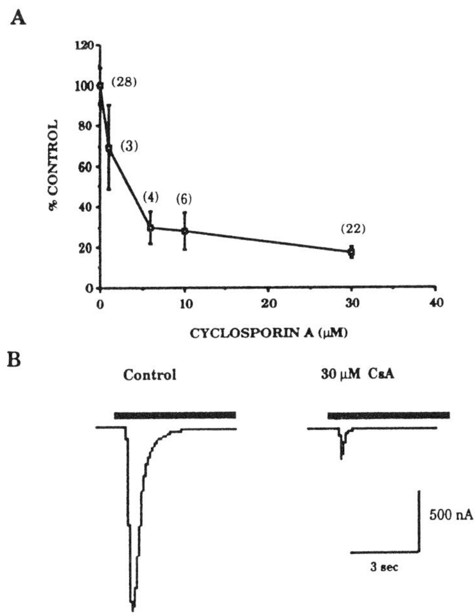

Figure 1. The functional expression of the α7 homo-oligomeric receptor in *Xenopus* oocytes is reduced by CsA. A. Each point on the graph is a mean of normalized peak current amplitude measured in 3-28 (in parentheses) oocytes from 6 separate experiments. Error bars represent standard errors of the mean. B. Traces show maximal currents induced by ~5 sec (horizontal bars above traces) applications of 500 µM nicotine in a pair of oocytes incubated in 30 µM CsA (right) and vehicle (left) containing Barth's saline. The membrane potential of these oocytes was held at -60 mV. From ref. 14.

Figure 2. Overexpression of exogenous CyP causes a reversal of CsA-induced blockade of functional α7 receptor expression. A. The bar graph shows normalized mean maximal nicotine-induced currents from oocytes in which CyPA cDNA (right pair) or luciferase cDNA (left pair) was co-expressed with α7 cDNA. Closed bars represent responses obtained in the presence (n = 5 for α7/luc, n = 5 for α7/CyP), and open bars represent those in the absence (n = 7 for α7/luc, n = 9 for α7/CyP) of 30 μM CsA. B. Traces of nicotine-induced α7 receptor-mediated currents obtained from a pair of CsA-treated oocytes. The left trace is from an oocyte co-injected with α7 and luciferase cDNAs (1:1 ratio) and the right trace is from one co-injected with α7 and CyPA cDNAs (1:1 ratio). From ref. 14.

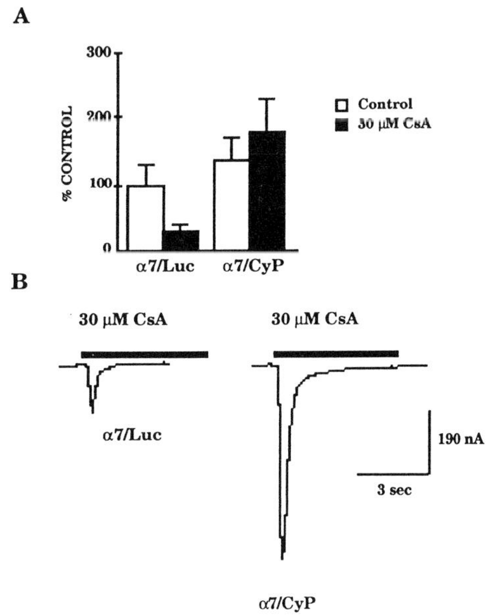

The CsA effect appears to be most likely mediated by the blockade of CyP because overexpression of exogenous rat brain CyP reversed this effect (Fig.2A and B). Blockade of another class of PPIases, namely FK506-binding proteins (15) with FK506, did not produce a CsA-like reduction in receptor expression (14). Since CsA and FK506 are known to produce immunosuppression through a common mechanism the above results indicate that the mechanism of CsA-induced reduction in receptor expression is distinct from that involved in immunosuppression (13, 16).

The Cyclosporin A Effect is not due to Blockade of Protein Synthesis

A simple explanation for the observed CsA-induced reduction in receptor expression is that CsA might be blocking protein synthesis. Two pieces of evidence rule out this possibility. The first piece of evidence comes from the observation that the functional expression of hetero-oligomeric muscle nAChRs expressed from mouse muscle receptor subunits that were subcloned in the same expression vector as α7 was not reduced by CsA (15). The second piece of evidence is that the synthesis of α7 polypeptide detected by immunoblotting techniques was not reduced in CsA-treated compared to untreated oocytes (Fig. 3). This result also rules out

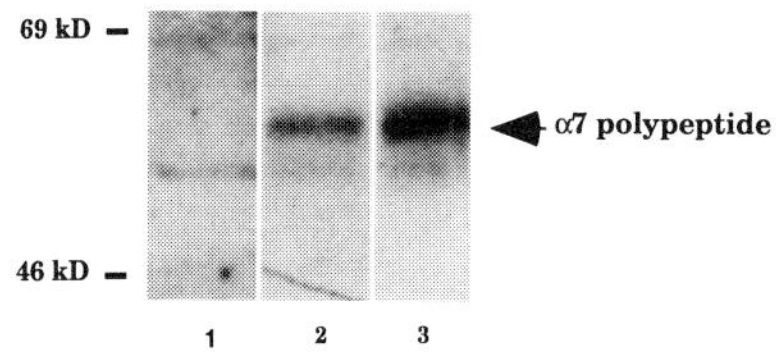

Figure 3. CsA does not block the synthesis of α7 polypeptide. Western blot with an antibody against an α7 C-terminal proton pump epitope tag shows a 56 - 58 kD immunoreactive band (α7-tag1 construct) in lane 2 (injected control oocyte lysate) and lane 3 (injected CsA-treated oocyte lysate), but not in lane 1 (uninjected oocyte lysate). From ref. 14.

the idea that an increased α7 subunit degradation in CsA-treated oocytes is responsible for the reduction in receptor expression. These data strongly indicate that the step at which CsA acts is downstream of the synthesis of α7 polypeptide. The two most likely CsA-affected steps, therefore, are receptor assembly and receptor targeting to the cell surface.

α7 Receptors are Rescued from CsA Blockade by Muscle Non-alpha Subunits

The observation that hetero-oligomeric muscle nAChR expression is not blocked by CsA, and the inference that the CsA-affected step is subsequent to synthesis of the α7 polypeptide lead us to the following two candidate mechanisms that may be affected by CsA: 1) In the absence of CyP the α7 polypeptide might fold into a conformation that is incapable of forming a ligand-gated ion channel. 2) In the absence of CyP, the α7 polypeptide might fold into a conformation that is incapable of forming a homo-oligomer.

To find out if one of these alternatives is right we co-expressed muscle non-alpha subunits, β1, γ and δ with α7 and tested if the sensitivity of the resulting α7 subunit containing receptor to CsA was reduced. When α7 cDNA was co-injected with β1, γ and δ cDNA at a ratio of 3:1:1:1 or 4:1:1:1 we observed large nicotine-induced currents that were not significantly reduced in 30 μM CsA-treated compared to untreated oocytes (Fig.4A and B). It is unlikely that these receptors are composed of β1, γ and δ subunits by themselves or in combination with an endogenous *Xenopus* alpha subunit because no nicotine-induced responses were observed in oocytes injected with β1, γ and δ subunit cDNA (10 oocytes). We are currently trying to determine which one(s) of the non-alpha subunits are sufficient to produce rescue from CsA blockade when co-expressed with α7 (17). These data strongly suggest that CyP is required only for the self-association of subunit monomers to form functional homo-oligomers, and not for the association of subunits forming hetero-oligomers. This idea was further

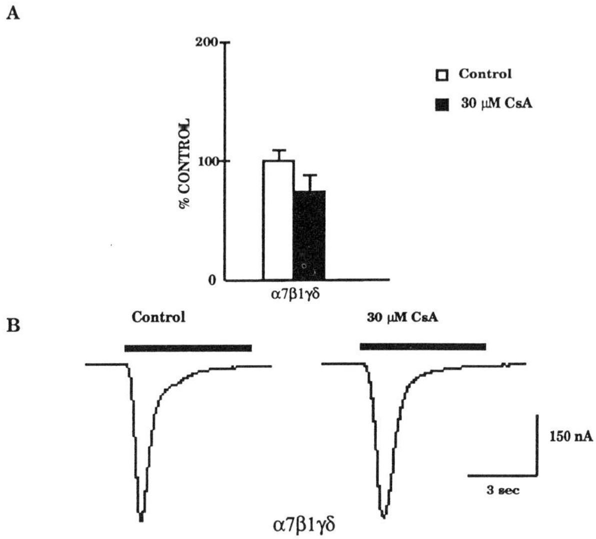

Figure 4. Co-expression of muscle non-alpha subunits rescues α7 receptors from CsA-blockade. cDNAs of α7, β1, γ and δ subunit were co-injected in a 3:1:1:1 or 4:1:1:1 ratio. A. The bar graph represents functional receptor expression in the presence (closed bars, n = 11) and absence (open bars, n = 10) of 30 μM CsA. B. Traces of nicotine-induced (500 μM, applied for ~5 sec) currents of putative α7 containing hetero-oligomeric receptors in CsA-treated (right) and untreated (left) oocytes. From ref. 14.

strengthened and generalized by the observation that the functional expression of mouse $5HT_3$ receptor was also reduced by CsA in a CyP-dependent manner (14).

The Role of Cyclophilin

The precise role that CyP plays in α7 receptor expression is unclear. Two possibilities appear likely. 1) CyP acts as a PPIase in the folding and assembly of this receptor. 2) CyP acts as a molecular chaperone to prevent α7 from misfolding. The large number of conserved proline residues within α7 subunit cause the first possibility to be of particular interest. Accordingly, we are currently systematically mutating these proline residues to study their effects on receptor expression and CsA sensitivity (18). It is reasonable to expect that *cis-trans* isomerization of one or more of these prolyl peptide bonds might be critical in the proper folding the α7 subunit. It is also possible that more than one stable prolyl isomers of α7 may exist, and participate in the formation of functional homo-oligomeric receptors.

Conclusion

In conclusion, we have provided evidence that the PPIase CyP may be required for the synthesis of α7 homo-oligomeric nAChR. This PPIase requirement has been generalized to the homo-oligomeric $5HT_3$ receptor, and may be extended to other ligand-gated ion channels that share many of the conserved prolines that are shared by these receptor subunits. Hetero-oligomeric receptors do not appear to be dependent on CyP for their synthesis. This finding might suggest structural differences between homo- and hetero-oligomeric receptors, especially with regard to the isomeric status of the prolines within their subunits.

***Acknowledgements**: We are grateful to Jacques Wadiche and Marietta Piattoni for technical assistance, Dr. Brian Gilbert for the gift of cyclosporin A, Drs. Andres Maricq and David Julius for the gift of 5HT$_3$ receptor subunit cDNA, and Dr. Mark Perin for the gift of tag1 antibody.*

References

1. Sargent PB. The diversity of neuronal nicotinic acetylcholine receptors. Ann. Rev. Neurosci. 1993; 16: 403-443.
2. Patrick JW. Neuronal nicotinic acetylcholine receptor diversity. NIDA Res. Monogr. 1992; 126: 4-13.
3. Clarke PB. The fall and rise of neuronal alpha-bungarotoxin binding proteins. Trends. Pharmacol. Sci. 1992; 13: 407-413.
4. Couturier S, Bertrand D, Matter JM, Hernandez MC, Bertrand S, Millar N, Valera-S, Barkas T, Ballivet M. A neuronal nicotinic acetylcholine receptor subunit (alpha 7) is developmentally regulated and forms a homo-oligomeric channel blocked by alpha-BTX. Neuron 1990; 5: 847-856.
5. Seguela P, Wadiche J, Dineley-Miller K, Dani JA, Patrick JW. Molecular cloning, functional properties, and distribution of rat brain $\alpha 7$: a nicotinic cation channel highly permeable to calcium. J. Neurosci. 1993; 13: 596-604.
6. Schoepfer R, Conroy WG, Whiting P, Gore M, Lindstrom J. Brain alpha-bungarotoxin binding protein cDNAs and MAbs reveal subtypes of this branch of the ligand-gated ion channel gene superfamily. Neuron 1990; 5: 35-48.
7. Bertrand D, Bertrand S, Ballivet M. Pharmacological properties of the homomeric alpha 7 receptor. Neurosci. Lett. 1992; 146: 87-90.
8. Galzi JL, Devillers-Thiery A, Hussy N, Bertrand S, Changeux JP, Bertrand D. Mutations in the channel domain of a neuronal nicotinic receptor convert ion selectivity from cationic to anionic. Nature 1992; 359: 500-505.
9. Chen D, Patrick J. Transient expression of rat neuronal nicotinic acetylcholine receptor subunit alpha 7 in COS cells. Soc. Neurosci. Abstr. 1993; 19: 466.
10. Lang K, Schmid FX, Fischer G. Catalysis of protein folding by prolyl isomerase. Nature 1987; 329: 268-270.
11. Fischer G, Bang H, Mech C. Nachweis einer enzymkatalyse fur die cis-trans-isomerisierung der peptidbindung in prolinhaltigen peptiden. Biomed. Biochim. Acta 1984; 43: 1101-1111.
12. Freskgard P-O, Bergenhem N, Jonsson B-H, Svensson M, Carlsson U. Isomerase and chaperone activity of prolyl isomerase in the folding of carbonic anhydrase. Science 1992; 258: 466-468.
13. Liu J, Farmer JD Jr, Lane WS, Friedman J, Weissman I, Schreiber SL. Calcineurin is a common target of cyclophilin-cyclosporin A and FKBP-FK506 complexes. Cell 1991; 66: 807-815.
14. Helekar SA, Char D, Neff S, Patrick P. Prolyl isomerase requirement for the expression of functional homo-oligomeric ligand-gated ion channels. Neuron 1994; 12: 179-189.
15. Siekierka JJ, Hung SHY, Poe M, Lin CS, Sigal NH. A cytosolic binding protein for the immunosuppressant FK506 has peptidyl-prolyl isomerase activity. Nature 1989; 341: 755-757.
16. Walsh CT, Zydowsky LD, McKeon FD. Cyclosporin A, the cyclophilin class of peptidylprolyl isomerases, and blockade of T cell signal transduction. J. Biol. Chem. 1992; 267: 13115-13118.
17. Colquhoun L, Le Pichon JB, Volrath M, Patrick J. $\alpha 7$ receptor function is rescued from cyclosporin block by co-expression with muscle d subunit. Soc. Neurosci. Abstr. 1994 (in press).
18. Dang H, Patrick J. Alterations of $\alpha 7$ nicotinic receptor by replacing the conserved proline residue within M-I to glycine. Soc. Neurosci. Abstr. 1993; 19: 280.

Effects of Nicotine on
Biological Systems II
Advances in Pharmacological Sciences

α-BUNGAROTOXIN RECEPTOR SUBTYPES

C. Gotti, W. Hanke[+], M. Moretti, R. Longhi[*], B. Balestra, L. Briscini and F. Clementi

CNR Center of Cytopharmacology, Department of Medical Pharmacology, University of Milano, 20129 Milano, [*]CNR Center Chimica degli Ormoni, 20133, Milano, Italy and [+]Institut für Zoophysiologie, Universitä't Hohenheim, Stuttgart 70, Germany

Summary: We used anti-α7 and anti-α8 subunit specific antibodies to study α-bungarotoxin (α-BTX) receptors at different developmental stages in chick optic lobe and retina. We found that only the α7 and α7-α8 subtypes are present at all developmental stages in chick optic lobe, whereas in addition to these two subtypes, chick retina contains an α8 subtype representing 50% of all α-BTX receptors at 1 day post hatching. By using an immunoaffinity column, we purified these subtypes and studied their biochemical and functional properties by reconstituting them in planar lipid bilayer. The α7 and α7-α8 receptor subtypes are made up of two subunits: a 57000 dalton subunit which binds α-BTX, and a 52000 dalton subunit whose function and structure is unknown. The three subtypes differ in pharmacological and biophysical properties.

Introduction

α-Bungarotoxin (α-BTX) is a potent competitive blocker of the nicotinic acetylcholine receptor (AChR) at the neuromuscular junction of most vertebrates, and it has played an essential role in the characterization of this neurotransmitter gated channel (1). α-BTX-binding nicotinic acetylcholine receptors are also present in the central and peripheral vertebrate nervous systems, but their structure and significance is still a matter of controversy (2).

Recent molecular cloning experiments have revealed a gene family encoding a number of neuronal AChR subunits (2). Of the genes so far isolated, only those encoding the α7 and α8 subunits seem to be related to the AChRs which bind and/or are blocked by α-BTX (3, 4). The peptides corresponding to sequences of both subunits bind α-BTX with high affinity, and when the α7 or α8 subunits alone are expressed in oocytes, they form homomeric channels activated by acetylcholine and nicotine, and blocked by α-BTX (3, 5, 6). Furthermore, immunoprecipitation experiments, with monoclonal antibodies raised against bacterially expressed α7 and α8 subunit fragments have demonstrated the presence of a family of α-BTX receptors in chick brain and retina (7, 8). Three subtypes of this family have so far been identified: the α7 subtype which contains the α7 subunit and is the major subtype present in chick brain; the α8 subtype which contains the α8 subunit and is the major subtype present in retina; and the α7-α8 subtype which contains both the α7 and α8 subunits (5, 7). However, neither the biochemical structure nor the functional characteristics of any of these subtypes are presently known. We generated anti-α7 and anti-α8 subunit specific antibodies in order to characterize the biophysical and pharmacological properties of chick brain α-BTX receptor subtypes further, and to establish their subunit composition. In addition, we studied the pattern of expression of these receptor subtypes in order to establish whether any changes occur during neurodevelopment.

Materials and Methods

Immunoprecipitation of the ^{125}I-α-BTX labeled receptors by different antibodies: The chick optic lobes (COLs) and retinas were dissected from in ovo chickens on embryonic days 7, 13 and 19 (E7, E13 and E19) and from 1-day old chickens (D1); they were then immediately frozen in liquid nitrogen and stored at -80°C for later use. The antibodies (Abs) were obtained against peptides with sequences that are unique to each subunit and purified by means of affinity chromatography (9). The antibodies from each serum were specific for their respective immunizing peptides in ELISA, the anti-α7 and anti-α8 antibodies only bound the corresponding antigens, each immunoprecipitation and immunolabeling was specifically inhibited by the peptide used for the immunization, but not by other peptides, and the elution of the receptor from the corresponding Ab could only be achieved by the peptide used for the production of that Ab (no elution of the receptor was obtained with other peptides). Quantitative immunoprecipitation assays were carried out as previously described (5, 9) using extracts prepared from tissue membranes dissected at the same times.

Immunopurification and characterization of the subtypes: All of the immunopurifications were done using receptors previously purified on an α-BTX-Sepharose column as previously described (10). For the purification of the α7 and α7-α8 subtypes, the COL toxin-purified receptor (50-100 pmol) was applied to a 4 ml column with bound anti-α8 antibodies (column B) for two hours at room temperature. The flowthrough of this column was collected and applied to a column with bound anti-α7 Abs (column A) for two hours. For the α8 subtype purified from the retina, the retina toxin-purified receptor was first incubated with column A and the flowthrough of this column was then incubated with column B. The columns were extensively washed with 80 ml of phosphate buffer saline containing 0.1% Triton X-100. The bound receptor subtypes were eluted using a concentration of 300 μM α7 peptides for three hours or a concentration of 200 μM α8 peptides for two hours; the immunoaffinity columns were then also washed at low pH to obtain complete receptor removal. The purified protein eluted from the toxin column or the subtypes eluted from the affinity columns were inserted in folded type planar lipid bilayers as previously described (5,10).

Results and Discussion

Determination of the α-bungarotoxin receptors containing the α7, α8, or both subunits during ontogenesis: In preliminary experiments, we found that the Ab doses necessary to immunoprecipitate the receptors were: 10 μg for the anti-α7 and 20 μg for the anti-α8 Abs. All of the receptors were labeled using a high concentration of α-BTX (30 nM) in order to label any possible low affinity receptor that may not have been detected by means of the binding assay.

Figure 1 shows the fmol of α-BTX receptors present in COL and chick retina at the indicated developmental stages and the amounts of the different subtypes. The values shown were calculated from immunoprecipitation experiments obtained using anti-α7, anti-α8 separately or both Abs together.

The number and type of receptor subtypes recognized by these Abs at the four developmental stages were different in COL and retina tissue. At E7, using the anti-α7 and anti-α8 Abs alone or together, we could only immunoprecipitate a maximum of 36% of the α-BTX receptors present in all three tissue solutions. This may indicate that the α-BTX receptor contains some unidentified a subunit (s) other than α7 and α8, although, the small amount of receptor at this developmental stage, may lead to technical problems in quantification.

Starting from E13, the percentage of receptors in COL extract precipitated by the anti-α7 Abs represents between 90% and 100% of all α-BTX receptors; the receptors immunoprecipitated by the anti-α8 Abs represent 15-25% of the receptors. In this tissue, the two Abs together immunoprecipitated the same percentage of receptors as the anti-α7 Abs alone, thus indicating that the α8 subunit is associated with the α7 subunit at all developmental stages.

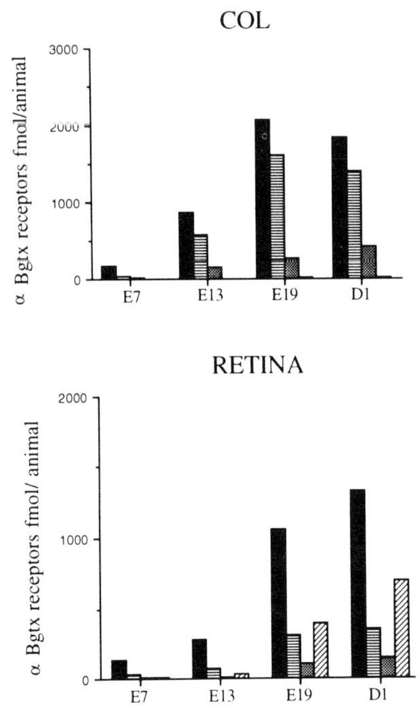

Figure 1. α-BTX receptor subtypes present in COL and chick retina at the indicated developmental age. The results are expressed as fmol /animal of total ^{125}I-α-BTX receptor (■), α7 subtype (≡), α7-α8 subtype (▩), and α8 subtype (▨). The values were obtained from immunoprecipitation experiments as described in Material and Methods, using saturating concentrations of anti-α7, anti-α8, or both Abs together. The ^{125}I-α-BTX receptors were calculated by binding assay in the solution before the immunoprecipitation.

Using the same Abs, the results were different in the extracts obtained from retina. At E13, anti-α7 and anti-α8 Abs together could only immunoprecipitate approximately 40% of the α-BTX receptors present in the solution. This suggests that some of the α-BTX receptors do not contain the α7 and α8 subunits at E13. Secondly, the percentage of receptors immunoprecipitated by the anti-α8 Abs increased during neurodevelopment to a greater extent than the percentage immunoprecipitated by the anti-α7 Abs.

Thirdly, the anti-α7 and anti-α8 Abs used together could immunoprecipitate more receptors than when the two Abs were used alone, but less than the sum of the two. This indicates that at least three α-BTX receptor subtypes, are present in retina: the α7, α8 and α7-α8 subtypes, which respectively represent the 26%, 52% and 11% of the α-BTX receptors at D1.

Purification and biochemical characterization of α-BTX receptors and their subtypes from chick retina and COL: Since our immunoprecipitation experiments have demonstrated that, at D1, the α7, α8 and α7-α8 subtypes in COL and retina are all expressed in sufficient amounts for careful biochemical characterization, we immunopurified the different subtypes from these tissues.

The α-BTX receptors purified from COL and retina have very similar biochemical characteristics: both are glycoproteins with an isoelectric point of 5.3 and, when run on a 5-20% sucrose gradient, they have a sedimentation coefficient of 10S. Fig. 2 shows the subunit composition of the purified receptors obtained after SDS-PAGE. Both receptors are made up of three polypeptides of Mr 67000, 57000, and 52000.

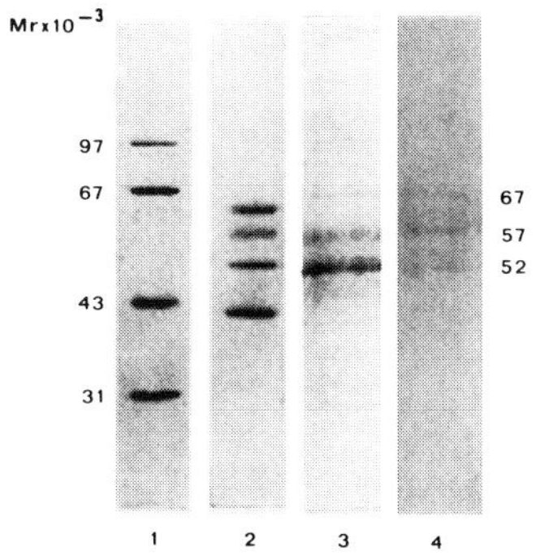

Figure 2. Analysis by SDS acrylamide gel electrophoresis of the subunit composition of α-BTX receptor purified from COL and retina. The receptors were separated on 9% acrylamide and stained with Coomassie blue. 1) Standard proteins. 2) 20 μg of purified Torpedo receptor. 3) 10 μg of α-BTX receptors purified from COL. 4) 7 μg of α-BTX receptors purified from retina.

In both purified receptors the 57000 Mr peptide is recognized by the anti-α7 and anti-α8 Abs. In order to quantify the number of the α-BTX binding subunits, we performed binding experiments on the receptors bound to the α-BTX-Sepharose column. The binding of ^{125}I-α-BTX to immobilized receptors detects approximately only 50% of the receptors present in solution after elution with the agonist carbamylcholine, thus indicating that the α-BTX binding subunit in both receptors is present in more than one copy per receptor.

The α-BTX receptors purified on the affinity column were then separated into their different subtypes by means of the purification on columns with bound anti-α7 or anti-α8 Abs. Table 1 shows the recovery obtained from the subtype immunopurification experiments.

Table 1. Immunoaffinity purification of α-BTX receptor subtypes from chick optic lobe and retina

Tissue	α-BTX receptor*	α7 subtype	α7-α8 subtype	α8 subtype
COL	100	32 ± 2	11.2 ± 2	-
RETINA	100	9.3 ± 3	2.8 ± 0.4	10.1± 2.9

* The values expressed are pmol and are the mean ± SEM of 6 experiments for COL and 4 experiments for retina

In order to study the subunit composition of the receptor subtypes, we labeled the COL toxin-purified receptor with ^{125}I-α-BTX and then immunopurified α7 and α7-α8 subtypes present in COL. Figure 3 shows the subunit composition of the ^{125}I-labeled affinity purified receptor subtypes.

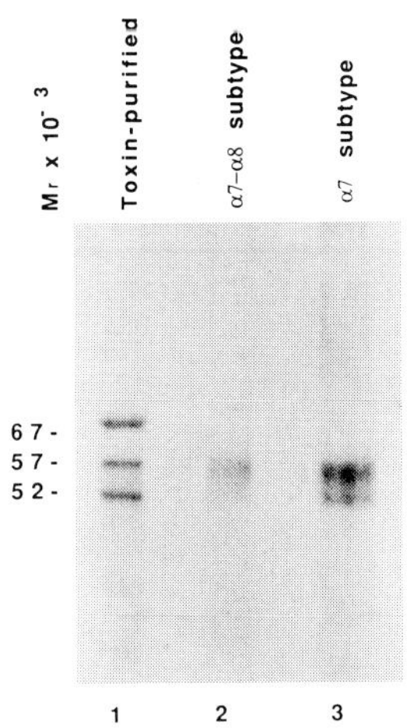

Figure 3. Subunit composition of COL α-BTX receptor subtypes. The receptors were separated on 9% acrylamide SDS gel electrophoresis, and autoradiographed. Lane 1) ^{125}I-labeled α-BTX receptor purified from COL; lane 2) ^{125}I-labeled α7-α8 subtype and 3) ^{125}I-labeled α7 subtype. Modified from ref. 5.

Upon SDS/PAGE, the ^{125}I-labeled COL toxin-purified receptor always contained three major components of Mr 67000, 57000 and 52000 (Fig.3, lane 1); the subtype α7-α8 only contained a major band of Mr 57000 and a faint band of Mr 52000 (Fig.3, lane 2); and the α7 subtype only two major bands of Mr 57000 and 52000 (Fig.3 lane 3). None of the receptors purified on the immunoaffinity columns contained the Mr 67000 band. These findings indicate that in vivo all the α-BTX receptors are heterooligomeric formed by at least two subunits, the 57 kD subunit, which binds α-BTX and the 52 kD subunit whose function is still unknown.

Subtype reconstitution in lipid bilayer: The toxin purified receptors and the α7 and α7-α8 COL subtypes and the α7 and α8 retina subtypes were inserted into the lipid bilayer. Both the α-BTX receptors and the subtypes form cationic channels activated by agonists in a dose-dependent manner.

The basic properties of these channels are similar. As shown in Table 2, they all are cation selective channels permeable to Ca^{++}; the α7 and α8 subtypes have very similar homogeneous conductances (45-50 pS) while the α7-α8 subtype has a much broader conductance (10-50 pS). They are all activated by agonists and blocked by d-tubocurarine.

The main difference between these subtypes is the concentration of agonists (EC_{50}) necessary to give 50% of maximal channels activation. These concentrations were calculated from the increase in integral open time with respect to the spontaneous activity of the reconstituted receptors.

The concentrations of both carbamylcholine and acetylcholine necessary to activate the α7 subtypes is 10-20 times higher than those necessary to activate the α8 subtype, whereas the α7-α8 subtype needs an agonist concentration which is intermediate between those of the α7 and α8 subtypes.

Table 2. Comparison of the electrophysiological characteristics of reconstituted α-BTX receptor subtypes

Channel properties	COL or retina* α7 subtype	COL α7-α8 subtype	Retina α8 subtype
Conductance (pS)	45-50*	10-50	50
Open state lifetime (ms)	5-10	1-3	20
Ion selectivity	Na≈K>>Cl^-	Na≈K>>Cl^-	Na≈K>>Cl^-
Ca^{++} permeability	++	++	N.D.
Agonist profile			
Acetylcholine EC_{50} (μM)	500-400*	150	50
Carbamylcholine EC_{50} (μM)	600-300*	200	30
Sensitivity to d-Tubocurarine	++	++	++

Comparison of the EC_{50} values obtained for the reconstituted subtypes are not identical, but similar to those obtained from the homomeric α7 and α8 channels expressed in oocytes. The EC_{50} for acetylcholine tested in the α8 homomer is 2.1 μM (5), but it is 110 μM for the α7 homomer (3). It seems that, both in the homomeric receptor expressed in oocytes and in the heterooligomeric receptors reconstituted in lipid bilayer, the presence of the α8 subunit gives receptors that are much more effectively gated by agonists than the α7 homomeric or hetero-oligomeric receptors.

A similar pattern of response was obtained when a number of agonists and antagonists were tested on the immunopurified subtypes in ^{125}I-α-BTX displacement experiments. The most relevant data were the higher affinity of α-BTX for the α7 than for the α8 subtype, and the higher affinity of agonists for the α8 subtype.

In conclusion we have found that three subtypes of the α-BTX receptor are present in chick nervous system. These subtypes have a different developmental pattern of expression and different biophysical and pharmacological properties; in addition, each tissue has a peculiar composition of subtypes.

The physiological role of these subtypes is presently unknown. It is possible that the neurons may vary their response to the environment by modifying the complement of different receptor subtypes in the membrane. This would give a significant degree of plasticity to neurons which use the same type of subunits, varying only their association and number. We are currently performing in vivo and in vitro experiments to test this hypothesis.

Acknowledgements: *We would like to thank Ida Ruffoni, Kevin Smart and Paolo Tinelli for their help with the manuscript. This work was supported in part by Consiglio Nazionale delle Ricerche grant "Chimica fine" and Telethon grant to C. G.*

References

1. Galzi J-L, Revah F, Bessis A, Changeux J-P. Functional architecture of the nicotinic acetylcholine receptor: from electric organ to brain. Ann. Rev. Pharmacol. 1991; 31: 37-72.
2. Sargent P. The diversity of neuronal nicotinic acetylcholine receptors. Annu. Rev. Neurosci. 1993; 16: 403-443.
3. Couturier S, Bertrand D, Matter J-M, Hernandez M-C, Bertrand S, Millar N, Valera S, Barkas T, Ballivet M. A neuronal nicotinic acetylcholine receptos subunit ($\alpha7$) is developmentally regulated and forms a homo-oligomeric channel blocked by a-BTX. Neuron 1990; 5: 847-856.
4. Schoepfer R, Conroy WG, Whiting P, Gore M, Lindstrom J. Brain. α-Bungarotoxin binding protein cDNAs and MAbs reveal subtypes of this branch of the ligand ion channel gene superfamily. Neuron 1990; 5: 35-48.
5. Gotti C. Hanke W, Maury K, Moretti M, Ballivet M, Clementi F, Bertrand D. Pharmacology and biophysical properties of $\alpha7$ and $\alpha7$-$\alpha8$ αBungarotoxin receptor subtypes immunopurified from chick optic lobe. Eur. J. Neurosci. 1994; 6: 1281-1291.
6. Gerzanich V, Anand R. Lindstrom J. Homomers of $\alpha7$ and $\alpha8$ subunits of nicotinic receptors exhibit similar channel but contrasting binding site properties. Mol. Pharm.1994; 45: 212-220.
7. Keyser KT, Britto LR, Schoepfer R, Withing P, Cooper J, Conroy W, Brozozowska-Prechtl A, Karten JH, Lindstrom J. Three subtypes of α-Bungarotoxin-sensitive nicotinic acetylcholine receptors are expressed in chick retina. J. Neurosci. 1992; 13: 442-454.
8. Anand R, Peng X, Ballesta J, Lindstrom J. Pharmacological characterization of α-Bungarotoxin-sensitive acetylcholine receptors immunoisolated from chick retina: contrasting properties of $\alpha7$ and $\alpha8$ subunit-containing subtypes. Mol. Pharm. 1993; 44: 1046-1050 .
9. Gotti C, Moretti M, Longhi R, Briscini L, Manera E, Clementi F. Anti-peptide specific antibodies for the characterization of different a subunits of α-Bungarotoxin binding acetylcholine receptors present in chick optic lobe. J. Rec. Res. 1993; 13: 453-465.
10. Gotti C, Esparis Ogando A, Hanke W, Schlue R, Moretti M, and Clementi F. Purification and characterization of an a-Bungarotoxin receptor that forms a functional nicotinic channel. Proc. Natl. Acad. Sci. USA 1991; 88: 3258-3262.

Effects of Nicotine on
Biological Systems II
Advances in Pharmacological Sciences

NEURONAL NICOTINIC RECEPTOR STRUCTURE AND FUNCTION

J. Lindstrom, R. Anand, X. Peng, V. Gerzanich

Department of Neuroscience, University of Pennsylvania Medical School
Philadelphia, PA 19104-6074

Summary: The best characterized subtype from the branch of the neuronal nicotinic receptor (nAChR) gene family which does not bind α-bungarotoxin has the subunit stoichiometry $(\alpha4)_2(\beta2)_3$ and accounts for >90% of the high affinity nicotine binding in mammalian brains. Chronic treatment of cells transfected with α4β2 AChRs causes an increase in the amount of AChR with pharmacological characteristics and a time course that parallel the nicotine-induced increase in brain α4β2 AChRs. This is the result of a decrease in turnover of surface AChRs. Many of these surface AChRs are permanently functionally inactive. The predominant subtype of the branch of the neuronal AChR gene family which binds α-bungarotoxin (α-BTX) contains α7 subunits. α7 AChRs predominate in brain, while α8 AChRs predominate in retina, and α7α8 AChRs are a minor component in both tissues. α7 and α8 homomers have similar channel properties, but α8 homomers and native α8 AChRs have lower affinity for α-BTX and higher affinity for small cholinergic ligands than do α7 homomers and α7 AChRs. The high Ca^{++} permeability and rapid desensitization of α7 and α8 AChRs may enable them to participate in unusual synaptic mechanisms.

Introduction

There are three branches of the nicotinic receptor (AChR) gene family: 1) muscle AChRs, 2) neuronal AChRs which do not bind α-bungarotoxin (α-BTX), and 3) neuronal AChRs which bind α-BTX. Muscle AChRs early in development have the subunit composition $(\alpha1)_2\beta1\gamma\delta$, whereas at mature neuromuscular junctions they have the subunit composition $(\alpha1)_2\beta1\varepsilon\delta$ (1,2). α-BTX binds to muscle AChRs with very high affinity. It is suspected that all subtypes of the nicotinic AChR family have sizes and shapes similar to those of muscle type AChRs, which, unlike the neuronal AChRs, have been characterized in some detail (1-3). Neuronal AChRs which do not bind α-BTX are thought to be formed from combinations of α2-α6 with β2-β4 subunits, but the subunit compositions, developmental changes, and functional roles of most potential subtypes *in vivo* are unknown (2,4). Neuronal AChRs which bind α-BTX are known to be composed of α7 or α8 subunits, or both, possibly in combination with additional structural subunits which remain unknown (5). Unlike other members of the gene family, α7 (6) and α8 (7) can function as homomers when expressed from cRNA in *Xenopus* oocytes, which has made them experimentally very useful, but the functional roles of α7 and α8 AChRs *in vivo* are not well known. In brain there are similar amounts of α4β2 AChRs (which represent the predominant subtype with high affinity for nicotine) and of α7 AChRs (which represent the predominant subtype with high affinity for α-BTX). Some recent experiments from our laboratory with α4β2 AChRs and α7 and α8 AChRs will be briefly reviewed.

Results and Discussion

α4β2 AChRs: From a structural point of view, the α4β2 AChR is the best characterized subtype of neuronal AChR. It accounts for >90% of the high affinity nicotine binding sites in mammalian brain (8,9). Its subunit composition was first determined by N-terminal amino acid sequencing of subunits of immunoaffinity purified AChRs (10-12) and then confirmed by studies with cholinergic ligand affinity purified AChRs (13), subunit-specific mAbs (9), and subunit-specific antisera (14). Its subunit stoichiometry was determined to be $(\alpha4)_2(\beta2)_3$ using AChRs expressed in *Xenopus* oocytes both by a biochemical (15) and an electrophysiological (16) approach. The biochemical method should be applicable to any multisubunit receptor expressed from cDNAs. It involves determining the relative amount of ^{35}S-methionine incorporated into each subunit of the purified receptor.

The physiological roles of α4β2 AChRs are not well characterized. Several lines of evidence from many laboratories suggest that many neuronal AChRs are found in presynaptic locations where they may modulate the release of transmitters. For examples, β2 subunits are transported along axons of retinal ganglion cells to the superior colliculus (17), much as AChRs from goldfish (18) or frog (19) retinas are transported to the optic tectum. Similar AChR transport occurs along the fasiculus retroflexus between the medial habenula and interpeduncular nuclei (20). AChRs on synaptosomes have been shown to enhance transmitter release (21).

Chronic exposure to nicotine in smokers (22) and in animals (23) treated with nicotine causes an increase in brain nicotine binding by as much as two fold. Nicotine upregulates brain α4β2 AChRs (14). This is not a result of upregulating α4 or β2 mRNA in brain (24). It has been suggested that this upregulation might be a specific response of neurons to a decrease in functional AChRs as a result of transient desensitization caused by nicotine. In fact, upregulation by nicotine is an intrinsic property of α4β2 AChRs (25). This is shown by the observation (25) that nicotine causes upregulation of α4β2 AChRs expressed in a permanently transfected mouse fibroblast cell line (26) or in *Xenopus* oocytes injected with α4β2 cRNAs. Concentrations of nicotine typical of those found in smokers are effective at causing upregulation in transfected cells ($EC_{50} = 2 \times 10^{-7}$M). A two fold increase in amount of α4β2 AChRs occurs over 3 days, which resembles the 2 fold increase in brain nicotine binding sites observed over 4 days (27). Other nicotinic agonists such as cytisine, carbamylcholine and DMPP also cause upregulation of α4β2 AChRs, and upregulation by nicotine is blocked by the competitive antagonist curare (25). Upregulation does not require ion flow through the cation channel as a result of agonist binding because the channel blocker mecamylamine also causes upregulation (25). Mecamylamine is synergistic with nicotine (25). This parallels additive effects of nicotine and mecamylamine on AChR upregulation in brain (28). Mecamylamine is an open channel blocker (29) which is more effective at producing a long lasting block in the presence of nicotine (25). Chronic exposure to nicotine of α4β2 AChRs expressed in *Xenopus* oocytes results in permanent inactivation of agonist-gated channel activity even though equilibrium ^{3}H-nicotine binding remains unaltered (25). Similar long lived desensitized states of AChRs after chronic exposure to nicotine have also been observed by others (30,31). Thus, a decrease in functional AChRs, despite an increase in total AChRs, can explain the tolerance to nicotine which is observed *in vivo*. The increase in α4β2 AChRs is due to a decrease in the rate of turnover of AChRs in the cell surface (25). The decrease in the rate of destruction of α4β2 AChRs appears to result from a change in the conformation of the AChRs induced by agonists and some noncompetitive antagonists.

α7 and α8 AChRs: α7 and α8 subunit cDNAs were initially cloned from chickens (5). They were shown to be subunits of neuronal AChRs which bind α-BTX by using mAbs to bacterially-expressed α7 and α8 sequences to immune precipitate ^{125}I-α-BTX labeled AChRs (5). These mAbs were made against chick sequences corresponding to a large cytoplasmic domain which in muscle AChR subunits usually provides epitopes in native and denatured

AChRs and is usually species-specific (32). Epitopes were mapped within this sequence using synthetic peptides (33). Some of these mAbs crossreact with mammalian α7 (34,35), especially after denaturation, and have been useful histologically on rat brain (35) as well as chick brain and retina (36-41). α7 has now been cloned also from rats (42) and humans (34,43,44). In chickens, α7 AChRs predominate in brain, while α8 AChRs predominate in retina, and α7α8 AChRs are a minor population in both tissues (39).

α7 homomers are efficiently expressed in *Xenopus* oocytes where they function as agonist-gated cation channels (6,45). However, *in vivo* α7 may exist in combination with as yet unknown structural subunits. All preparations of purified brain α-BTX binding proteins have exhibited multiple protein bands on SDS PAGE, some of which may be additional subunits (46). α8 homomers are also functional when expressed in *Xenopus* oocytes, but assemble much less efficiently, suggesting that they have a greater dependence on as yet unknown structural subunits than do α7 AChRs (7). α7 and α8 homomers have nearly identical sequences in the putative channel forming domains and adjacent putative transmembrane domains and consequently have essentially identical channel properties (7). Both α7 and α8 homomers desensitize very rapidly and exhibit high Ca^{++} permeability. In *Xenopus* oocytes the Ca^{++} which enters through the homomer channels triggers a Ca^{++}-dependent Cl^- channel which amplifies their effect. *In vivo*, Ca^{++} entering through α7 and α8 AChRs might act as a second messenger to trigger other ion channels or to affect cellular processes like neurite extension (47,48).

α7 and α8 AChRs exhibit pharmacological differences (39,49,50) which are reflected in the properties of α7 and α8 homomers (7,49). α7 AChRs and α7 homomers have higher affinity for α-BTX but lower affinity for small cholinergic ligands than do α8 AChRs or α8 homomers. The ligand binding properties of α7 homomers are similar but not identical to those of native α7 AChRs (34,49). This could reflect a difference in post translational modification in *Xenopus* oocytes, but more likely it reflects the presence of structural subunits in the native AChR. There are some species differences in pharmacological responses. For example DMPP is virtually an antagonist on chick α7 homomers (<3% partial agonist) (7), but on human α7 homomers it is a full agonist and more potent than ACh or nicotine (34). In order to determine which of the limited sequence differences between α7 and α8 account for their pharmacological differences a series of mosaics were made between α7 and α8 (Anand,

Gerzanich and Lindstrom, unpublished). This revealed that the amino acids responsible for the pharmacological differences between $\alpha7$ and $\alpha8$ are among the 5 which differ in the region 179-208.

In order to investigate possible physiological roles of $\alpha7$ and $\alpha8$ AChRs we studied cochlear hair cells (Anand, Cooper, Peng, Fuchs, and Lindstrom, unpublished). Paul Fuchs and coworkers (51) had shown that chick cochlear hair cells have an excitatory AChR which is blockable by α-BTX and which permits entry of Ca^{++} that then acts as a second messenger to trigger a Ca^{++}-sensitive K^{+} channel that causes a net prolonged inhibitory response. Like $\alpha7$ AChRs (7,34,50), this AChR was inhibited by the muscarinic antagonist atropine and the glycinergic antagonist strychnine. Cochleae were found to contain $\alpha7$ mRNA (Anand et al., unpublished). mAbs to $\alpha7$, but not to $\alpha8$, $\alpha1$, or $\beta2$, could immunoisolate ^{125}I-α-BTX labeled AChRs from cochleae. These $\alpha7$ AChRs were found to have pharmacological properties identical to brain $\alpha7$ AChRs. Single dissected hair cells bound mAbs to $\alpha7$ but not to $\alpha8$ or $\alpha1$. However, nicotine is an agonist on $\alpha7$ homomers, but an antagonist on AChRs in hair cells (Fuchs, unpublished). Thus, the AChR studied by Fuchs may include $\alpha7$ subunits in combination with structural subunits that affect their pharmacological properties, or in addition to $\alpha7$ AChRs another AChR may be present which accounts for all of the pharmacological properties reported by Fuchs et al. $\alpha9$ subunits initially reported at this meeting in the poster by Elgoyhen et al. might serve as this "structural" subunit (in the same sense that $\alpha8$ associates with $\alpha7$ (5,39)) or as an additional AChR subtype with pharmacological properties like those observed by Fuchs (51). In any case, the unusual sort of synaptic mechanism observed in hair cells, where Ca^{++} acts as a second messenger, provides an example of the potentially novel and as yet unknown sorts of mechanisms in which $\alpha7$ and $\alpha8$ AChRs may participate.

Acknowledgments: *Research in the laboratory of J.L. is supported by grants from the National Institutes of Health, the Muscular Dystrophy Association, Council for Tobacco Research, USA, Inc., and Smokeless Tobacco Research Council, Inc. R.A. is supported in part by a National Research Service Award.*

References

1. Karlin A. Structure of nicotinic acetylcholine receptors. Curr. Opin. Neurobiol. 1993; 3:299-309.
2. Lindstrom J. Nicotinic acetylcholine receptors in CRC Handbook of Receptors. Alan North, ed. CRC Press: in press.

3. Unwin N. Nicotinic acetylcholine receptor at 9Å resolution. J. Mol. Biol. 1993; 229:1101-1124.
4. Sargent P. The diversity of neuronal nicotinic acetylcholine receptors. Annu. Rev. of Neurosci. 1993; 16:403-443.
5. Schoepfer R, Conroy WG, Whiting P, Gore M, Lindstrom J. Brain α-bungarotoxin-binding protein cDNAs and mAbs reveal subtypes of this branch of the ligand-gated ion channel superfamily. Neuron 1990; 5:35-48.
6. Courturier S, Bertrand D, Matter J, Hernandez M, Bertrand S, Millar N, Valera S, Barkas T, Ballivet M. A neuronal nicotinic acetylcholine receptor subunit (α7) is developmentally regulated and forms a homomeric channel blocked by α-bungarotoxin. Neuron 1990; 5:847-856.
7. Gerzanich V, Anand R, Lindstrom J. Homomers of α8 subunits of nicotinic receptors functionally expressed in Xenopus oocytes exhibit similar channel but contrasting binding site properties compared to α7 homomers. Mol. Pharmacol. 1994; 45:212-220.
8. Whiting PJ, Lindstrom JM. Purification and characterization of a nicotinic acetylcholine receptor from rat brain. Proc. Natl. Acad. Sci. USA 1987; 84:595-599.
9. Whiting PJ, Lindstrom JM. Characterization of bovine and human neuronal nicotinic acetylcholine receptors using monoclonal antibodies. J. Neurosci. 1988; 8:3395-3404.
10. Whiting P, Esch F, Shimasaki S, Lindstrom J. Neuronal nicotinic acetylcholine receptor b subunit is coded for by the cDNA clone α4. FEBS Letters 1987; 219:459-463.
11. Schoepfer R, Whiting P, Esch F, Blacher R, Shimasaki S, and Lindstrom J. cDNA clones coding for the structural subunit of a chicken brain nicotinic acetylcholine receptor. Neuron 1988; 1:241-248.
12. Whiting P, Schoepfer R, Conroy WG, Gore MJ, Keyser K, Shimasaki S, Esch F, Lindstrom J. Differential expression of nicotinic acetylcholine receptor subtypes in brain and retina. Mol. Brain Res. 1991; 10:61-70.
13. Nakayama H, Shirase M, Nakashima T, Kurogochi Lindstrom J. Affinity purification of nicotinic acetylcholine receptor from rat brain. Mol. Brain Res. 1990; 7:221-226.
14. Flores C, Rogers S, Pabreza L, Wolfe B, Kellar K. A subtype of nicotinic cholinergic receptor in rat brain is composed of α4 and β2 subunits and is upregulated by chronic nicotine treatment. Mol. Pharmacol. 1992; 41:31-37.
15. Anand R, Conroy WG, Schoepfer R, Whiting P, Lindstrom J. Chicken neuronal nicotinic acetylcholine receptors expressed in Xenopus oocytes have a pentameric quaternary structure. J. Biol. Chem. 1991; 266:11192-11198.
16. Cooper E, Couturier S, Ballivet M. Pentameric structure and subunit stoichiometry of a neuronal nicotinic acetylcholine receptor. Nature 1991; 350:235-238.
17. Swanson L, Simmons D, Whiting P, Lindstrom J. Immunohistochemical localization of neuronal nicotinic receptors in the rodent central nervous system. J. Neurosci. 1987; 7:3334-3342.
18. Henley J, Lindstrom J, Oswald R. Acetylcholine receptor synthesis in retina and transport to the optic tectum in goldfish. Science 1986; 1627-1629.
19. Sargent P, Pike S, Nadel S, Lindstrom J. Nicotinic acetylcholine receptor-like molecules in the retina, retinotectal pathway, and optic tectum of the frog. J. Neurosci. 1989; 9:565-573.
20. Clarke P, Hamill G, Nadi N, Jacobowitz D, Pert A. ^{3}H-Nicotine and ^{125}I-α-bungarotoxin-labeled nicotinic receptors in the interpeduncular nucleus of rats. II. Effects of habenular deafferentation. J. Comp. Neurol. 1986; 251:407-413.
21. Rapier C, Lunt G, Wonnacott S. Stereoselective nicotine-induced release of dopamine from striatal synaptosomes: Concentration dependence and repetitive stimulation. J. Neurochem. 1988; 50:1123-1130.
22. Benwell M, Balfour D, Anderson J. Evidence that tobacco smoking increases the density of (-)-[^{3}H] nicotine binding sites in human brain. J. Neurochem. 1988; 50:1243-1247.
23. Schwartz R, Kellar K. *in vivo* regulation of [^{3}H] acetylcholine recognition sites in brain by nicotinic cholinergic drugs. J. Neurochem. 1985; 45:427-433.
24. Marks M, Pauly J, Gross D, Deneris E, Hermans-Borgmeyer I, Heinemann S, Collins A. Nicotine binding and nicotinic receptor subunit RNA after chronic nicotine treatment. J. Neurosci. 1992; 12:2765-2784.
25. Peng X, Gerzanich V, Anand R, Whiting P, Lindstrom J. Nicotine-induced increase in neuronal nicotinic receptors results from a decrease in the rate of receptor turnover. Mol. Pharmacol., in press.
26. Whiting P, Schoepfer R, Lindstrom J, Priestley T. Structural & pharmacological characterization of the major brain nicotinic acetylcholine receptor subtype stably expressed in mouse fibroblasts. Mol. Pharmacol. 1991; 40:463-472.

27. Marks M, Stetzel J, Collins A. Time course study of the effects of chronic nicotine infusion on drug response and brain receptors. J. Pharmacol. Exp. Ther. 1985; 235:619-628.
28. Collins A, Bhat R, Pauly J, Marks M. Modulation of nicotine receptors by chronic exposure to nicotinic agonists and antagonists. Ciba Foundation Symposium 152 The Biology of Nicotine Dependence, G. Bock and J. Marsh, editors, John Wiley and Sons, Chichester, 1990: 68-82.
29. Varanda W, Aracova A, Sherby S, Van Meter W, Eldefrawi M, Albuquerque E. The acetylcholine receptor of the neuromuscular junction recognizes mecamylamine as a noncompetitive antagonist. Mol. Pharmacol. 1985; 28:128-137.
30. Lukas R. Effects of chronic nicotinic ligand exposure on functional activity of nicotinic acetylcholine receptors expressed by cells of the PC12 rat pheochromocytoma or the TE671/RD human clonal line. J. Neurochem. 1991; 56:1134-1145.
31. Marks M, Grady S, Collins A. Downregulation of nicotinic receptor function after chronic nicotine infusion. J. Pharmacol. Exp. Ther. 1993; 266:1268-1275.
32. Das M, Lindstrom J. Epitope mapping of antibodies to acetylcholine receptors. Biochemistry 1991; 30:2470-2477.
33. McLane K, Wu X, Lindstrom J, Conti-Tronconi B. Epitope mapping of polyclonal and monoclonal antibodies against two α-bungarotoxin binding subunits from neuronal nicotinic receptors. J. Neuroimmunol. 1992; 38:115-128.
34. Peng X, Katz M, Gerzanich V, Anand R, Lindstrom J. Human α7 acetylcholine receptor: cloning of the α7 subunit from the SH-SY5Y cell line and determination of pharmacological properties of native receptors and functional α7 homomers expressed in Xenopus oocytes. Mol. Pharmacol. 1994; 45:546-554.
35. Del Toro E, Juiz J, Peng X, Lindstrom J, Criado M. Immunocytochemical localization of the α7 subunit of the nicotinic acetylcholine receptor in the rat central nervous system. J. Comp. Neurol. (in press).
36. Hamassaki-Britto D, Brzozowska-Prechtl A, Karten H, Lindstrom J, Keyser K. GABA-like immunoreactive cells containing nicotinic acetylcholine receptors in the chick retina. J. Comp. Neurol. 1991; 313:394-408.
37. Britto L, Hamassaki-Britto D, Ferro E, Keyser K, Karten H, Lindstrom J. Neurons of the chick brain and retina expressing both α-bungarotoxin-sensitive and α-bungarotoxin-insensitive nicotinic acetylcholine receptors: an immunohistochemical analysis. Brain Res. 1992; 590:193-200.
38. Britto L, Keyser K, Lindstrom J, Karten H. Immunohistochemical localization of nicotinic acetylcholine receptor subunits in the mesencephalon and diencephalon of the chick (Gallus gallus) J. Comp. Neurol. 1992; 317:325-340.
39. Keyser K, Britto L, Schoepfer R, Whiting P, Cooper J, Conroy W, Brozozowska-Prechtl A, Karten H, Lindstrom J. Three subtypes of α-bungarotoxin-sensitive nicotinic acetylcholine receptors are expressed in chick retina. J. Neurosci. 1993; 13:442-454.
40. Hamassaki-Britto D, Brzozowska-Prechtl A, Karten H, Lindstrom J. Bipolar cells of the chick retina containing α-bungarotoxin-sensitive nicotinic acetylcholine receptors. Vis. Neurosci. 1994; 11:63-70.
41. Hamassaki-Britto D, Gardino P, Hokoc J, Keyser K, Karten H, Lindstrom J, Britto L. Differential development of α-bungarotoxin-sensitive and α-bungarotoxin-insensitive nicotinic acetylcholine receptors in the chick retina. J. Comp. Neurol. (in press).
42. Séguéla P, Wadiche J, Dinelly-Miller K, Dani J, Patrick J. Molecular cloning, functional properties, and distribution of rat brain α7: a nicotinic cation channel highly permeable to calcium. J. Neurosci. 1993; 13:596-604.
43. Chini B, Raimond E, Elgoyhen A, Moralli D, Bolzaretti M, Heinemann S. Molecular cloning and chromosomal localization of the human α7 nicotinic receptor subunit gene (CHRNA7). Genomics 1994; 19:379-381.
44. Doucellestamm L, Monteggia L, Donnellyroberts D, Wang M, Lee J, Tian J, Giordano T. Cloning and sequence of the human α7 nicotinic acetylcholine receptor. Drug Devel. Res. 1993; 30:252-256.
45. Lindstrom J, Anand R, Peng X, Gerzanich V, Wang F, Li Y. Neuronal nicotinic receptor subtypes. Abel Lajtha and Leo Abood, editors. New York Academy of Science volume on Functional Diversity of Interacting Receptors, in press.
46. Gotti C, Moretti M, Longhi R, Briscini L, Manera E, Clementi F. Anti-peptide specific antibodies for the characterization of different a subunits of α-bungarotoxin binding acetylcholine receptors present in chick optic lobe. J. Recept. Res. 1993; 13:453-465.
47. Lipton S, Frosch M, Phillips M, Tauck D, Aizenman E. Nicotinic antagonists enhance process outgrowth by rat retinal ganglion cells in culture. Science 1988; 239:1293-1296.

48. Pugh P, Berg D. Neuronal acetylcholine receptors that bind α-bungarotoxin mediate neurite retraction in a calcium-dependent manner. J. Neurosci. 1994 14:889-896.
49. Anand R, Peng X, Lindstrom J. Homomeric and native α7 acetylcholine receptors exhibit remarkably similar but nonidentical pharmacological properties suggesting that the native receptor is a heteromeric protein complex. FEBS Lett. 1993; 327:241-246.
50. Anand R, Peng X, Ballesta J, Lindstrom J. Pharmacological characterization of α-bungarotoxin-sensitive AChRs immunoisolated from chick retina: contrasting properties of α7 and α8 subunit-containing subtypes. Mol. Pharmacol. 1993; 44:1046-1050.
51. Fuchs P, Murrow B. A novel cholinergic receptor mediates inhibition of chick cochlear hair cells. Proc. R. Soc. Lond. B. 1992; 248:35-40

DETERMINANTS REGULATING NEURONAL NICOTINIC RECEPTOR FUNCTION

S. Bertrand, B. Buisson, I Forster* and D. Bertrand

*Dept of Physiology, Zürich, and Dept of Physiology, Geneva Medical Faculty, 1211 Geneva 4, Switzerland

Summary: The structure function relationship of ligand-gated channels was investigated at the molecular level using the α7 homomeric nicotinic acetylcholine receptor as a model of the superfamily of allosteric proteins. Chimera between the neuronal nicotinic acetylcholine receptor (nAChR) and the serotoninergic ($5HT_3$) receptor revealed functional protein domains, and also allowed the demonstration of the allosteric modulation induced by extracellular calcium. Rectification, another key feature of ligand-gated channels, could be attributed to intracellular magnesium block.

Introduction

Neuronal nicotinic receptors are members of the superfamily of ligand-gated channels which encompass receptors permeable to cations or anions. Investigations using high resolution electron microscopy as well as atomic force microscopy have shown that, as expected from earlier hypotheses, nAChRs are formed by the assembly of five subunits (1, 2). Concomitantly, progress in molecular biology has allowed the isolation and characterization of the genes encoding a number of ligand-gated channels and the determination of their protein sequences (3-5).

However despite all this progress, the structure-function relationship of the ligand-gated channels still remains to be elucidated. As a first step toward this understanding we used a combination of site-directed mutagenesis and electrophysiological recordings to investigate determinants regulating the receptor function.

Materials and Methods

Oocyte preparation and injection: The method used throughout these experiments is derived from previously described work (6). *Chemicals and solutions.* All chemicals were obtained from Sigma. Agonists and antagonists were prepared in stock solution of 0.1 M in H_2O and kept frozen. Dilutions were made fresh in OR2 medium for every experiment. OR2 contained in mM: 82.5 NaCl, 2.5 KCl, 2.5 $CaCl_2$, 1 $MgCl_2$ and 15 Hepes, and was adjusted to pH 7.4 with NaOH. To prevent a possible activation of endogenous muscarinic receptors, 1 μM atropine was added. Recordings, data acquisition and treatment: One to four days following intranuclear injection, oocytes were placed in the recording chamber and their properties assessed using dual electrode voltage clamp (GeneClamp from AXON instrument). Cells were maintained at a steady potential of -100 mV and tested for their responses by application of short pulses of ACh (typically 1s) using a concentration near to the receptor EC_{50}. Data were captured on-line and stored on a personal computer PC 386 or 486. Data acquisition and analysis were performed using DATAC (7). *Mutagenesis:* Mutants were prepared as described (8, 9) and their coding sequence was always completely checked.

Results and Discussion

Functional domains of ligand-gated channels: Ligand-gated channels are integral membrane proteins which display several properties with at least the agonist binding site and the ionic pore whose opening leads to modification of the cell membrane potential. Binding of the agonist, or closely related pharmacological agents, induces the allosteric transformation of the protein which, eventually, results in the modification of conformation and opening of the pore (10). It is now widely accepted that ligand-gated channels are composed of five subunits each of which spans the membrane four times (reviewed in 3). At the center of this pentameric assembly, a minute aqueous pore is formed through which ions can flow. Several lines of evidence have demonstrated that the second transmembrane segment of each of the five subunits constitute the walls of the ionic pore (3, 11). Other experimental evidence, such as labeling of amino acid residues that form the putative binding site with photo affinity ligands (12), has shown that the agonist binding pocket is formed by a small segment of the first hydrophilic domain of the protein. This hypothesis was further supported by point mutation experiments in which exchanges of some the previously labeled amino acids resulted in a

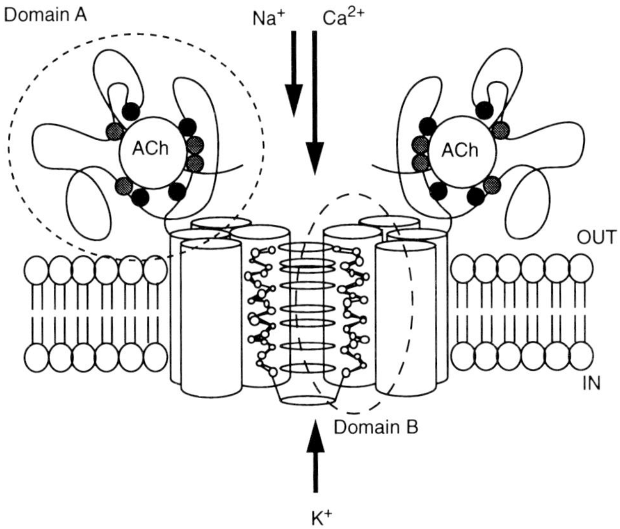

Figure 1. Schematic representation of the functional domains of the neuronal nAChR. For clarity, only two of the five subunits of the neuronal nAChR are represented.

large modification of the physiological EC_{50} (13). All these data point to the existence of distinct functional domains with a clear separation between the binding site and aqueous pore that is schematically represented as shown in Figure 1.

The postulate that functional domains are distinct and might be conserved across members of the ligand-gated channels superfamily implies that it should be possible to produce a fusion protein containing one domain of one receptor type and the other from another member of the family.

To test this hypothesis we have produced a fusion protein between the neuronal nAChR $\alpha7$ and the serotoninergic ($5HT_3$) receptor. We found that functional chimera can indeed be obtained and, as described below, this has lead to a new technique for investigating the receptor structure-function relationship (14).

Calcium, a potent allosteric modulator of the $\alpha7$ receptor: Recordings of some neuronal nAChR responses done in media containing different calcium concentrations showed that the amplitude of the ACh-evoked current strongly correlates with the concentration of these ions. Careful analysis done on native receptors, expressed by neurons from the medial habenula (15), revealed that calcium acts as an allosteric modulator of these neuronal nAChRs. To verify if these observations related to a common feature which applies to all the members of the neuronal nicotinic receptors we have explored the effect of extracellular calcium on $\alpha7$. In $\alpha7$ homooligomers, however, the high calcium permeability of the channel precludes any good correlation between extracellular calcium and the ACh-evoked current, because it is almost impossible to dissociate the contribution of putative allosteric receptor modulation from the calcium and calcium induced currents (16, 17). Thus, in the α7 wild type receptor, determination of the ACh dose-response curves in different extracellular calcium conditions constitutes the only evidence of putative allosteric modulation.

A better approach would be to make use of proteins in which the channel domain has been modified to suppress calcium permeability. In this context, the chimeric receptor that was produced between the wild type α7 and $5HT_3$ receptor is, up to now, one of the best tools available. This chimera was produced by fusing the extracellular domain of α7 (from the N-terminal up to position 201, domain A Figure 1) with the transmembrane domains of the $5HT_3$ receptor. The main advantages of this construct arise from the fact that the $5HT_3$ channel is impermeable to calcium and that in the wild type $5HT_3$ receptor, reduction of the extracellular calcium produces no detectable shift of the dose response curve.

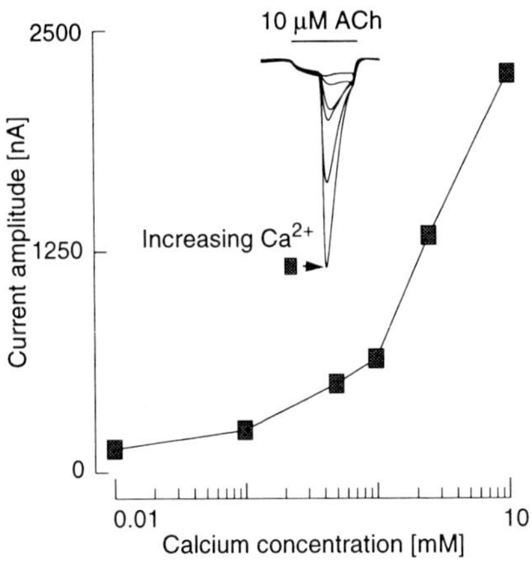

Figure 2. Calcium allosteric modulation is revealed on the chimeric α7-$5HT_3$ receptor. The inset illustrates the experimental paradigm. ACh-evoked currents were recorded either at a low calcium concentration (0.01 mM) or during an abrupt pulse of calcium applied once the agonist evoked current is established. Cells were maintained at -100 mV during the entire experiment. Differences between the peak and steady evoked current are plotted in the lower panel as a function of the calcium concentration.

Thus, if extracellular calcium induces any effect on the ACh-evoked current of the chimera this must be attributed to the interaction between these ions and the protein extracellular domain. A special experimental protocol illustrated in Figure 2 was designed to assess this putative allosteric calcium modulation. The potentiation of ACh-evoked currents induced by the stepwise jumps in calcium concentrations supports the hypothesis of the modulatory role of calcium on the α7 receptor and suggests that this divalent cation might act as a co-agonist. Moreover, determination of the peak induced current as a function of the extracellular calcium demonstrates that this allosteric modulation occurs at physiological concentrations.

Role of divalent cations in channel rectification: One of the key features of many neuronal nAChRs is the voltage dependency of the ACh-evoked current (18, 19, 20). For these receptors rectification results in a reduction or suppression of the neurotransmitter-evoked current at depolarized voltages. Thus, to be active, neuronal nicotinic receptors need both the presence of the neurotransmitter and an adequate resting membrane potential. Therefore, neuronal nAChR can be viewed as coincidence detectors that might play an important role in synaptic transmission and thereby information processing.

Determination of the voltage dependence properties of the $\alpha 7$ receptor under steady state and voltage jump conditions revealed significant differences between the two methods the latter showing little or no rectification. However, the absence of rectification in voltage step experiments can be attributed to the contribution of endogenous calcium dependent chloride current. This was confirmed by intracellular injection of the calcium chelating agent BAPTA (21). Another positive maneuver was the substitution of the extracellular calcium ions with equimolar barium which is known to be less effective in activating chloride currents (22). In previous work (16), we have observed that mutation of a single amino acid at the channel inner mouth, the glutamate residue at location 237 (E237), into an alanine uncharged residue is sufficient to abolish calcium permeability and to suppress rectification(17). Furthermore, we found that mutation of this glutamate by the equivalent charged aspartate residue, which has a smaller lateral chain, significantly decreases rectification.

These data therefore indicate that both charge and size of residues located at the channel inner mouth play an important role in determining the voltage dependency. A possible explanation that could account for these observations is that intracellular divalent cations block the channel by steric hindrance in a way comparable to the effect of extracellular Mg^{2+} described for the NMDA receptors (23). According to this hypothesis magnesium ions would enter the ionic pore under the force exerted by the transmembrane electrical field and block the channel even if it was previously opened by activation of the receptor. This hypothesis and mode of action was later validated by intracellular injection of magnesium chelating agents. Using a number of substances we observed first a correlation between the chelating capacity of the employed compounds and second that rectification could be progressively reduced depending upon the concentration of the chelating agent.

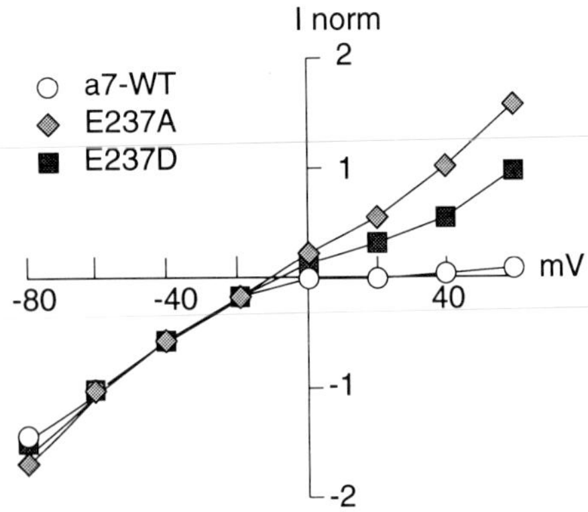

Figure 3. Current-voltage relationship of α7 wild type and mutants at position 237. Mutation of the glutamate residue into asparate reduces rectification.

Conclusions

The α7 nAChR is highly permeable to calcium (more than NMDA) (11, 17), displays fast activation and inactivation kinetics (Buisson and Bertrand, unpublished), and is blocked by intracellular magnesium at positive potentials. Furthermore, α7 was shown to be potently modulated by the extracellular calcium concentration within the physiological range.

Acknowledgments: *We are indebted to J.L. Galzi and A. Devillers-Thiéry for the preparation of most of the gene constructions used in this work and to J.P. Changeux for his collaboration and constructive suggestions. The α7 cDNA was kindly provided by M. Ballivet. This work was supported by grants from the Swiss National Foundation to I.F. and to D.B., and by grants from Human Frontier and the EEC to D.B.. B.B. was a research fellow from the Jules Thorn Foundation.*

Abbreviations: ACh acetylcholine, nAChR nicotinic acetylcholine receptor, NMDA N-methyl-D-aspartate, 5HT serotonin, BAPTA 1,2-bis (2-aminophenoxy) ethane-*N, N,N',N'*-tetraacetic acid.

References

1. Unwin N. The nicotinic acetylcholine receptor at 9 Å resolution. J. Mol. Biol. 1993; 229: 1101-1124.
2. Lal R, Yu L. Atomic force microscopy of cloned nicotinic acetylcholine receptor expressed in *Xenopus* oocytes. Proc. Natl. Acad. Sci. USA 1993; 90: 7280-7284.
3. Devillers-Thi éry A, et al. Functional architecture of the nicotinic acetylcholine receptor: a prototype of ligand-gated ion channels. J. Membrane Biol. 1993; 136: 97-112.
4. Betz H. Ligand-gated ion channels in the brain: the amino acid receptor superfamilly. Neuron 1990; 5: 383-392.
5. Gasic GP, Heinemann S. Determinants of the calcium permeation of ligand-gated cation channels. Current Biology 1992; 4: 670-677.
6. Bertrand D, Cooper E, Valera S, Rungger D, Ballivet M. Electrophysiology of neuronal nicotinic acetylcholine receptors expressed in *Xenopus* oocytes following nuclear injection of genes or cDNA. Conn, M. 1991: 174-193.
7. Bertrand D, Bader CR, DATAC. A multipurpose biological data analysis program based on a mathematical interpreter. Int. J. Bio-medical Computing 1986; 18: 193-202.
8. Revah F, et al. Mutations in the channel domain alter desensitization of a neuronal nicotinic receptor. Nature 1991; 353: 846-849.
9. Bertrand D, et al. Unconventional pharmacology of a neuronal nicotinic receptor mutated in the channel domain. Proc. Natl. Acad. Sci. USA. 1992; 89: 1261-1265.

10. Changeux JP. In: Fidia Research Foundation Neuroscience Award Lectures. Changeux JP, Llinas RR, Purves D, Bloom FE, editors. Raven Press, Ltd., 1990: 21-168.
11. Bertrand D, Galzi J-L, Devillers-Thi éry A, Bertrand S, Changeux J-P. Stratification of the channel domain in neurotransmitter receptors. Current Opinion in Cell Biol. 1993; 5: 688-693.
12. Galzi JL, et al. Allosteric transitions of the acetylcholine receptor probed at the amino acid level with a photolabile cholinergic ligand. Proc. Natl. Acad. Sci. USA. 1991; 88: 5051-5055.
13. Galzi JL, et al. Functional significance of aromatic amino acids from three peptide loops of the α7 neuronal nicotinic receptor site investigated by site-directed mutagenesis. FEBS Lett. 1991; 294: 198-202.
14. Eiselé, JL, et al. Chimearic nicotinic-serotoninergic receptor combines distinct ligand binding and channel specificities. Nature 1993; 366: 479-483.
15. Mulle C, Léna C, Changeux JP. Potentiation of nicotinic receptor response by external calcium in rat central neurons. Neuron 1992; 8: 937-945.
16. Galzi JL, et al. Mutations in the ion channel domain of a neuronal nicotinic receptor convert ion selectivity from cationic to anionic. Nature 1992; 359: 500-505.
17. Bertrand D, Galzi JL, Devillers-Thiery A, Bertrand S, Changeux JP. Mutations at two distinct sites within the channel domain M2 alter calcium permeability of neuronal α7 nicotinic receptor. Proc. Natl. Acad. Sci. USA 1993; 90: 6971-6975.
18. Bertrand D, Ballivet M, Rungger D. Activation and blocking of neuronal nicotinic acetylcholine receptor reconsituted in *Xenopus* oocytes. Proc. Natl. Acad. Sci. USA 1990; 87: 1993-1997.
19. Mathie A, Colquhoun D, Cull-Candy SG. Rectification of currents activated by nicotinic acetylcholine receptors in rat sympathetic ganglion neurons. J. Physiol. (London) 1990; 427: 625-655.
20. Couturier S, et al. A neuronal nicotinic acetylcholine receptor subunit (alpha 7) is developmentally regulated and forms a homo-oligomeric channel blocked by alpha-BTX. Neuron 1990; 5: 845-856.
21. Forster I, Bertrand D. Molecular determinants of inward rectification in α-bungarotoxin sensitive neuronal nicotinic ACh receptors. 1994; in press.
22. Miledi R, Parker I. Chloride current induced by injection of calcium into *Xenopus* oocytes. J. Physiol. 1984; 357: 173-183.
23. Ascher P, Nowak L. The role of divalent cations in the responses of central neurons to N-methyl-D-aspartate. J. Physiol. London. 1988; 399: 247-266.

Effects of Nicotine on
Biological Systems II
Advances in Pharmacological Sciences

EXPRESSION, FUNCTION, AND REGULATION OF NEURONAL ACETYLCHOLINE RECEPTORS CONTAINING THE α7 GENE PRODUCT

Darwin K. Berg, Zhong-wei Zhang, William G. Conroy, Phyllis C. Pugh, Roderick A. Corriveau, Suzanne J. Romano, Margaret M. Rathouz, Bo Huang, and Sukumar Vijayaraghavan

Department of Biology, 0357; University of California, San Diego; La Jolla, CA 92093-0357

Summary: Native acetylcholine (ACh) receptors that contain the α7 gene product function as ligand-gated ion channels that are cation-selective, prefer nicotine over ACh, and rapidly desensitize. Activation of the receptors elevates intracellular calcium levels in neurons, triggering calcium-dependent events such as neurite retraction and the release of arachidonic acid. Arachidonic acid, in turn, reversibly inhibits the receptors, suggesting negative feedback regulation. The α7 gene product is expressed not only in neurons but also in developing muscle, raising the possibility that such receptors mediate cholinergic regulation of calcium-dependent events in a variety of cells.

Introduction

Neuronal nicotinic acetylcholine receptors (AChRs) that bind α-bungarotoxin (α-BTX) are widely distributed throughout the central and peripheral nervous sytems (1,2). Subunit analysis with monoclonal antibodies has revealed that the predominant forms of the receptors (α-BTX-AChRs) contain either α7 subunits, α8 subunits, or both together (3,4). None of the other five neuronal AChR α-type genes (α2-α6) nor the three neuronal β-type genes (β2-β4) identified to date appear to be present in α-BTX-AChRs. Expression studies in *Xenopus* oocytes demonstrate that the α7 gene alone can produce functional AChRs and that the receptors are blocked by α-BTX (5). Despite the discovery of α-BTX-AChRs on neurons over two decades ago and despite the characterization of electrophysiological responses from α7-containing receptors in oocytes, the functional properties of native α-BTX-AChRs remained controversial until recently. Evidence is presented here demonstrating that the receptors function as ligand-gated ion channels that can elevate intracellular calcium levels in neurons and, as a result, have the potential for mediating cholinergic regulation of calcium-dependent events in cells.

Materials and Methods

Whole-cell patch clamp recordings (6) were performed on dissociated ciliary ganglion neurons from 14- to 15-day old chick embryos using rapid application of agonist from large bore multibarrel micropipettes under gravity flow as previously described (7). Exposure of individual cells to arachidonic acid (AA) and other compounds prior to challenging cells with agonist was achieved by using similar large bore micropipettes. Data were collected, analyzed, and compiled as previously described (7).

Drug effects on neurite length were examined by growing 8-day embryonic ciliary ganglion neurons in dissociated cell culture and applying the drugs to individual neurites by pressure from a nearby micropipette as previously described (8). The drugs were applied for 3 ten second intervals 6 minutes apart and neurite length was monitored over a 1 hour period. α-BTX, when present, was bath applied at 50 nM for ≥ 1 hour at 37°C.

Release of AA from neurons in culture was measured by loading the cells overnight with 1 μCi [^{3}H]AA, rinsing, treating with nicotine, collecting the released radioactivity, and quantifying by scintillation counting.

In situ hybridizations were performed on skeletal myotube cultures prepared from embryonic day 11 pectoral muscle (9) using α1 and α7 RNA probes labeled with α[^{35}S]UTP as previously described (10). RNAse protection experiments were performed as previously described (11).

Results and Discussion

Current responses from native α-BTX-AChRs: Chick ciliary ganglion have two major classes of AChRs. One class, termed α-BTX-AChRs, binds α-BTX, is located primarily in non-synaptic membrane on the neurons, and contains α7 subunits (12). The other class, termed mAb 35-AChRs, binds the monoclonal antibody mAb 35, is concentrated primarily in postsynaptic membrane, and collectively contains α3, β4, and α5 gene products. When nicotinic agonists are applied to ciliary ganglion neurons by pressure from a micropipette, whole-cell patch clamp recording reveals relatively long-lasting responses that can be attributed to mAb 35-AChRs. Rapid application of agonist elicits a large, rapidly decaying response in addition to slower components (Fig. 1A). The rapidly decaying response can be completely blocked by preincubation of the neurons in 60 nM α-BTX (Fig. 1B), indicating that the response arises from α-BTX-AChRs.

Dose-response curves yielded an apparent EC_{50} of about 10 μM for nicotine; much higher concentrations of ACh were required to induce equivalent currents that could be blocked by α-BTX. The rapidly-decaying responses were completely and reversibly abolished by 20 μM d-tubocurarine and by 20 μM strychnine, as previously reported for α7-containing AChRs expressed in *Xenopus* oocytes (13). Current-voltage plots revealed a reversal potential of about 0 mV for currents arising from α-BTX-AChRs. The receptors were cation-selective since subsitution of sodium isethionate (130 mM) for sodium chloride in the extracellular solution had no effect on the reversal potential. Internal perfusion of the cells with 10 mM BAPTA had no effect on the nicotine-induced currents measured, indicating that calcium-activated currents are not likely to make a significant contribution to the signal. The results show that α-BTX-AChRs are ligand-gated, cation-selective, rapidly-desensitizing ion channels. Previous failures to detect currents from α-BTX-AChRs on neurons probably resulted from their rapid desensitization and their relatively poor affinity for ACh which would have permitted other classes of AChRs to dominate the response.

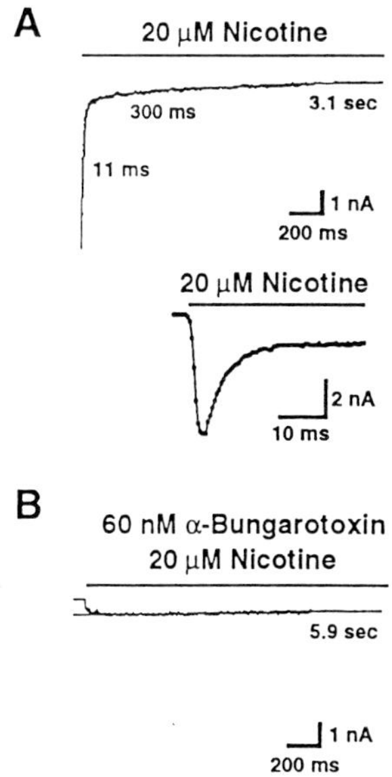

Figure 1. Selective blockade of the rapid nicotinic response by α-BTX. A. Whole-cell current from a cell held at -60 mV and challenged with 20 μM nicotine. Nicotine produced an inward current with a peak of -6.37 nA which decayed rapidly (see inset for detail). The decay of nicotine-evoked current was best fit with three exponentials (solid line) having time constants of 11 ms, 300 ms, and 3.1 sec, respectively. B. Nicotine-induced whole-cell current from a cell incubated 1 hour with 60 nM α-BTX prior to testing. The response shows a peak current of -0.67 nA, and the decay is adequately fit with a single exponential (solid line) having a time constant of 5.9 sec. α-BTX completely abolished the fast-decaying current seen in panel A. Reprinted from (7).

Neurite retraction induced by α-BTX-AChRs: Activation of α-BTX-AChRs on ciliary ganglion neurons can substantially elevate intracellular calcium levels (14). This, together with the fact that the receptors are located on pseudodendrites on the neurons in vivo (15), suggested that the receptors might influence morphogenic events such as neurite extension and retraction which are known to depend on intracellular calcium levels (16). This was tested by growing dissociated 8-day embryonic ciliary ganglion neurons in cell culture, and focally applying brief pulses of 20 μM nicotine directly to the tips of individual neurites. Nicotine reversed the forward growth. On average the neurites retracted about 35% during the one hour test period rather than growing the net 20% observed in control cultures (Fig. 2). Application of nicotine in the absence of calcium had no effect on growth. Pretreatment of the neurons with 50 nM α-BTX largely blocked the nicotine effect, showing that it was dependent on α-BTX-AChRs. d-Tubocurarine also blocked it as did ω-conotoxin, the latter implicating voltage-gated calcium channels. Apparently α-BTX-AChRs depolarize the membrane sufficiently to recruit contributions from voltage-gated channels that help induce the neurite retraction.

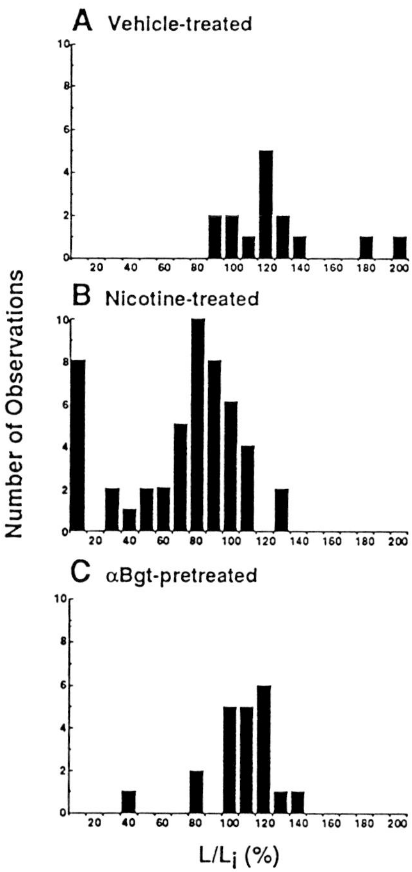

Figure 2. Histogram of neurite lengths following treatments to induce retraction. Neurites were stimulated with vehicle alone (A), vehicle plus nicotine (B), or vehicle plus nicotine after blockade with α-BTX (C). The relative change in length for each neurite was calculated by expressing the length at the sustained maximum retraction (L) as a percent of the initial length (L_i) at the beginning of stimulation. When neurite length continually changed during the test period and did not undergo a sustained retraction, the relative length was calculated by expressing the final length after 1 hour as a percent of the initial length. The values have been pooled into bins of 10%,. e.g. the 20% bin included all neurites that retracted to 11-20% of their initial lengths. The results show that nicotine treatment shifts the distribution of neurite lengths to shorter values and that α-BTX treatment partially blocks the effect. Reprinted from (8).

Arachidonic acid and possible negative feedback regulation: The ability of α-BTX-AChRs to elevate intracellular calcium in neurons presents an opportunity for regulatory cascades. One such cascade may involve arachidonic acid as a second messenger. Nicotine induces AA release from neurons preloaded with [^{3}H]AA. The nicotine-induced release is blocked by α-BTX, d-tubocurarine, and removal of extracellular calcium (Fig. 3A). Maximal release is induced by 50 μM nicotine and occurs within 2 minutes.

AA released by nicotine may serve as a negative feedback regulator because it can reversibly inhibit the receptors. Quantifying only the rapidly decaying response elicited by nicotine indicates that a 10 second application of AA at 5 μM reversibly inhibits about 90% of the α-BTX-AChR response (Fig. 3B). Inhibition of the slower response attributed to mAb 35-AChRs is less pronounced, required 20 μM AA to achieve a 50-60% inhibition. α7-Containing AChRs expressed in *Xenopus* oocytes behave like α-BTX-AChRs; 10 μM AA induces an 80% inhibition of the response. The results demonstrate that activation of α-BTX-

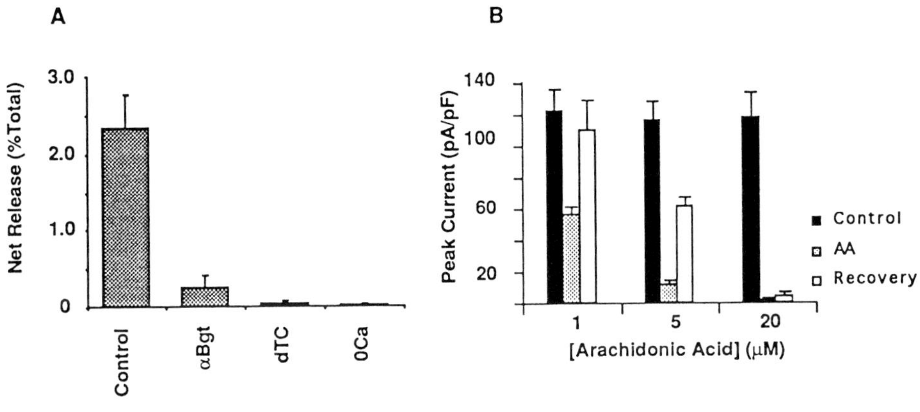

Figure 3. A. Nicotine-induced release of [^{3}H]AA from neurons in culture. Ciliary ganglion neurons from 8-day embryos were grown for 6 days in culture and then incubated overnight in 1 μCi [^{3}H]AA, rinsed, and incubated in the indicated test conditions prior to and during a 5 minute exposure to 50 μM nicotine. The amount of radioactivity released was determined and expressed as a percent of the total initially associated with the cells. Values represent the mean ± SEM of 6-11 cultures pooled from 3 separate experiments. Control, no blockers; α-BTX, 60 nM α-BTX for 40 minutes; d-TC, 20 μM d-tubocurarine for 5 minutes; 0 Ca, 0.5 mM EGTA in place of the calcium normally present in the medium. Total mean radioactivity associated with the neurons at the outset of the release period was 3.7 ± 0.1 x 10^5 cpm/culture (mean ± SEM, n = 32 cultures). B. Concentration dependence of AA inhibition of α-BTX-AChR responses. Whole cell currents induced by 10 μM nicotine in neurons before, during, and after application of the indicated concentrations of AA were recorded and compared. The rapidly decaying peak current that arises solely from α-BTX-AChR activation was measured in each of the conditions and was found to be reversibly inhibited by AA.

AChRs on neurons can trigger second messenger cascades and, in the case of AA, the second messenger can act back on the receptors to inhibit them, potentially exerting negative feedback control.

Expression of α7 gene product in developing muscle: The α7 gene has been designated neuronal because it is thought to be expressed only in neurons. RNAse protection assays reveal, however, that the α7 gene is expressed both in skeletal myotubes in culture and in embryonic muscle including pectoral and leg samples. Embryonic day 8 pectoral muscle contains about 20% as many α7 transcripts as α1 transcripts. The levels peak at embryonic day 11 and decline almost to background by embryonic day 17. In situ hybridizations with α7 antisense probe confirm that α7 mRNA is present in skeletal myotubes in cell culture (Fig. 4). Grains are found over multinucleated cells as well as over mononuclear cells that may represent myocytes. The level of labeling varies significantly among nuclei even within the same myotube. Similar results were obtained with tissue sections prepared from embryonic day 11 leg and pectoral muscles. Grains were detected over myotubes as well as presumptive myoblasts among the interstitial cells present. Only background levels of grains were observed over extracellular spaces and over cells hybridized with α7 sense probe.

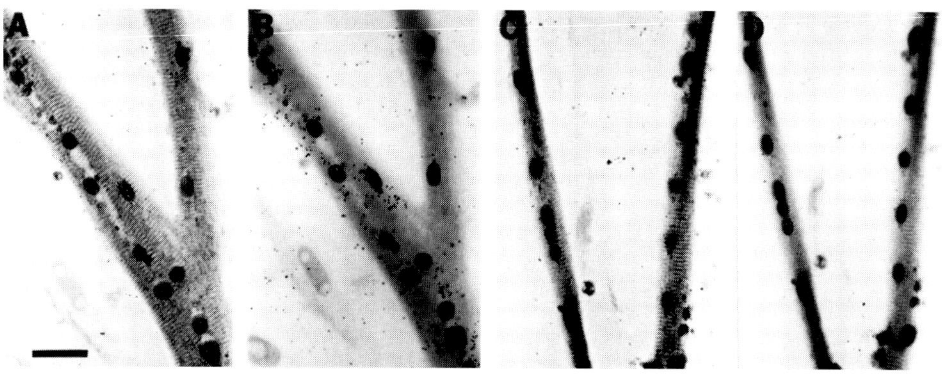

Figure 4. In situ hybridization showing α7 transcripts in skeletal myotubes in culture. Cultures of skeletal muscle were hybridized with [^{35}S]labeled probe, coated with emulsion, exposed for four weeks, developed, and stained with hematoxylin and eosin. (A, C) Phase contrast optics with the plane of focus on the cell layer; (B, D) bright field optics with the plane of focus on the silver grains. (A,B) α7 antisense probe hybridized to myotubes. Heterogeneity of α7 expression among nuclei within the same myotube was also evident. Background levels of silver grains were observed over extracellular spaces and (C, D) over cells hybridized with α7 sense probe. Scale bar, 30 μm.

The α7 transcript present in muscle is translated into protein. Solid phase immunoprecipitation assays using mAbs 318 and 319, which are specific for the α7 protein (3), confirm the present of α7 protein in embryonic leg and pectoral muscle. Sucrose gradient sedimentation followed by solid phase assays on the fractions indicate that the α7-containing species in muscle has the size expected for a fully assembled receptor (9-10S). Cross-precipitations with anti-α7 versus anti-α1 mAbs demonstrate that the two kinds of α-type gene products are not co-assembled in muscle. Competition binding studies using cholinergic ligands to compete against $[^{125}I]$α-BTX for binding to α7-containing muscle components tethered with mAbs 318 and 319 indicate a pharmacological profile similar to that of ciliary ganglion α-BTX-AChRs and clearly different from conventional α1-containing muscle AChRs. The results are consistent with α7-containing receptors being expressed at early stages in embryonic muscle and suggest a role for them in muscle development.

The ability of α-BTX-AChRs to elevate intracellular calcium provides a mechanism by which cholinergic signalling could achieve pleiotropic effects through second messenger cascades. The presence of α-BTX-AChRs both in neurons and in non-neuronal cells implies the cascades may have general significance.

Acknowledgements: *We thank Lynn Ogden for expert technical assistance, and Everado Gutierrez, Michael Silverberg, Cynthia Brent, and Leticia Oliva for dissections and preparation of cell cultures. Grant support was provided by the National Institutes of Health (NS12601 and NS25916) and by the California Tobacco-Related Disease Research Program.*

References

1. Clarke PBS. The fall and rise of neuronal α-bungarotoxin-binding proteins. Trends in Pharmacol Sci 1992; 13:407-413.
2. Sargent PB. The diversity of neuronal nicotinic acetylcholine receptors. Annu Rev Neurosci 1993; 16:403-443.
3. Schoepfer R, Conroy WG, Whiting P, Gore M, Lindstrom J. Brain α-bungarotoxin binding protein cDNAs and MAbs reveal subtypes of this branch of the ligand-gated ion channel gene superfamily. Neuron 1990; 5:35-48.
4. Anand R, Peng X, Ballesta J, Lindstrom J. Pharmacological characterization of α-bungarotoxin-sensitive acetylcholine receptors immunoisolated from chick retina: contrasting properties of α7 and α8 subunit-containing subtypes. Molec Pharm 1993; 44:1046-1050.
5. Couturier S, Bertrand D, Matter J-M, Hernandez M-C, Bertrand S, Millar N, Valera S, Barkas T, Ballivet M. A neuronal nicotinic acetylcholine receptor subunit (α7) is developmentally regulated and forms a homo-oligomeric channel blocked by α-BTX. Neuron 1990; 5:847-856.
6. Hamill OP, Marty A, Neher E, Sakmann B, Sigworth FJ. Improved patch clamp techniques for high-resolution current recordings from cells and cell-free membrane patches. Pfluegers Arch 1981; 291:85-100.

7. Zhang Z-w, Vijayaraghavan S, Berg DK. Neuronal acetylcholine receptors that bind α-bungarotoxin with high affinity function as ligand-gated ion channels. Neuron 1994; 12:167-177.
8. Pugh PC, Berg DK. Neuronal acetylcholine receptors that bind α-bungarotoxin mediate neurite retraction in a calcium-dependent manner. J Neurosci 1994; 14:889-896.
9. Fischbach GD. Synapse formation between dissociated nerve and muscle cells in low density cell cultures. Devel Biol 1972; 28:407-429.
10. Simmons DM, Arriza JL, Swanson LW. A complete protocol for *in situ* hybridization of messenger RNAs in brain and other tissues with radiolabeled single-stranded RNA probes. J Histotechnol 1989; 12: 169-181.
11. Corriveau RA, Berg DK. Coexpression of multiple acetylcholine receptor genes in neurons: quantification of transcripts during development. J Neurosci 1993; 13: 2662-2671.
12. Vernallis AB, Conroy WG, Berg DK. Neurons assemble acetylcholine receptors with as many as three kinds of subunits while maintaining subunit segregation among receptor subtypes. Neuron 1993; 10:451-464.
13. Seguela P, Wadiche J, Dineley-Miller K, Dani JA, Patrick JW. Molecular cloning, functional properties, and distribution of rat brain α7: a nicotinic cation channel highly permeable to calcium. J Neurosci 1993; 13:596-604.
14. Vijayaraghavan S, Pugh PC, Zhang Z-w, Rathouz MM, Berg DK. Nicotinic receptors that bind α-bungarotoxin on neurons raise intracellular free Ca^{2+}. Neuron 1992; 8:353-362.
15. Jacob MH, Berg DK. The ultrastructural localization of α-bungarotoxin binding sites in relation to synapses on chick ciliary ganglion neurons. J Neurosci 1983; 3:260-271.
16. Kater SB, Mills LR. Regulation of growth cone behavior by calcium. J Neurosci 1991; 11:891-899.

Nicotinic receptor regulation and tolerance

Effects of Nicotine on
Biological Systems II
Advances in Pharmacological Sciences

BIOCHEMICAL MEASURES OF NICOTINIC RECEPTOR DESENSITIZATION

Michael J. Marks, Sharon R. Grady, Scott F. Robinson,
Amy E. Bullock, and Allan C.Collins

Institute for Behavioral Genetics, University of Colorado, Boulder, CO, USA

Summary: Exposure to either nonactivating (nM) or activating (μM) concentrations of nicotine desensitized nicotine-stimulated [^{3}H]dopamine release from striatal synaptosomes and $^{86}Rb^+$ efflux from thalamic synaptosomes, but the kinetics of desensitization were not identical. Although nicotine-stimulated $^{86}Rb^+$ efflux was completely inhibited by exposure to either low or high concentrations of nicotine, a significant fraction of [^{3}H]dopamine release persisted. The kinetics of functional desensitization was similar to that for high affinity [^{3}H]nicotine binding.

Introduction

It has been nearly 40 years since Katz and Thesleff (1) observed that prolonged exposure of motor endplate to nicotine desensitized the response. Their proposal of a cyclical model of activatable and desensitized receptor forms, with the desensitized form of the receptor possessing higher affinity for agonists, has been a remarkably durable first approximation of nicotinic receptor desensitization. High affinity binding of agonists to the desensitized state of the receptor was invoked to explain why nicotine and other agonists bind to putative nicotinic receptors at concentrations much lower than those required to elicit a response to the agonist (2). Chronic receptor desensitization has also been offered as the explanation for the paradoxical increase in the number of nicotinic binding sites after chronic treatment with nicotine and other nicotinic agonists (3,4).

The assumption that brain nicotinic receptors will desensitize after chronic exposure to agonists, in analogy to the desensitization observed with muscle-type nicotinic receptors, is not unreasonable. However, until relatively recently, methods to measure the desensitization of CNS nicotinic receptors have not been available. The results reported in this manuscript used two methods recently described to measure nicotinic receptor function in mouse brain ([^{3}H]dopamine release from striatal synaptosomes (5) and $^{86}Rb^+$ efflux from midbrain synaptosomes (6)). The results presented below compare the properties of receptor desensitization determined on a sec to min time scale.

Methods

Crude synaptosomes (P2 fractions) were prepared from defined brain regions of female C57BL/6 mice, loaded with radioactive tracers, and superfused as described previously (5,6). Desensitization following prolonged stimulation was measured using 0.1 - 100 μM nicotine. Samples were exposed to 1 - 200 nM nicotine before stimulation with 10 μM nicotine to determine the effects of nonstimulating concentrataions. The time course of [^{3}H]nicotine binding at 22°C was measured using the method of Marks et al. (7).

Primary results from each individual experiment were analyzed using nonlinear least squares curve fitting with SigmaPlot or Ligand. In addition, the kinetics of desensitization of $^{86}Rb^+$ efflux and [^{3}H]nicotine binding were analyzed using SimuSolve with the assistance of J.-Y. Yang and P.M. Lippiello.

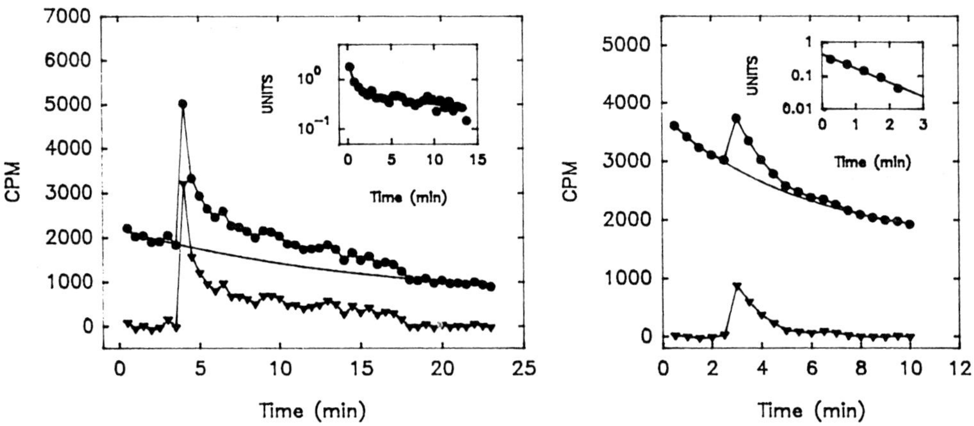

Figure 1. Time courses for the release of [^{3}H]dopamine (left) and efflux of $^{86}Rb^+$ (right) were determined for synaptosomes exposed to 3 μM nicotine for 12 and 5 min, respectively. The upper trace in the main panels shows actual data (•) while the lower trace shows data after subtraction of baseline (▾). Insets of each figure are semilogarithmic plots for nicotine-stimulated response. Data derived from ref. 13.

Results

Comparison of Desensitization Properties of Nicotine-Stimulated [³H]Dopamine Release and ⁸⁶Rb⁺ Efflux. Comparison of a representative time course for [³H]dopamine release from mouse striatal synaptosomes (Figure 1, left panel) with one for ⁸⁶Rb⁺ efflux from midbrain synaptosomes (Figure 1, right panel) after stimulation with 3 μM nicotine indicates that both of these responses diminish with time of exposure to nicotine. However, while complete desensitization of nicotine-induced ⁸⁶Rb⁺ efflux occurred within 5 min, a substantial fraction of nicotine-stimulated [³H]dopamine release persisted even after 10 min of exposure.

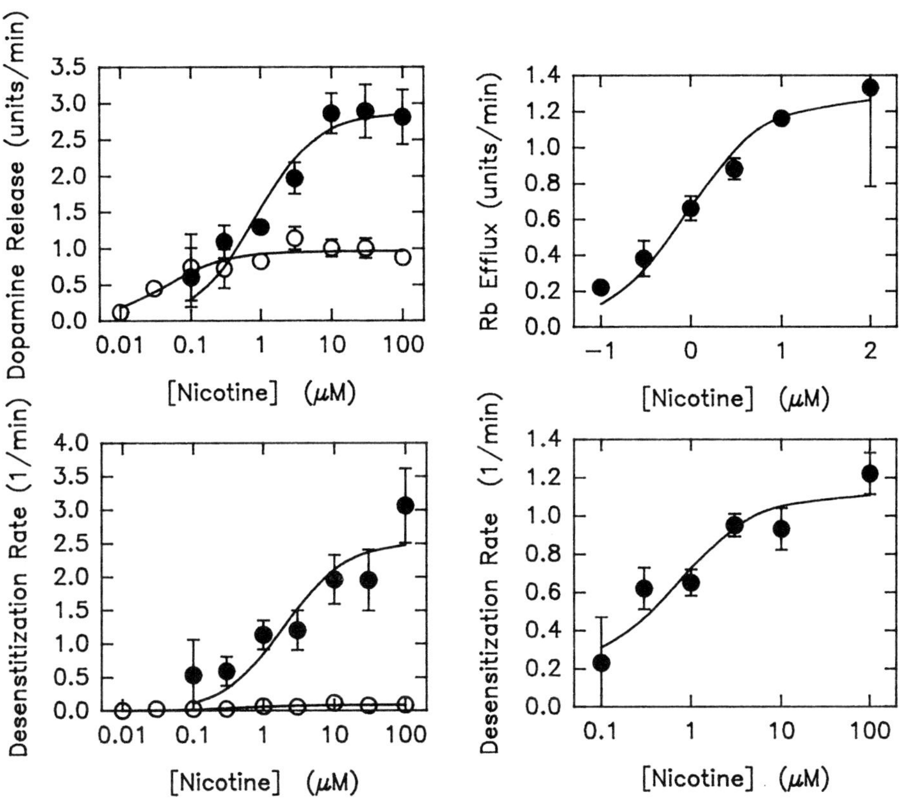

Figure 2. The concentration dependence of stimulation of response (upper panels) and desensitization of response (lower panels) for nicotine-stimulated [³H]dopamine release (left panels) and ⁸⁶Rb⁺ efflux (right panels) are shown. Curves for both the persistent (○) and transient (•) phases of [³H]dopamine release are displayed. Data derived from refs. 13 and 14.

The desensitization properties for both [^{3}H]dopamine release and $^{86}Rb^+$ efflux at several nicotine concentrations were measured. Concentration-dependent increases in initial response and desensitization rate were observed. Two different curves were indicated for the nicotine-stimulated [^{3}H]dopamine release: one corresponding to the persistent (very slowly desensitizing) release that displayed an EC_{50} about 30 nM and a second corresponding to the transient (desensitizing) release that displayed an EC_{50} of about 2 μM. Nicotine-stimulated $^{86}Rb^+$ efflux displayed only the transient process with an EC_{50} for activation of 0.8 μM. The desensitization rate for the persistent phase of [^{3}H]dopamine release was low and virtually unchanged when the nicotine concentration was increased. However, the apparent desensitization rates for the transient component of [^{3}H]dopamine release and $^{86}Rb^+$ efflux increased with nicotine concentration (EC_{50} = 0.8 μM for both).

The effect of the low concentrations on the subsequent response to a stimulating concentration of nicotine is summarized in Figure 3. A concentration-dependent decrease in response was observed for both [^{3}H]dopamine release and $^{86}Rb^+$ efflux after exposure to nM concentrations of nicotine, with IC_{50} values of 7 nM and 21 nM, respectively. While an maximal inhibition of 97% was estimated for $^{86}Rb^+$ efflux, the maximal inhibition of [^{3}H]dopamine release was 75%.

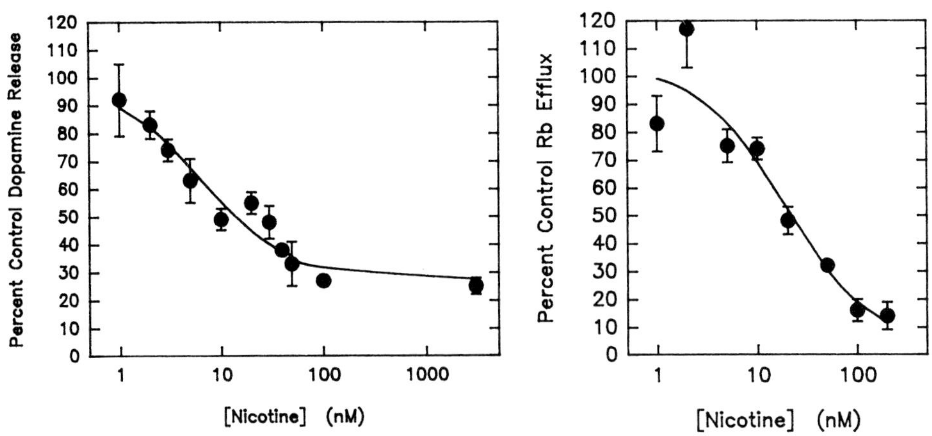

Figure 3. Concentration-effect curves for inhibition of the response to a 1-min test stimulation by with 10 μM nicotine by prior exposure to nM concentrations of nicotine were determined for both [^{3}H]dopamine release (left panel) and $^{86}Rb^+$ efflux (right panel). Data derived from refs. 13 and 14.

Comparison of Kinetic Properties of Nicotine-Stimulated $^{86}Rb^+$ Efflux and [^3H]Nicotine Binding. Initial characterization of nicotine-stimulated $^{86}Rb^+$ efflux from mouse brain synaptosomes indicated that the relative potency of several nicotinic agonists and the variation among several brain regions of the efflux process were very similar to those of [^{3}H]nicotine binding. These relationships suggested that under the conditions of the experiments that the $^{86}Rb^+$ efflux may be a functional expression of those receptors measured by high affinity [^{3}H]nicotine binding. The kinetics of [^{3}H]nicotine binding to either rat (8) or mouse brain (9) deviates from the properties expected for a simple bimolecular process. The time courses for nicotine association of 5 nM or 50 nM [^{3}H]nicotine to mouse midbrain membranes are illustrated in Figure 4. The binding measured using 50 nM [^{3}H]nicotine is higher and occurs more rapidly than that measured at 5 nM. The representation of the data as semilogarithmic plots (Figure 4, inset) illustrates the biphasic nature of the binding in that both a fast and slow component of the process is observed. Note that fast phase binding observed at 50 nM is virtually complete at the first time point (15 s). Also note that slow phase binding is more rapid at 50 nM nicotine (k_{app} = 0.88 s^{-1}) than at 5 nM (k_{app} = 0.26 s^{-1}). A monophasic dissociation of bound [^{3}H]nicotine is observed independent of the concentration used to achieve the equilibrium binding. Results for the time courses for [^{3}H]nicotine binding at concentrations between 1 nM and 200 nM and for nicotine-stimulated $^{86}Rb^+$ efflux at 6 concentrations were analyzed to calculate the 8 rate constants proposed by the two-state Katz-Thesleff model. The model and the estimated rate constants are summarized in Table 1. Constants calculated for the two processes are comparable.

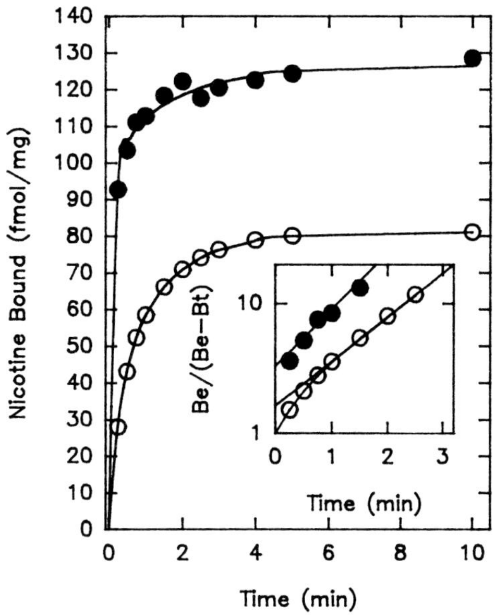

Figure 4. Time course for the binding of [^{3}H]nicotine to midbrain fractions measured using 5 μM (○) or 50 nM (•) is shown. Data derived from ref. 14.

Table 1. Two-state Katz-Thesleff model and the kinetic parameters derived from the nonlinear curve fitting of the time courses for $^{86}Rb^+$ efflux (0.1 - 100 μM) and the time courses for [^{3}H]nicotine binding (1 nM - 200 nM). Data derived from ref. 14.

MODEL

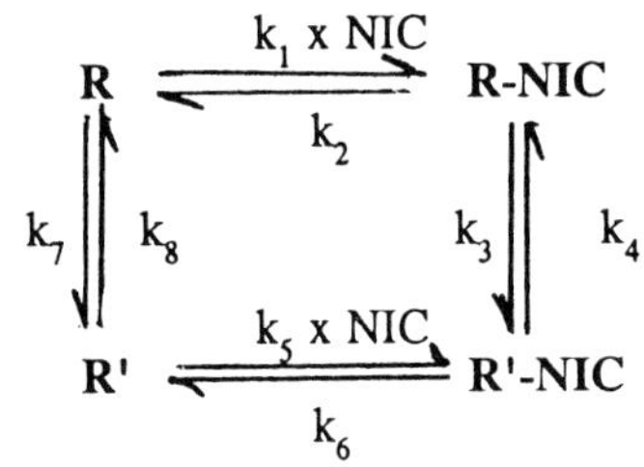

CALCULATED CONSTANTS

Parameter	$^{86}Rb^+$ Efflux	[^{3}H]Nicotine Binding
k_1(nM^{-1}min^{-1})	0.57	0.51
k_2(min^{-1})	199.	235.
k_3(min^{-1})	1.68	1.26
k_4(min^{-1})	0.003	0.003
k_5(nM^{-1}min^{-1})	0.31	0.32
k_6(min^{-1})	0.72	0.77
k_7(min^{-1})	0.30	0.34
k_8(min^{-1})	0.15	0.15

Discussion

Initial characterization of nicotine-stimulated [^{3}H]dopamine release from mouse striatal and $^{86}Rb^+$ efflux from mouse midbrain synaptosomes indicated that these processes displayed pharmacological properties consistent with those expected for nicotinic receptors. However, differences in the relative potency of several agonists and in the inhibition by neuronal bungarotoxin ([^{3}H]dopamine release sensitive, $^{86}Rb^+$ efflux insensitive) suggest that the nicotinic receptor subtypes mediating these two responses may be different. Several properties measured for these processes are very similar. The EC_{50} for both receptor activation (about 1 μM) and desensitization by activating concentrations (about 1 μM) are indistinguishable, as are rates for desensitization after exposure to 30 nM nicotine (about 0.35 min^{-1}). However, some of the kinetic parameters measured for [^{3}H]dopamine release and $^{86}Rb^+$ are different: maximal desensitization rate (2.5 min^{-1} and 1.1 min^{-1}, respectively), IC_{50} for desensitization by low nicotine concentrations (6.5 nM and 19 nM), and maximal inhibition after exposure to either nonstimulating or stimulating concentrations of nicotine (75% and 100%).

Kinetic modeling of the desensitization of $^{86}Rb^+$ efflux and [^{3}H]nicotine binding as summarized in Table 1 indicates that the simple two-state model of Katz and Thesleff adequately explains the data and generates estimates of the rate constants that are virtually identical for the two measurements. This result is consistent with previously reported similarities between nicotine binding and $^{86}Rb^+$ efflux (6).

The persistence of substantial nicotine-stimulated [^{3}H]dopamine release after prolonged exposure to nicotine may have interesting consequences, especially if such actions are important in mediating the reinforcing properties of nicotine. The reason for the persistence of [^{3}H]dopamine response after desensitization is not yet known. Presuming that the response is receptor mediated, several hypotheses for the persistent response can be advanced. A first hypothesis would propose that at least two receptor subtypes are being measured. One subtype would exhibit properties similar to those measured for $^{86}Rb^+$ efflux for which complete desensitization is observed after exposure to either low or high concentrations of nicotine. A second subtype would have the relatively unusual properties of being activated at low nicotine concentrations (EC_{50} about 40 nM) and undergoing little desensitization. A second hypothesis would propose that a single receptor subtype is involved, that the Katz-Thesleff model holds, but that the desensitized receptor form (R'-NIC in Table 1) can stimulate release, but at a lower rate than that of the activated receptor (R-NIC). A third hypothesis would propose that only the activated receptor stimulates release, but that a significant fraction of desensitized liganded receptor (R'-NIC) can be converted to activated receptor (R-NIC). In the context of the two-state model, this hypothesis predicts that k_3 and k_4 are of similar magnitude. (Note that k_3 calculated for $^{86}Rb^+$ efflux was approximately 500 fold larger than k_4.) Additional data will be required before detailed kinetic modeling of the desensitization of [^{3}H]dopamine release can be attempted. Unpublished studies examining the effects of nicotinic antagonists on both the transient and persistent components of nicotine-stimulated [^{3}H]dopamine release have not identified any substantial differences in the K_i values for either competitive or noncompetitive inhibitors. This result is consistent with, but surely does not prove the hypothesis that the two components are different kinetic manifestations of the same receptor. Considering that many nicotinic receptor subtypes may be expressed in striatum, any mechanism proposing the action of a single receptor subtype in modulating [^{3}H]dopamine release must be regarded cautiously.

Rigorous examination of the role of nicotinic receptor subtypes in the regulation of functional responses will be required to establish which subtypes contribute to each of the functional responses. Initial pharmacological examination and the similarity of the desensitization kinetics would suggest that the $^{86}Rb^+$ efflux is primarily mediated by the same receptor that is measured with high affinity [^{3}H]nicotine binding, presumably the α4ß2 subtype (10,11). The sensitivity of [^{3}H]dopamine release to neuronal bungarotoxin suggests that α3 subunits may be contained in the receptor(s) that mediate this response (12). However, the relatively low EC_{50} for stimulation of the release, as well as the ability to induce desensitization with low nicotine concentrations suggests that α4 receptor subunits may be present. Evaluation of the functional responses in mice that have altered expression, for example using anti-sense oligonucleotide treatment strategies, will be extremely useful for the evaluation of the relationship between nicotinic receptor subunit expression and nicotinic receptor function in the central nervous system.

Acknowledgments: *The support of grant DA-03194 from NIDA is much appreciated.*

References

1. Katz B, Thesleff S. A study of the 'desensitization' produced by acetylcholine at the motor end-plate. J Physiol (Lond) 1957; 138:63-80.
2. Romano C and Goldstein A. Stereoselective nicotine receptors in rat brain. Science (Wash DC) 1980; 210:647-649.
3. Marks MJ, Burch JB, Collins AC. Effects of chronic nicotine infusion on tolerance development and cholinergic receptors. J Pharmacol Exp Ther 1983; 226:806-816.
4. Schwartz RC, Kellar KJ. In vivo regulation of [^{3}H]acetylcholine recognition sites in brain by cholinergic drugs. J Neurochem 1985; 45:427-433.
5. Grady SR, Wonnacott S, Marks MJ, Collins AC. Characterization of nicotinic receptor-mediated [^{3}H]dopamine release from synaptosomes prepared from mouse striatum. J Neurochem 1992; 59:848-856.
6. Marks MJ, Farnham DA, Grady SR, Collins AC. Nicotinic receptor function determined by stimulation of rubidium efflux from mouse brain synaptosomes. J Pharmacol Exp Ther 1993; 264:542-552.
7. Marks MJ, Stitzel JA, Romm E, Wehner JM, Collins AC. Nicotinic binding sites in rat and mouse brain: Comparison of acetylcholine, nicotine and α-bungarotoxin. Mol Pharmacol 1986; 30:427-436.
8. Lippiello PM, Sears SB, Fernandez KG. Kinetics and mechanism of L-[^{3}H]nicotine to putative high affinity receptor sites in rat brain. Mol Pharmacol 1987; 31:392-400.
9. Bhat RB, Marks MJ, Collins AC. Effects of chronic nicotine infusion on kinetics of high-affinity nicotine binding. J Neurochem 1994; 62:574-581.
10. Lindstrom J, Schoepfer R, Conroy W, Whiting P, Das M, Saedi M, Anand R. The nicotinic acetylcholine receptor gene family: Structure of nicotine receptors from muscle and neurons and neuronal α-bungarotoxin binding proteins. Adv Exp Med Biol 1991; 287:255-278.
11. Flores CM, Rogers SW, Pabreza LA, Wolfe BB, Kellar KJ. A subtype of nicotinic cholinergic receptor in rat brain is composed of α4 and ß2 subunits and is up-regulated by chronic nicotine treatment. Mol Pharmacol 1992; 41:31-37.
12. Luetje CW, Wada K, Rogers S, Abramson SN, Tsuji K, Heinemann S, Patrick J. Neurotoxins distinguish between different neuronal nicotinic acetylcholine receptor subtype combinations. J Neurochem 1990; 55:632-640.
13. Grady SR, Marks MJ, Collins AC. Desensitization of nicotine-stimulated [^{3}H]dopamine release from mouse striatal synaptosomes. J Neurochem 1994; 62:1390-1398.
14. Marks MJ. Grady SR, Yang J-M, Lipiello PM, Collins AC. Desensitization of nicotine-stimulated $^{86}RB^+$efflux from mouse striatal synaptosmomes. J Neuochem (in press).

Effects of Nicotine on
Biological Systems II
Advances in Pharmacological Sciences

THE ROLE OF DESENSITIZATION IN CNS NICOTINIC RECEPTOR FUNCTION

P.M. Lippiello, M. Bencherif and R.J. Prince*

R&D Department, R.J. Reynolds Tobacco Co., Winston-Salem, NC 27102 USA
*Integrated Toxicology Program, Duke University, Durham, NC 27710 USA

Summary: Kinetic parameters describing [^{3}H]-nicotine binding to high affinity nicotinic receptors in rat brain were very good predictors of activation and desensitization of [^{3}H]-dopamine release from striatal synaptosomes. Equilibrium binding affinities of nicotine and other agonists were not good predictors of their efficacy or potency in evoking dopamine release, but their potencies for desensitization of release were highly correlated with binding. A simple quaternary amine (tetramethylammonium chloride) was found to fully activate and desensitize dopamine release. We conclude that high affinity receptor binding is a good predictor of relative agonist potencies for functional desensitization and a positive charge center is structurally sufficient to elicit both activation and desensitization.

Introduction

Nicotinic cholinergic receptors serve as the prototypical representative of the family of ligand-gated ion channels. Recent studies using cloned subtypes of these receptors point to the potential structural and functional diversity of these receptors, particularly in the central nervous system (1,2). One of the most abundant and thoroughly studied of the brain subtypes of nicotinic receptor, now identified as having an $\alpha 4\beta 2$ subunit configuration (3), is unique in that it binds nicotine and many other agonists with high affinity (4,5). As such, it has been relatively straightforward to characterize the equilibrium and kinetic binding properties of this particular subtype (6,7). The kinetics of ligand binding to the high affinity nicotinic receptors in rat brain can be described by a two-state model whereby nicotinic agonists can bind to either of two pre-existing receptor conformations, one (low affinity) presumed to be activatable and the other (high affinity) desensitized (8,9). The findings indicate that nicotine tends to stabilize the receptors in the high affinity conformation, prompting the suggestion that high affinity binding is related to the process of functional desensitization (10). In the present studies we have tested the predictive value of the two-state model with respect to activation and desensitization of ligand-evoked release of dopamine from rat brain striatal synaptosomes and probed the relationship of ligand structure to these processes

Methods

The kinetics of binding of [^{3}H]-nicotine to rat brain membrane preparations were measured and analyzed as previously described by Lippiello et al. (9). Ligand-evoked release of [^{3}H]-dopamine from rat brain striatal synaptosomes was monitored using an open superfusion system according to the method of Grady et al. (11), except that higher flow rates (4 ml/min) and more rapid fraction collection (10 sec) were utilized. Ligands utilized in these studies were either commercially available or were synthesized according to published methods.

Results

The association and dissociation kinetics for [^{3}H]-nicotine binding to high affinity nicotinic receptors in rat brain homogenates were measured to obtain estimates of the eight rate constants (k1...k8) which describe the two-state model shown below:

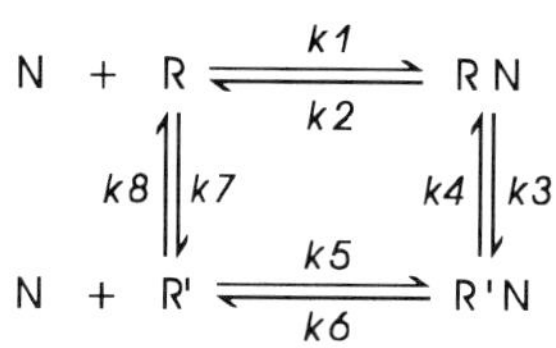

$k1 = 0.3\ \text{min}^{-1}\text{nM}^{-1}$ $k5 = 0.3\ \text{min}^{-1}\text{nM}^{-1}$

$k2 = 45\ \text{min}^{-1}$ $k6 = 0.4\ \text{min}^{-1}$

$k3 = 1.6\ \text{min}^{-1}$ $k7 = 0.12\ \text{min}^{-1}$

$k4 = 0.02\ \text{min}^{-1}$ $k8 = 0.18\ \text{min}^{-1}$

where R and R' are the activatable and desensitized states of the receptors, respectively, and N=nicotine. According to the model, nicotine (or other agonists) can bind to R, leading to receptor activation and related responses (e.g. neurotransmitter release), or to spontaneously desensitized receptors (R'), with no functional consequences. In rat brain we have estimated that around 60% of the receptors are in the R state and 40% in the R' state (9). At saturating concentrations of nicotine essentially all of the receptors are converted to the high affinity state (R'N) within approximately 5 min at 22°C. This is predictive of a concomitant decrease in activated receptors, RN. To test the ability of the model to predict functional desensitization, nicotine-evoked dopamine release was monitored during continuous exposure to ligand (Fig. 1).

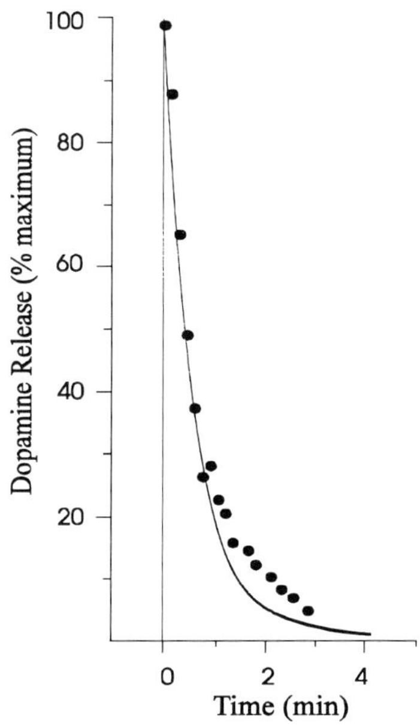

Figure 1. Desensitization of dopamine release by nicotine. Rat brain striatal synaptosomes were loaded with [^{3}H]-dopamine and superfused with 10 μM S-(-)-nicotine. Closed circles represent the amount of dopamine released in each successive 10 sec fraction over a period of 3 min, normalized to the peak response. The solid line is a theoretical curve for the amount of activated receptor (RN), generated from an integrated solution of the two-state model. The rate constants used were determined from nicotine receptor binding kinetics at 22°C (see above), assuming a constant concentration of 10 μM nicotine.

The amount of [^{3}H]-dopamine released decreased with time and at a nicotine concentration which gave a maximal peak response (ca. 10 μM) the release process was fully desensitized within 5 minutes. After a washout with buffer for 20 minutes 80-90% of the initial response could be restored (data not shown), showing that desensitization was reversible and that the decreased response was probably not due to a significant rundown of the neurotransmitter.

In order to study the processes of receptor activation and desensitization separately we took advantage of an important prediction of the model. From nicotine binding kinetics it was found that the affinity of the activatable state (R) for ligand is around 150 nM and for the high affinity state (R') around 1 nM. Thus the model predicts that low nicotine concentrations should desensitize the receptors without activating them and that activation should require higher concentrations as shown in Figure 2. The EC50 for dopamine release was in fact found to be around 120 nM, in good agreement with the affinity of activatable receptors predicted from binding kinetics. To test the prediction of predesensitization, synaptosomes were exposed to non-activating levels of ligand (low nM) for 10 min before evoking the dopamine response with activating (μM) concentrations of ligand (Fig. 2). The results show that predesensitization significantly decreases subsequent dopamine release evoked by maximally activating ligand levels.

In order to determine the relationship of ligand binding affinity to receptor activation and desensitization the Ki values for inhibition of [^{3}H]-nicotine binding by a number of nicotinic

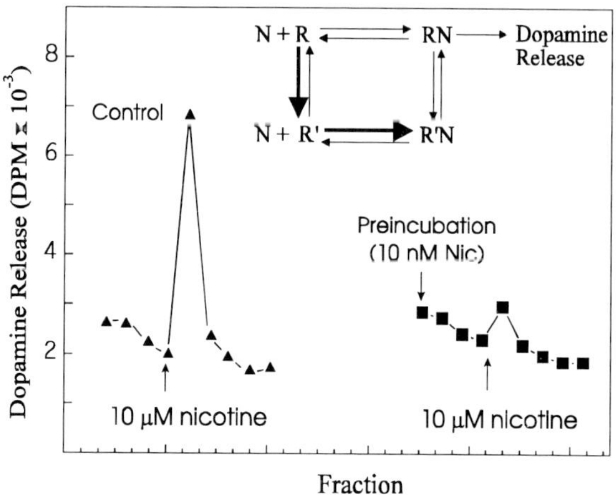

Figure 2. Predesensitization of nicotinic receptors by nicotine. Synaptosomes were superfused for 10 min either with buffer (control) or 10 nM nicotine before evoking dopamine release with a maximally activating nicotine concentration (10 μM). The large arrows show the branches of the two-state model which predict the route of predesensitization at low ligand concentrations.

agonists were compared to their potency and efficacy in eliciting dopamine release and to their potency in predesensitizing the receptors. The results in Figure 3 (panel A) indicate that there is very little correlation, if any, between binding affinity and efficacy (Emax) for activation of dopamine release. A similar lack of correlation was found for Ki vs potency (EC50). By comparison, initial predesensitization studies with a number of ligands indicate a very high correlation between binding affinity and potency of desensitization (Fig. 3, panel B).

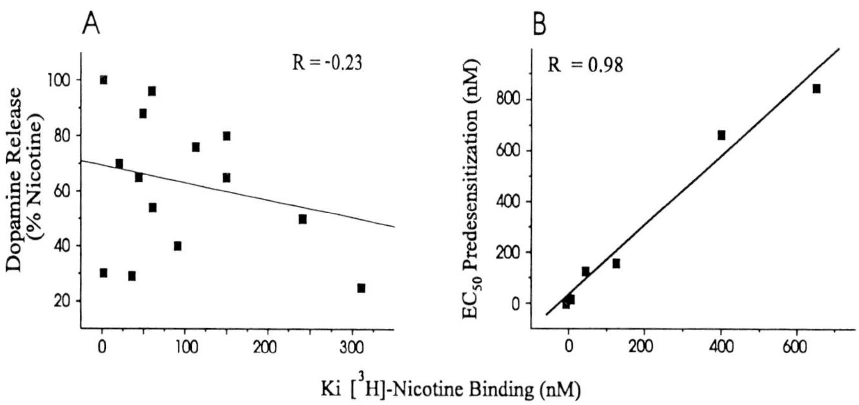

Figure 3. Correlation of nicotine binding affinity with activation and desensitization of dopamine release. *Panel A*: Correlation of inhibition binding constants (Ki) for various nicotinic agonists with efficacy for evoked dopamine release. *Panel B*: Correlation of Ki values determined for six nicotinic agonists with their potency in predesensitizing the dopamine release response.

To assess the role of ligand structure in receptor activation and desensitization, the effects of the simple quaternary amine TMA (tetramethylammonium chloride) on dopamine release were compared to those of nicotine. TMA contains the positive charge center of the putative nicotinic pharmacophore but lacks a hydrogen bonding moiety analogous to the pyridine nitrogen of nicotine or the carbonyl oxygen of acetylcholine. The results shown in Table 1 indicate that TMA is a full agonist relative to nicotine and can fully predesensitize the receptors as judged by inhibition of the dopamine response. Although TMA was less potent than nicotine in evoking dopamine release and in desensitizing the response, the rates of desensitization by nicotine and TMA were essentially the same.

Table 1. Activation and desensitization of dopamine release by nicotine and TMA.

Agonist	Ki or Kd (nM)	Dopamine Release		Predesensitization	Desensitization Rate (min^{-1})
		EC50 (nM)	Emax (% baseline)	EC50 (nM)	
Nicotine	3	117	86	6	1.6
TMA	650	18000	107	837	1.4

Discussion

In general, the results suggest that the two-state model is a good predictor of receptor activation and desensitization by nicotine and other nicotinic agonists. For example, the affinity of nicotine for the activatable state of the receptors (k2/k1=150 nM), determined from receptor binding kinetics, agreed quite well with the EC50 for dopamine release (ca. 120 nM), demonstrating that the functional response correlates with the extent of receptor activation (RN). Another key prediction of the model supported by the present results is the ability of low concentrations of agonists to desensitize the receptors without causing activation. This occurs by the route R-> R'-> R'N. In the case of nicotine, this results from the large difference between the affinities for the activated (RN) and desensitized (R'N) states (ca. 150-fold) revealed by ligand binding kinetics. The close correspondence between equilibrium binding affinity and EC50 for predesensitization of dopamine release (3 nM vs 6 nM, respectively for nicotine) supports the notion that high affinity binding of nicotinic agonists is a direct indicator of their relative potency for functional desensitization. This idea is further supported by the lack of correlation between high affinity binding and either functional potency or efficacy (see Figure 3).

Exploration of the structure-activity relationships for activation and desensitization led to some unexpected findings. In particular, the simple quaternary amine TMA acted as a full agonist in evoking dopamine release and was able to fully desensitize the response. This result calls into question the necessity of having the full nicotinic cholinergic pharmacophore to achieve receptor activation and/or desensitization and suggests that the positive charge center is sufficient in this regard. Based on these findings, one could also conclude that the additional hydrogen bonding center present in acetylcholine and other nicotinic ligands acts primarily to modulate agonist potency with respect to both activation and desensitization. The fact that the rates of desensitization at maximally activating concentrations of nicotine and TMA are the same further suggests that the process of desensitization which occurs through functional activation, i.e. the upper branch of the two-state model (R-> RN), is an intrinsic attribute of the receptor which is less sensitive to ligand structure. At the mechanistic level, the full agonist nature of TMA indicates that bridging of the binding site by different parts of the putative nicotinic pharmacophore does not play an essential role in inducing the conformational change which leads to channel opening. Likewise, the suggestion put forth by others (12) that the shift to the high affinity (desensitized) form of the receptor is due to the formation of a hydrogen bond between ligand and receptor needs to be reexamined. Further studies will obviously be required to fully elucidate the role of ligand structure in the processes of receptor activation and desensitization.

Acknowledgments: *The authors wish to thank Elisa Lovette, Joanna Gregory, and Ian Martyn for their technical contributions to this work.*

References

1. Lukas RJ, Bencherif M. Heterogeneity and regulation of nicotinic acetylcholine receptors. Int Rev Neurobiol 1992; 34: 25-131.
2. Luetje CW, Patrick J. Both α- and β-subunits contribute to the agonist sensitivity of neuronal nicotinic acetylcholine receptors. J Neurosci 1991; 11(3): 837-845.

3. Flores CM, Rogers SW, Pabreza LA, Wolfe BB, Kellar KJ. A subtype of nicotinic cholinergic receptor in rat brain is composed of $\alpha4$ and $\beta2$ subunits and is up-regulated by chronic nicotine treatment. Mol Pharmacol 1991; 41: 31-37.
4. Schwartz RD, McGee R, Kellar KJ. Nicotinic cholinergic receptors labeled by [^{3}H]-acetylcholine in rat brain. Mol Pharmacol 1982; 22: 56-62.
5. Anderson DJ, Arneric SP. Nicotinic receptor binding of [^{3}H]-cytisine, [^{3}H]-nicotine and [^{3}H]-methylcarbamylcholine in rat brain. Eur J Pharmacol 1994; 253: 261-267.
6. Lippiello PM, Fernandes KG. The binding of L-[^{3}H]-nicotine to a single class of high affinity sites in rat brain membranes. Mol Pharmacol 1986; 29: 448-454.
7. Martino-Barrows AM, Kellar KJ. [^{3}H]-Acetylcholine and [^{3}H](-)-nicotine label the same recognition site in rat brain. Mol Pharmacol 1987; 31: 169-174.
8. Katz B, Thesleff S. A study of the "desensitization" produced by acetylcholine at the motor end-plate. J Physiol (Lond) 1957; 138: 63-80.
9. Lippiello PM, Sears SB, Fernandes KG. Kinetics and mechanism of L-[^{3}H]-nicotine binding to putative high affinity receptor sites in rat brain. Mol Pharmacol 1987; 31: 392-400.
10. Marks MJ, Burch JB, Collins AC. Effects of chronic nicotine infusion on tolerance development and nicotinic receptors. J Pharm Exp Ther 1983; 226: 817-825.
11. Grady S, Marks MJ, Wonnacott S, Collins AC. Characterization of nicotinic receptor-mediated [3H]-dopamine release from synaptosomes prepared from mouse striatum. J Neurochem 1992; 59: 848-856.
12. Cockcroft VB, Osguthorpe DJ, Barnard EA, Lunt GG. Modeling of agonist binding to the ligand-gated ion channel superfamily of receptors. Proteins 1990; 8: 386-397.

Effects of Nicotine on
Biological Systems II
Advances in Pharmacological Sciences

PRESYNAPTIC NICOTINIC AUTORECEPTORS AND HETERORECEPTORS IN THE CNS

S. Wonnacott, G. Wilkie, L. Soliakov and P. Whiteaker
School of Biology & Biochemistry, University of Bath, Bath, BA2 7AY, U.K.

Summary: Four agonists were compared for their abilities to release [^{3}H]dopamine from striatal synaptosomes and [^{3}H]acetylcholine from hippocampal synaptosomes in the rat, to see if there are differences between the presynaptic nicotinic receptors presumed to mediate these effects. Dose-response profiles for [^{3}H]acetylcholine release were bell-shaped, with a rank order of potency: anatoxin-a>cytisine=(-)nicotine>>isoarecolone. A similar rank order was determined for [^{3}H]dopamine release, but the dose-response profiles were typically sigmoidal. More detailed analysis of nicotine-evoked [^{3}H]dopamine release suggested a biphasic response consistent with two components. These data are discussed with respect to receptor subtypes. Attempts to extend the study to [^{3}H]glutamate release failed to disclose a specific nicotinic effect.

Introduction

Several lines of evidence support a presynaptic localisation for a considerable proportion of nicotinic acetylcholine receptors (nAChR) in mammalian brain (1). The most extensively studied example is the nicotinic modulation of dopamine release from striatal nerve terminals (2-4). This is consistent with the presence of presynaptic nicotinic heteroreceptors on dopamine terminals, the physiological context of which may be provided by axo-axonic synapses made by cholinergic interneurons onto these terminals (5). Other neurotransmitters whose release may be influenced by nicotine include GABA in the rat hippocampus (6), interpeduncular nucleus (7) and avian lateral geniculate nucleus (8). Electrophysiological studies have also provided evidence for the nicotinic potentiation of glutamate release in the rat prefrontal cortex (9).

Anatomically, the simplest concept of presynaptic receptor regulation is via an autoreceptor, where no physical arrangement between different transmitter-containing neurons is implied. The nicotinic modulation of acetylcholine (ACh) release has been described in cortex and hippocampus (10,11), consistent with a physiological feedback regulation by ACh, although this could also arise through axo-axonic synapses between neighbouring cholinergic neurons.

With the exception of [^{3}H]dopamine release, quantitative pharmacological comparisons have not been undertaken in biochemical preparations. However, the emergence of a plethora of nAChR subunits expressed in the CNS, and the potential for an enormous variety of nAChR subtypes (12) demands a closer examination of nAChR pharmacology to see if distinctions (that might reflect subtype differences) are demonstrated by, for example, hetero- versus autoreceptors. As an initial approach to this objective, we have compared 4 nicotinic agonists for their nAChR-mediated effects on [^{3}H]dopamine and [^{3}H]ACh release. Attempts to extend this study to the nicotinic regulation of [^{3}H]glutamate release have failed to disclose such a phenomenon; the possible explanations of this negative outcome are discussed.

Materials & Methods

[^{3}H]Dopamine release. Synaptosomes were prepared from rat striatum as a P2 pellet, resuspended to a concentration of 5 mg protein/ml in Kreb's bicarbonate buffer (NaCl, 118 mM; $NaHCO_3$, 25 mM; KCl, 2.4 mM; KH_2PO_4, 1.2 mM; $CaCl_2$, 2.4 mM; $MgSO_4$, 1.2 mM; glucose, 10 mM; gassed with 95% O_2/5% CO_2, pH 7.4), supplemented with 10 μM pargyline and 1 mM ascorbic acid. Synaptosomes were loaded with [^{3}H]dopamine (0.1 μM; 50 Ci/mmol; 25 min) and superfused at a rate of 0.5 ml/min in open, homemade chambers essentially as previously described (13). Uptake and release were carried out within a hood maintained at 37°C. Following a 30 min washout, 40 s agonist or KCl (20 mM) pulses were administered via the perfusion line, separated from the bulk flow by 10 s air bubbles.

[^{3}H]Glutamate and [^{3}H]aspartate release. Excitatory amino acid release was investigated in cortex, using purified synaptosomes prepared on Percoll gradients (14), and in striatum using P2 preparations. Superfusion was carried out exactly as described for [^{3}H]dopamine, except that synaptosomes were loaded with amino acid by incubation with [^{3}H]glutamate (0.2 μM; 57 Ci/mmol) or [^{3}H]aspartate (0.8 μM; 12 Ci/mmol) for 25 min at 37°C.

[^{3}H]ACh release. Synaptosomes were prepared from rat hippocampus on Percoll gradients (14) and resuspended to a concentration of 1 mg protein/ml in Kreb's bicarbonate buffer. The synaptosomes were loaded with [^{3}H]choline (0.8 μM; 32 Ci/mmol; 30 min) and aliquots (150 μl) introduced into chambers of a Brandell superfusion apparatus. Superfusion conditions

were essentially as described for [^{3}H]dopamine release.

Data analysis. Evoked release was measured as the area under the peak above basal release (see Fig. 2b, insert). Inter-assay variation was normalised by reference to a 1 μM-nicotine standard incorporated into every experiment.

Results

[3H]ACh release. Nicotine evoked the release of [^{3}H]ACh from hippocampal synaptosomes in a dose-dependent, dihydroβerythroidine-sensitive and Ca^{2+} dependent manner. Each of the 4 agonists examined gave "bell shaped" dose-response curves, showing marked attenuation of the responses at higher agonist concentrations (Fig. 1a). Anatoxin-a was the most potent agonist, nicotine and cytisine were equipotent, while isoarecolone was considerably less potent. From Fig. 1a it is evident that these 4 agonists are equally efficacious in evoking [^{3}H]ACh release.

[^{3}H]Dopamine release. Nicotine-evoked [^{3}H]dopamine release from striatal synaptosomes was robust and reproducible, as previously reported (2-4). At concentrations above 50 μM there was a large non-specific (mecamylamine-insensitive) component of release. Consequently all agonists were examined in the presence and absence of 20 μM-mecamylamine. Total and mecamylamine-insensitive nicotine-evoked [^{3}H]dopamine release are shown in Fig. 1b, together with the difference curve representing specific release. This is complex, consistent with two components with approximate EC_{50}'s of 1 μM and 40 μM. Dose-response curves for mecamylamine-sensitive release elicited by anatoxin-a, cytisine and isoarecolone appear monophasic, with a sustained plateau. However the number of concentrations studied was less than for nicotine. The rank order of potencies is anatoxin-a>cytisine>nicotine>isoarecolone, similar to that for [^{3}H]ACh release, but the efficacies of anatoxin-a and cytisine are clearly lower than that of nicotine.

[^{3}H]Glutamate release. Initial experiments compared the uptake and release of [^{3}H]glutamate and [^{3}H]aspartate from cortex synaptosomes (Fig 2). Both amino acids gave clear responses to KCl stimulation, but nicotine failed to elicit any release except at high concentrations. Whereas KCl-evoked [^{3}H]glutamate release was partially Ca^{2+}-dependent, release elicited by 100 μM nicotine was not (Fig 2a, insert). Furthermore, the release evoked by 100 μM nicotine could not be blocked by 10 μM mecamylamine (Fig 2a,b), indicating that it is non-

specific. A similar pattern of results was observed for striatal synaptosomes (not shown).

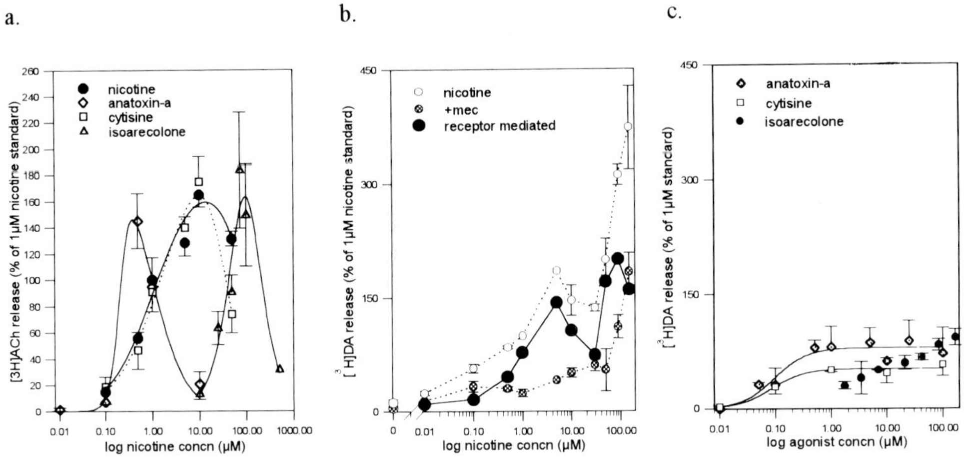

Figure 1. Agonist evoked neurotransmitter release from perfused synaptosomes. (a) [^{3}H]ACh release from hipocampal synaptosomes; (b) nicotine evoked [^{3}H]dopamine release from striatal synaptosomes; (c) mecamylamine-sensitive agonist evoked release of [^{3}H]dopamine.

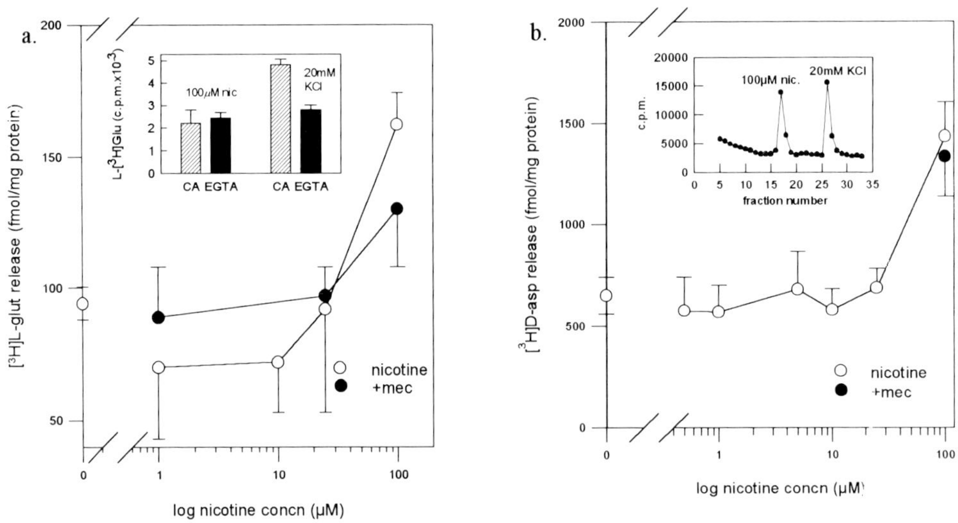

Figure 2. Nicotine evoked release of (a) [^{3}H]L-glutamate and (b) [^{3}H]D-aspartate, from purified cortical synaptosomes superfused in the absence and presence of 20 μM mecamylamine. Insert: (a) Ca^{2+}-dependence of nicotine- and KCl-evoked [^{3}H]L-glutamate release; (b) A typical superfusion profile for [^{3}H]D-aspartate release. The nicotine response could not be blocked by mecamylamine.

Discussion

Similar concentrations of nicotine elicit the release of [^{3}H]ACh and [^{3}H]dopamine but the shapes of the dose-response profiles differ between the two transmitters (Fig. 1). This could simply reflect differences in the complex release processes involved. However, nicotinic responses measured biochemically over a similar time course (40 s to a few minutes) typically result in inverted-"U" dose responses (15,16), as seen for [^{3}H]ACh release here. Indeed this is a common phenomenon in pharmacology (17), attributed to desensitisation of some step in the process being measured; in this case desensitisation or inactivation of the nAChR itself is likely (15). The sustained release of [^{3}H]dopamine that has been documented previously (3) prompted its re-examination over a more detailed concentration range. This has disclosed a distinct inflection in the profile (Fig. 1b) compatible with two nAChR of differing affinities. Overlap of their agonist sensitivities could combine to give the seemingly monophasic curves derived for the other agonists.

Another difference between agonist-evoked [^{3}H]dopamine and [^{3}H]ACh release is agonist efficacy. The 4 agonists examined were equi-efficacious in evoking [^{3}H]ACh release, whereas anatoxin-a and cytisine attained only 54% and 35% of the maximum [^{3}H]dopamine release elicited by nicotine. Cytisine has been reported to be a full agonist with respect to [^{3}H]dopamine release from <u>mouse</u> striatum (3) but was less efficacious than nicotine in a rat study (4), so species differences in nAChR properties or composition may be important.

Table 1. Approximate EC_{50} values (μM) and % maximum nicotine response for 4 agonists

	Striatal [^{3}H]dopamine release		Hippocampal [^{3}H]ACh release		M10 cells (α4,β2) ^{86}Rb influx
(+)anatoxin	0.05±0.02	(54%)	0.14±0.04	(90%)	0.05±0.02
(-)nicotine	1.1±0.5 {≈40	(100%) (159%)}	0.99±0.23	(100%)	2.7±0.7
cytisine	≈0.1	(35%)	1.06±0.15	(103%)	3.7±1.1
isoarecolone	≥10	(>68%)	43±7	(106%)	nd

Which nAChR subtypes are presynaptic? Several pieces of evidence support the notion that nAChR composed of α4 and β2 subunits (defined by tritiated agonist binding) are present on nerve terminals (1). For example, [^{3}H]nicotine binding sites are lost from striatum following

6-hydroxydopamine lesions of the nigro-striatal pathway (18), and these binding sites decline in parallel with presynaptic markers in hippocampus and cortex during degeneration of cholinergic axons in Alzheimer's disease (19). Agonist-evoked Rb^+ influx into mouse fibroblasts transfected with α4 and β2 subunits (20) results in sharply "bell-shaped" curves, similar to those seen for [^{3}H]ACh release (21). Comparison of agonist potencies (Table 1) shows reasonable concordance between the two release systems with this defined nAChR, although [^{3}H]dopamine release is unique in displaying variable agonist efficacies.

It is regrettable that more subtype-selective probes are not available. [^{3}H]ACh release and both phases of [^{3}H]dopamine release are insensitive to α-bungarotoxin and low concentrations of methyllycaconitine, excluding the α7 subunit. However, it is difficult to distinguish α2, α3 and α4-containing nAChR. Neuronal bungarotoxin (nBgt) offered some hope in this direction (22), although its complex kinetics, determined by the β subunit (23), complicate interpretation. We previously observed a 50% inhibition of [^{3}H]dopamine release by 100 nM nBgt (24). This result contrasts with the total block of striatal dopamine release by 100 nM nBgt reported by others (3,25). Whereas we examined 3 μM nicotine, the other studies used 50 or 100 μM nicotine. From the dose response curve in Fig. 1b, we can rationalise this discrepancy: 3 μM nicotine would act principally on the higher affinity component of release that may have low sensitivity to nBgt, whereas 50 and 100 μM nicotine would activate the lower affinity component which may have high sensitivity to nBgt. Thus the higher affinity component is more consistent with α4 and β2 subunits while the low affinity component might reflect α3 and β2 subunits (22). To address this proposition nBgt should be examined across the full nicotine dose-response range, but the current unavailablility of nBgt precludes this experiment.

It must be borne in mind that additional subunits might also contribute to the nAChR, thereby changing their properties. This could account for differences in agonist efficacy between [^{3}H]dopamine and [^{3}H]ACh release: for example, the β subunit can determine the effectiveness of cytisine as an agonist (23).

Are there nAChR on glutamate terminals? We were unable to demonstrate mecamylamine-sensitive nicotine-evoked [^{3}H]glutamate release from cortical or striatal synaptosomes, despite electrophysiological evidence for such an action (9). The possibility that the accumulated radiolabel did not enter the vesicular pool of glutamate but was released by reversal of the carrier (26) is likely for [^{3}H]aspartate. However, the partial Ca^{2+} dependence of K^+-evoked

[^{3}H]glutamate release (Fig. 2b) argues that in this case at least some release occurred by exocytosis. The electrophysiological evidence for the presynaptic nicotinic modulation of glutamate release is derived from the prefrontal area of the neocortex, where only 14% of neurons responded to nicotine (9). Therefore assay of radiolabelled glutamate from synaptosome populations necessarily derived from relatively large brain regions may lack sufficient resolution to reveal a modest effect of nicotine that is restricted in its localisation. Evidently the major excitatory transmitter glutamate is not subject to widespread modulation via presynaptic nAChR.

Acknowledgments: *This work was supported by a grant from the SERC and by postgraduate studentships from SERC and MRC to GW and PW respectively. We are grateful to SmithKline Beecham Pharmaceuticals, Harlow, Essex UK for the isoarecolone and to Dr. T. Gallagher, University of Bristol, U.K., for the anatoxin-a.*

References

1. Wonnacott S, Drasdo A, Sanderson E, Rowell P. Presynaptic nicotinic receptors and the modulation of transmitter release. In: Marsh J Ciba Foundation Symposium 152: The biology of nicotine dependence Chichester: John Wiley and Sons pp 87-101, 1990.
2. Rapier C, Lunt GG, Wonnacott S. Nicotinic modulation of [^{3}H]dopamine release from striatal synaptosomes: Pharmacological characterisation. J Neurochem 1990; 54: 937-945.
3. Grady S, Marks M, Wonnacott S, Collins AC. Characterization of nicotinic receptor mediated [^{3}H]dopamine release from synaptosomes prepared from mouse striatum. J Neurochem 1992; 59: 848-856.
4. El-Bizri H, Clarke, PBS. Blockade of nicotinic receptor-mediated release of dopamine from striatal synaptosomes by chlorisondamine and other nicotinic antagonists administered *in vitro*. Brit J Pharmacol 1994; 111: 406-413.
5. McGeer EG, McGeer PL, Grewaal DS, Singh VK. Striatal cholinergic interneurons and their relation to dopaminergic nerve endings. J Pharmacol (Paris) 1975; 6: 143-152.
6. Wonnacott S, Fryer L, Irons J, Lunt GG. Presynaptic sites of nicotine's action in the brain. In: Rand MJ and Thurau K The pharmacology of nicotine Oxford: IRL Press, 1988.
7. Lena C, Changeux J-P, Mulle C. Evidence for "preterminal" nicotinic receptors on GABAergic axons in the rat interpeduncular nucleus. J Neurosci 1993; 13: 2680-2688.
8. McMahon LL, Yoon K-W, Chiappinelli VA. Electrophysiological evidence for presynaptic nicotinic receptors in the avian ventral lateral geniculate nucleus. J Neurophysiol 1994; 71: 826-829.
9. Vidal C, Changeux J-P. Nicotinic and muscarinic modulations of excitator synaptic transmission in the rat prefrontal cortex. Neurosci 1993; 56: 23-32.
10. Araujo DM, Lapchak PA, Collier B, Quirion R. Characterisation of [^{3}H]N-methylcarbamylcholine binding sites and effect of N-methylcarbamylcholine on acetylcholine release in rat brain. J Neurochem 1988; 51: 292-299.
11. Wonnacott S, Irons T, Rapier C, Thorne B, Lunt GG. Presynaptic modulation of transmitter release by nicotinic receptors. In: Nordberg A, Fuxe K, Holmstedt B and Sundwall A Nicotinic receptors in the CNS: their role in synaptic transmission Amsterdam: Elsevier, 1989: 157-163.
12. Sargent PB. The diversity of neuronal nicotinic acetylcholine receptors. Ann Rev Neurosci 1993; 16: 403-443.
13. Rowell PP, Wonnacott S. Evidence for functional activity of up-regulated nicotine binding sites in rat striatal synaptosomes. J Neurochem 1990; 55: 2105-2110.

14. Thorne B, Wonnacott S, Dunkley PR. Isolation of hippocampal synaptosomes on Percoll gradients: cholinergic markers and ligand binding sites. J Neurochem 1990; 56: 479-484.
15. Marley PD. Desensitisation of the nicotinic secretory response in adrenal chromaffin cells. Trends Pharmacol Sci 1988; 9: 102-107.
16. Boyd ND. Two distinct kinetic phases of acetylcholine receptors of clonal rat PC12 cells. J Physiol (London) 1987; 389: 45-67.
17. Pliska V. Models to explain dose-response relationships that exhibit a downturn phase. Trends Pharmacol Sci 1994; 15: 178-181.
18. Clarke PBS, Pert A. Autoradiographic evidence for nicotine receptors on nigrostriatal and mesolimbic dopaminergic neurons. Brain Res 1985; 348: 355-358.
19. Perry EK, Perry RH, Smith CJ, Purohit D, Bonham J, Dick DJ, Candy JM, Edwardson JA, Fairbairn A. Cholinergic receptors in cognitive disorders. Can J Neurol Sci 1987; 13: 521-527.
20. Whiting P, Schoepfer R, Lindstrom J, Priestley T. Structural and pharmacological characterisation of the major brain nicotinic acetylcholine receptor subtype stably expressed in mouse fibroblasts. Mol Pharmacol 1991; 40: 463-472.
21. Thomas P, Stephens M, Wilkie G, Amar M, Lunt GG, Whiting P, Gallagher T, Pereira E, Alkondon M, Albuquerque EX, Wonnacott S. (+)Anatoxin-a is a potent agonist at neuronal nicotinic acetylcholine receptors. J Neurochem 1993; 60: 2308-2311.
22. Luetje CW, Wada K, Rogers S, Abramson SN, Tsuji K, Heinemann S, Patrick, J Neurotoxins distinguish between different neuronal nicotinic acetylcholine receptors. J Neurochem 1990; 55: 632-640.
23. Papke RL, Duvoisin RM, Heinemann SF. The amino terminal half of the nicotinic beta subunit extracellular domain regulates the kinetics of inhibition by neuronal-bungarotoxin. Proc Roy Soc Lond B 1993; 252: 141-148.
24. Wonnacott S, Drasdo A. Presynaptic actions of nicotine in the CNS. In: Adlkofer F Effects of nicotine on biological systems Basel: Birkhauser Verlag pp 295-306, 1991.
25. Schulz DW, Zigmond RE. Neuronal bungarotoxin antagonises nicotinic function in rat caudate-putamen. Neurosci Lett 1989; 98: 310-316.
26. Attwell D, Barbour B, Szatkowski M. Nonvesicular release of neurotransmitter Neuron 1993; 11: 401-407.

Effects of Nicotine on
Biological Systems II
Advances in Pharmacological Sciences

REGULATION OF NEURONAL NICOTINIC RECEPTORS IN PRIMARY CULTURES

M.I. Dávila-García, Y. Xiao, R.A. Houghtling,
S.S. Qasba, C.M. Flores and K.J. Kellar

Department of Pharmacology, Georgetown University School of
Medicine, Washington D.C. 20007, USA

Summary: We used [^{3}H]cytisine to characterize nicotinic cholinergic receptor binding sites in rat brain neurons grown in culture. Primary cells were obtained from several different areas of fetal rat brain at 18 days gestation and were kept in culture for up to 15 days. [^{3}H] Cytisine binds with high affinity to primary cultures of neurons from cerebral cortex, hippocampus and striatum; but the highest binding (≈ 50 fmol/mg protein) was found in neurons from a region containing thalamus, midbrain and brainstem, which we refer to here as the TMB. Culturing cortical neurons in the presence of nicotine for 7 days increases the number of nicotinic receptor binding sites labeled by [^{3}H]cytisine. Primary neuronal cultures should be useful as model systems for studies of subunit composition and molecular mechanisms involved in the regulation of neuronal nicotinic receptors.

Introduction

The density of nicotinic receptors in rat and mouse brain is increased by chronic administration of nicotine (1-5) and an increase in nicotinic receptors has been found in autopsied human brain from people who smoked cigarettes (6), suggesting that these receptors are involved in processes underlying addiction to nicotine. The molecular and cellular mechanisms leading to the increase in brain nicotinic receptors by chronic administration of nicotine are not understood; however, based on earlier studies of nicotine's effects on autonomic ganglia (7,8), prolonged blockade or desensitization of receptors by nicotine has been postulated as an important and possibly the initiating factor in this increase (1-4).

In fact, certain brain-mediated responses to nicotine administration *in vivo*, such as the release of prolactin and ACTH from the pituitary, desensitize rapidly and nearly completely. For example, after a single injection of nicotine these neuroendocrine responses to a subsequent injection of nicotine are markedly diminished or absent for several hours (9,10). Furthermore, after chronic administration of nicotine, the nicotine-induced release of prolactin is absent for more than 8 days (11). *In vitro*, nicotine-induced release of dopamine from striatal tissues also desensitizes rapidly (12-14), and recently Collins and colleagues have attempted to relate the kinetics of desensitization of this response to the kinetics of the high affinity nicotinic binding sites (14). Interestingly, nicotine appears to desensitize the brain nicotinic receptors that mediate the prolactin response *in vivo* (10) and the release of dopamine *in vitro* (14) more readily than it activates them. This is similar to nicotine's effects in adrenal PC12 cells, where it was found to be more potent in desensitizing the sodium ion flux response than in activating it (15). Thus, although nicotine acts initially as an agonist, its time-averaged effect on at least some neuronal nicotinic receptors appears to be that of an antagonist.

One major hinderance to a better understanding of the cellular and molecular processes underlying desensitization of nicotinic receptors and the increase in nicotinic binding sites after chronic administration of nicotine is the absence of a clonal cell line containing neuronal nicotinic receptors with high affinity for nicotinic ligands. However, Lippiello et al. (16, 17) have found that the pharmacological properties of nicotinic receptor binding sites labeled by $[^3H]$(-)nicotine in primary cultures of fetal rat cortical neurons are similar to those in adult rat brain. We have examined cultured primary neurons from several fetal rat brain regions to try to develop cell models suitable for studying the subunit composition and regulation of brain nicotinic receptors.

Materials and Methods

Primary cultures of neurons from embryonic day 18 rat brain were prepared by standard procedures. The brains were dissected and cells were cultured from several brain areas, including cerebral cortex (CTX), hippocampus (HIPP), caudate (CA) and a subcortical region containing the septum, thalamus, midbrain and brainstem (TMB). The cells were plated in 6 well tissue culture plates previously coated with poly-D-lysine (10 µg/ml) and grown in DMEM/F12 containing 10% fetal calf serum for 48 to 72 hr before changing the medium to serum-free N2 defined medium containing 10 µM cytosine arabinoside for 24 to 48 hr to stop glial proliferation. The medium was then changed to N2 medium without cytosine arabinoside. Cells were usually grown for 15 days at 37°C with 5% CO_2 before harvesting. In some cases, 100 µM nicotine bitartrate was added to the cells during the last 7 days in culture.

Binding of [^{3}H]cytisine to nicotinic receptors with high affinity for agonists was measured in homogenates of primary neuronal cultures using slight modifications of the procedures of Pabreza et al. (18).

Results and Discussion

Tissues from four regions of embryonic day 18 rat brain were maintained in culture for 15 days. The cerebral cortical neuronal cultures contained several cell types, including pyramidal, fusiform and multipolar neurons. The hippocampal and caudate cultures contained primarily bipolar neurons. The TMB cultures contained a variety of neuronal cell types, with the largest number being small bipolar cells.

High affinity binding sites for [^{3}H]cytisine were present in cultured cells from all 4 brain areas examined, but the highest binding was found in cells from the TMB, intermediate binding was found in cells from the hippocampus and caudate, and the lowest binding was found in cells from the cerebral cortex (Fig. 1).

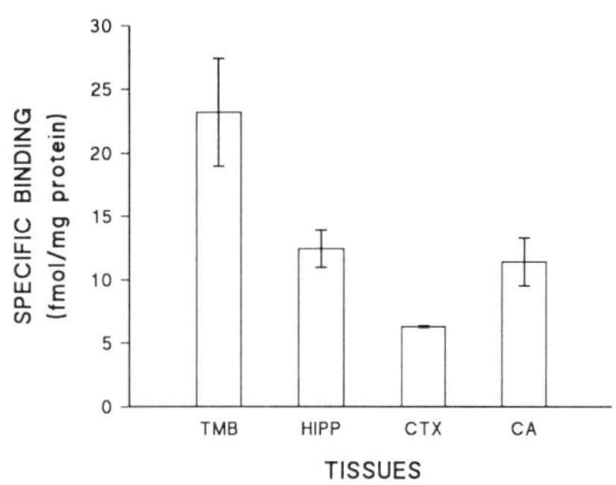

Figure 1. *Binding of 2 nM [^{3}H]cytisine to homogenates from primary cultures of neurons from several brain areas*. Primary cultures were prepared from TMB, the hippocampus, the cerebral cortex and the caudate nucleus. The cells were kept in culture for 15 days before assay. Values are the mean ± SEM of 7 - 12 determinations.

[^{3}H]Cytisine binding in cultures from the TMB was more than 3-times higher than in cultures from the cerebral cortex. Therefore, the binding of [^{3}H]cytisine in TMB cells was examined further. In the TMB, [^{3}H]cytisine binds with a K_d of approximately 0.4 nM and a density (B_{max}) of 50 fmol/mg protein (Fig. 2). This density of nicotinic binding sites is as high as in many areas of adult rat brain. In addition, [^{3}H]cytisine binding in the TMB displays a very favorable ratio of specific to non-specific binding (Fig. 2).

The pharmacological characteristics of [^{3}H]cytisine binding sites in TMB cells were examined in competition assays (Table 1). Epibatidine, a nicotinic agonist with very high affinity for [^{3}H]nicotine binding sites in mouse and rat brain (19,20), and nicotine compete potently for [^{3}H]cytisine binding sites in TMB, with IC_{50} values of 2 and 16 nM, respectively.

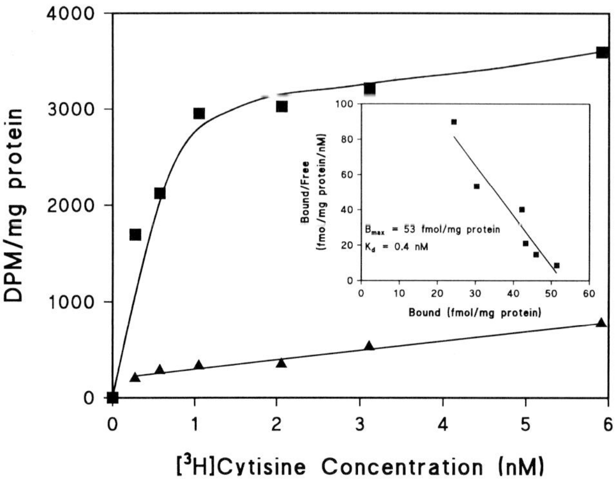

Figure 2. *Saturation binding of [^{3}H]cytisine to homogenates from TMB cells grown in culture for 15 days.* Specific binding (■) was determined by subtracting nonspecific binding (▲) from total binding. Inset shows Scatchard plot of specific binding.

The nicotinic agonist carbachol and the competitive antagonist dihydro-ß-erythroidine (DßE) compete with IC_{50} values of less than 1 µM, while α-bungarotoxin (α-BTX), at concentrations up to 2 µM, did not compete effectively for [^{3}H]cytisine binding. This pharmacological profile of [^{3}H]cytisine binding sites in cells from the TMB is very similar to that found in adult rat brain (18), suggesting that the nicotinic receptor labeled by [^{3}H]cytisine in cells from the TMB is comprised of α4 and ß2 subunits, as it is in adult rat brain (21); however, the subunit composition of these receptors in the TMB remains to be determined.

Table 1. *Competition by nicotinic drugs for [^{3}H]cytisine binding in TMB cell homogenates.* Drugs competed against 4 nM [^{3}H]cytisine. The IC_{50} values are the mean ± SEM of three determinations, except for epibatidine which is from a single determination.

Drug	IC_{50} (nM)
Epibatidine	2
Nicotine	16 ± 9
Carbachol	850 ± 50
DßE	850 ± 50
α-BTX	> 2000

The nicotinic receptors increased by chronic administration of nicotine in adult rat brain are comprised of α4 and ß2 subunits (21). To determine whether chronic exposure to nicotine can affect nicotinic receptors in primary neuronal cultures, we measured [^{3}H]cytisine binding in cells from the cerebral cortex and TMB grown in the presence of 100 μM nicotine for 7 days. [^{3}H]Cytisine binding in cortical cells grown in the presence of nicotine was markedly increased compared to control cells (Fig. 3). In contrast, [^{3}H]cytisine binding in TMB cells was not significantly increased by the presence of nicotine for 7 days (Fig. 3).

The cellular mechanisms underlying the nicotine-induced increase in nicotinic receptor binding sites in adult brain are not known; however, mRNA for α4 or ß2 subunits does not appear to increase (22, 23), suggesting that the mechanisms involved in the increase in binding sites are post-transcriptional. It is not yet known whether the nicotine-induced increase in [^{3}H]cytisine binding sites in these primary cultures is accompanied by changes in mRNA. It is possible, however, that the nicotine-induced change in nicotinic binding sites in both the adult brain *in vivo* and in these primary cultures reflects a decrease in the rate of degradation of the receptor, rather than an increase in its synthesis. In this regard, nicotine's effect to increase nicotinic receptor binding sites is not uniform throughout the rat brain (24, 25); the regional differences in the degree of receptor increase following chronic nicotine may reflect differences in the turnover rate of the receptor in different brain regions. Studies in cell models for neuronal nicotinic receptors may contribute to answers to this question as well as to other questions related to the signal transduction mechanisms underlying nicotinic receptor regulation.

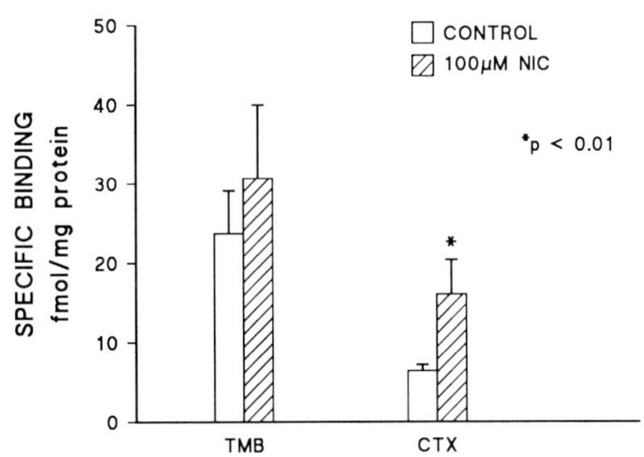

Figure 3. *[^{3}H]Cytisine binding to nicotinic receptors in primary neuronal cultures grown in the presence of nicotine for 7 days.* Cells from the TMB and the cerebral cortex (CTX) were grown from day 8 to day 15 in the presence of 100 μM nicotine bitartrate. Values are the mean ± SEM from 5 to 9 independent determinations.

Acknowledgements: *These studies were supported by NIH grants DA06486, AG09884 and DA05417.*

References

1. Schwartz RD, Kellar KJ. Nicotinic cholinergic receptor binding sites in brain: regulation in vivo. Science 1983; 220:214-216.
2. Schwartz RD, Kellar KJ. In vivo regulation of [^{3}H]acetylcholine recognition sites in brain by nicotinic cholinergic drugs. J Neurochem 1985; 45:427-433.
3. Marks MJ, Burch JB, Collins AC. Effects of chronic nicotine infusion on tolerance development and nicotinic receptors. J Pharmacol Exp Ther 1983; 226:817-825.
4. Marks MJ, Stitzel JA, Collins AC. Time course study of the effects of chronic nicotine infusion on drug response and brain receptors. J Pharmacol Exp Ther 1985; 235:619-628.
5. Nordberg A, Wahlstrom G, Arnelo U, Larsson C. Effect of long-term nicotine treatment on [^{3}H]nicotine binding sites in rat brain. Drug Alcoh Depend 1985; 16:9-17.
6. Benwell MEM, Balfour DJK, Anderson JM. Evidence that tobacco smoking increases the density of [^{3}H](-)nicotine binding sites in human brain. J Neurochem 1988; 50:1243-1247.
7. Eccles JC. The action potential of the superior cervical ganglion. J Physiol (London) 1935; 85:179-206.
8. Ginsborg BL, Guerrero S. On the action of depolarizing drugs on sympathetic ganglia of the frog. J Physiol (London) 1964; 172:189-206.
9. Sharp BM, Beyer HS, Rapid desensitization of the acute stimulatory effects of nicotine on rat plasma adrenocorticotropin and prolactin. J Pharmacol Exp Ther 1986; 238:486-491.
10. Hulihan-Giblin BA, Lumpkin MD, Kellar KJ. Acute effects of nicotine on prolactin release in the rat: agonist and antagonist effects of a single injection of nicotine. J Pharmacol Exp Ther 1990; 252:15-20.
11. Hulihan-Giblin BA, Lumpkin MD, Kellar KJ. Effects of chronic administration of nicotine on prolactin release in the rat: inactivation of prolactin response by repeated injections of nicotine. J Pharmacol Exp Ther. 1990; 252:21-25.
12. Rapier C, Lunt GG, Wonnacott S. Stereoselective nicotine-induced release of dopamine from striatal synaptosomes: concentration dependence and repetitive stimulation. J Neurochem 1988; 50:1123-1130.
13. Schulz DW, Zigmond RE. Neuronal bungarotoxin blocks the nicotinic stimulation of endogenous dopamine release from rat striatum. Neurosci Lett 1989; 98:310-316.
14. Grady S R, Marks MJ, Collins AC. Desensitization of nicotine-stimulated [^{3}H]dopamine release from mouse striatal synaptosomes. J Neurochem 1994; 62:1390-1398.
15. Boyd ND. Two distinct kinetic phases of desensitization of acetylcholine receptors of clonal rat PC12 cells. J Physiol (London) 1987; 389:45-67.
16. Lippiello PM, Fernades KG. Identification of putative high affinity nicotinic receptors on cultured cortical neurons. J Pharmacol Exp Ther 1988; 246:409-416.
17. Lippiello PM, Fernandes KG, Langone JJ, Bjercke RJ. Characterization of nicotinic receptors on cultured neurons using anti-idiotypic antibodies and ligand binding. J Pharmacol Exp Ther 1991; 257:1216-1224.
18. Pabreza L A, Dhawan S, Kellar KJ. [^{3}H]Cytisine binding to nicotinic cholinergic receptors in brain. Mol Pharmacol 1991; 39:9-12.
19. Badio B, Daly JW. Epibatidine, a potent analgetic and nicotinic agonist. Mol Pharmacol 1994; 45:563-569.
20. Houghtling RA, Dávila-García MI, Kellar KJ. In preparation.
21. Flores CM, Rogers SW, Pabreza LA, Wolfe BB, Kellar KJ. A subtype of nicotinic cholinergic receptor in rat brain is composed of α4 and ß2 subunits and is up-regulated by chronic nicotine treatment. Mol Pharmacol 1992; 41:31-37.
22. Marks MJ, Pauly JR, Gross SD, Deneris ES, Hermans-Borgmeyer I, Heinemann SF, Collins AC. Nicotine binding and nicotinic receptor subunit RNA after chronic nicotine treatment. J Neurosci 1992; 12:2765-2784.
23. Xiao Y, Flores CM, Houghtling RA, Riegel AT, Kellar KJ. In preparation.
24. Kellar KJ, Giblin BA, Lumpkin MD. Regulation of brain nicotinic cholinergic recognition sites and prolactin release by nicotine. Progress in Brain Research 1989; 79:209-216.
25. Sanderson EM, Drasdo AL, McCrea, Wonnacott S. Upregulation of nicotinic receptors following continuous infusion of nicotine is brain region-region specific. Brain Res 1993; 617:349-352.

Effects of Nicotine on
Biological Systems II
Advances in Pharmacological Sciences

CONDITIONED TOLERANCE TO NICOTINE IN RATS

A.R. Caggiula, L.H. Epstein, S.M. Antelman, S. Knopf, K.A. Perkins,
S. Saylor, E. Donny and R. Stiller

Departments of Psychology, Psychiatry and Anesthesiology,
University of Pittsburgh, Pittsburgh, PA, 15260 and Pittsburgh Cancer Institute, USA

Summary. Evidence is reviewed for the role of conditioning in tolerance to nicotine. Possible mechanisms are discussed, including the hypothesis that a conditioned release of adrenal steroids may mediate conditioned tolerance to some of nicotine's effects.

Repeated exposure to drugs can result in enhanced (sensitization) or reduced (tolerance) responsiveness to future drug exposure. These changes reflect neuroadaptations which may be critical in the development and maintenance of drug dependence. Considerable effort is being made to identify the variables controlling such adaptations and their underlying mechanisms. Our laboratory is engaged in studies of sensitization (1) and tolerance (2-5) to several therapeutic and addictive drugs. Here we will discuss research on tolerance to intermittent nicotine exposure suggesting that learning plays an important role in its development and manifestation.

Conditioned drug tolerance

Chronic tolerance is thought to reflect kinetic modifications in drug distribution and metabolism or cellular nervous system changes in direct response to repeated drug exposure. However, in the late 1960s and early 1970s, evidence began to accumulate that behavioral demands and learning also could affect the development and manifestation of tolerance (see 6). For example, tolerance that develops when morphine is given in conjunction with predictive, environmental cues, becomes dependent, at least in part, on learned associations with those cues. Similar results for other therapeutic and addictive drugs suggest that chronic tolerance consists of "associative" or environment-dependent, as well as "nonassociative" components (6).

Several studies have considered the influence of environmental variables or behavioral demands present in the drug-testing situation on nicotine-tolerance, but with mixed results (7-9). However these studies did not address the potential importance of environmental stimuli present during drug administration. In 1989 we reported evidence for associative tolerance to nicotine, using an external inhibition paradigm (2). In this paradigm (6), tolerance is tested after either the addition of an extraneous stimulus or the omission of part of the normal drug-administration ritual (including a change in physical context). During the development of tolerance to the antinociceptive effects of nicotine -- delay in tail-withdrawal from a hot water bath -- the same physical and temporal cues consistently predicted drug delivery. Giving nicotine (1.0 mg/kg free base, sc) without those cues disrupted tolerance and completely reinstated the original nicotine effect. Evidence for associative tolerance was also obtained for anorectic (4, 5) and corticosterone-elevating effects (5), using lower doses of nicotine (0.33 and 0.66 mg/kg) and an appetitive behavior that differed from the tail-withdrawal response in the likely involvement of CNS or peripheral sites in nicotine's action (10).

Our initial studies focused on the reversal of tolerance by external inhibition. Conditioning theory also predicts that tolerance *develops* faster if the drug is experienced in a consistent, distinctive environment than in a less consistent and distinctive one (e.g., a familiar setting). Using a discriminative conditioning paradigm, one group received nicotine (0.66 mg/kg) in a distinctive environment, and saline in the home environment. A second group received the opposite (saline/distinctive, nicotine/home). Both received saline in the home environment on alternate days. Tail-withdrawal tests were after the first, third and fifth nicotine injections. Several control groups received saline in different contexts. Rats consistently receiving nicotine in the distinctive environment developed tolerance faster than those given nicotine in a less distinctive environment (colony room) in which they also received saline injections (Figure 1).

The above research suggests that tolerance develops faster in a consistent, predictable environment than in a changing one. In our first attempt to apply this principle to humans (3), male smokers received five trials of smoking in either a predictable context, or in a context that changed each trial. Initially, smoking increased heart rate for both groups. Tolerance developed in the predictable environment but not for the changing group. These differences could not be explained by the environmental manipulation per se or differences in alveolar CO levels.

In our most recent study, rats were given 7 injections of 0.5 mg/kg nicotine in either the distinctive or home environment, using a discriminative conditioning paradigm (see above), and then tested for antinociception after receiving one of 3 doses of nicotine in the distinctive environment. Saline groups received saline or nicotine before the test. This study used hind-paw withdrawal from a hot plate, which is a more sensitive measure and more likely to be mediated entirely by central actions of nicotine than is tail-withdrawal (8). The single measurement also eliminated the possible influence of repeated-testing experience (9). Tolerance was exhibited only by the distinctive environment group at the two lowest test doses (Figure 2). The fact that both groups showed partial tolerance at the highest dose may reflect a greater proportion of nonassociative tolerance at this dose (9).

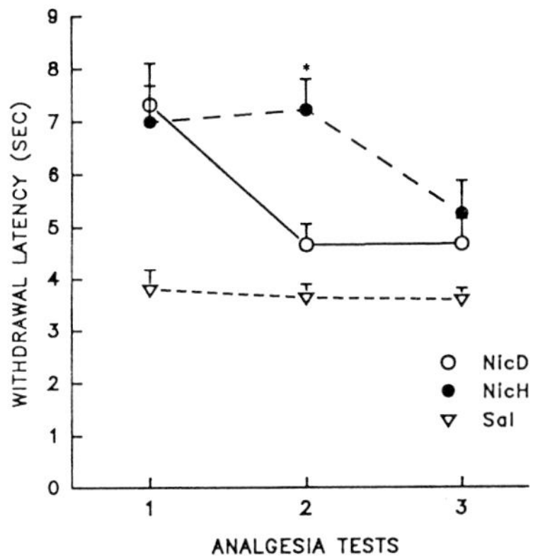

Figure 1. Tolerance to the tail-withdrawal effects of nicotine (0.66 mg/kg) in either a distinctive (NicD) or the home environment (NicH; N=10). *=P<0.001 *vs* NicH (ANOVA + linear contrasts).

Mechanisms of conditioned tolerance

Investigators differ regarding whether the mechanisms mediating learned tolerance are the same as those for cellular tolerance (10). Similar mechanisms might involve, for example, a nicotine-induced decrease in receptor sensitivity, which may underlie nonassociative tolerance to nicotine (11). However, such changes would have to be extremely rapid and easily reversible to accommodate conditioned tolerance, the expression of which is temporally-linked to the presence of a conditioned stimulus (CS). Alternative mechanisms include homeostatic counter-responses which become conditioned to the CS and either directly oppose, or indirectly antagonize the drug response (6). One such response under consideration by us and others is conditioned activation of the hypothalamic-pituitary-adrenocortical (HPA) system and release of corticosterone (CORT; 12,13). According to this hypothesis, as conditioned tolerance to repeated nicotine injections develops, predictive environmental cues (CS) come to evoke a con-

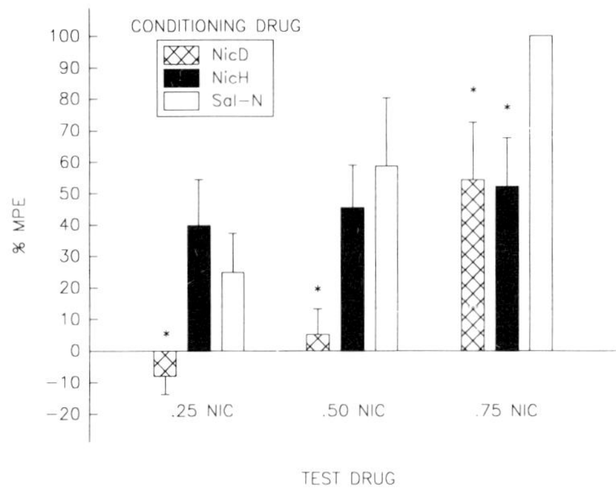

Figure 2. Tolerance to paw-withdrawal effects of nicotine (0.5 mg/kg) in a distinctive (NicD) or a home environment (NicH; N=8). MPE (maximum possible effect) = 60s cutoff - saline controls (17.4±2.7s SEM). *=P<0.001 vs respective Sal-N (acute nicotine).

ditioned release of CORT, which reduces responsiveness to nicotine. In the absence of the CS, the conditioned response is not elicited and the reduction in responsiveness, due to conditioning, does not occur. Support for this hypothesis comes from several laboratories.

Stimulation of CORT is an unconditioned response to nicotine. There is mounting evidence that environmental cues (CS) associated with nicotine delivery evoke a conditioned CORT response (CR). The development of tolerance to nicotine in mice was accompanied by a pre-injection elevation in CORT (12). We have reported evidence for a similar elevation, in rats, temporally-related to the presentation of environmental events associated with nicotine (13). In a direct test of this hypothesis, nicotine (0.66 mg/kg) was paired either with a distinctive environment 100% of the time or with the home environment 50% of the time. Also included were saline-injected controls receiving either saline or nicotine on the test day, and home cage controls. Significant tolerance developed to the tail-withdrawal effects of nicotine for distinctive-environment, but not for home-environment rats after 9 injections (data not shown). The critical test occurred on the next day. If conditioned tolerance is mediated by a CORT response to drug-related cues, then those cues should elicit a CORT response in anticipation of nicotine administration. The animals in each group were placed into the distinctive environment (with or without a saline injection) and sacrificed for CORT measurement 5 minutes later. Rats that experienced nicotine in the home environment tended to show an increase in CORT when injected with saline in the distinctive environment, perhaps reflecting a generalization of handling and injection cues (Figure 3). However, only rats with a history of nicotine in the distinctive environment responded to that environment with a CORT response significantly higher than saline controls, indicating that the onset of the CORT response was more rapid in those rats.

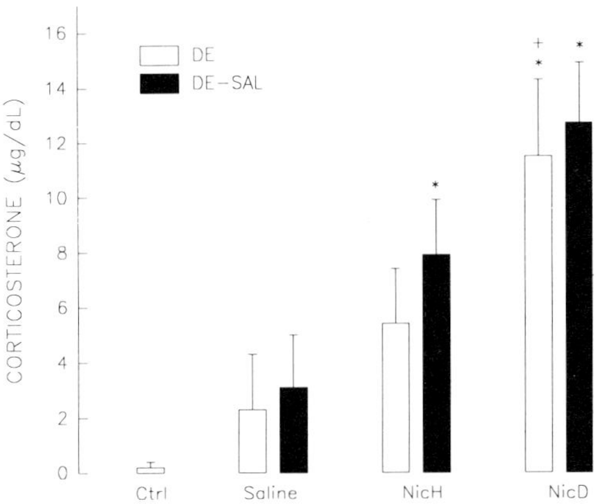

Figure 3. CORT response to a distinctive environment with (DE-SAL) or without (DE) saline after a history of nicotine in the DE (NicD) or in the home environment (NicH; N=10). Saline & Ctrl = saline injected (N=5) or untreated controls (N=10). *=P<0.01 vs respective Saline; +=P<0.05 vs NicH(DE).

The hypothesis that a conditioned CORT response participates in conditioned tolerance to nicotine depends on the effectiveness of CORT to acutely and rapidly reduce responsiveness to nicotine. Chronic administration of CORT to mice reduces several effects of nicotine, whereas adrenalectomy increases nicotine responsiveness (see 11). However these studies dealt with chronic, not acute effects. We have recently shown that *acute* administration of CORT, or stressors that release CORT, reduce nicotine-responsiveness within the 5-10 minute time-frame corresponding to our usual conditioning ritual, i.e. CS-drug interval (13). CORT injections 10 minutes before nicotine attenuated its antinociceptive effects (tail-withdrawal response) and accelerated the development of tolerance. Of particular interest is the finding that exposure to a distinctive environment elevated CORT after it had been paired with nicotine on 5 previous trials, but not in rats that were placed in this environment for the first time. Similarly, preexposure to the distinctive environment significantly decreased nicotine responsiveness only after the first trial. There was no evidence that repeated nicotine elevated baseline levels of circulating CORT. One possible mechanism by which CORT decreases nicotine-responsiveness is by reducing nicotine's actions at nicotinic-cholinergic receptors (11). The onset of this effect must be extremely rapid for it to be considered a plausible mediator of associative tolerance.

The demonstration that stress or CORT-administration alters nicotine sensitivity raises the question of whether the disruption of tolerance after environmental change, seen in our earlier studies, was an *unconditioned* stress effect. We have discussed elsewhere the reasons for rejecting this hypothesis (13). For example, in some cases, changing the physical environment involved switching the injections from the testing to the home environment, which was not accompanied by a CORT response or change in the behavioral effect of nicotine in control animals. Moreover, stress decreases, whereas environmental change increases nicotine responsiveness.

Concluding remarks

It is worthwhile to review some of what we do not know about conditioned tolerance to nicotine. When nicotine is administered to rats by discrete injections, tolerance to some of its effects can include an associative component. We do not know whether this applies to all of nicotine's actions, to species other than rats, and, perhaps most importantly, when the drug is self-administered. The fact that CORT decreases nicotine responsiveness and that a conditioned release of CORT parallels tolerance development does not prove a causative relationship. While we have focused on CORT in our initial work, other conditionable neural or neuroendocrine responses, such as CRF and ACTH, cannot be ruled out as important mediators in addition to, or instead of CORT, and must be considered in future work (13).

Acknowledgements: *This research was supported by DA07546 and MH24114*

References

1. Antelman SM, Caggiula AR, Kocan D, Knopf S, Meyer D, Edwards D, Barry III H. One experience with "lower" or "higher" intensity stressors, respectively enhances or diminishes responsiveness to haloperidol weeks later: Implications for understanding drug variability. Brain Res 1991; 566: 276-283.
2. Epstein LH, Caggiula AR, Stiller RL. Environment-specific tolerance to nicotine. Psychopharm 1989; 97: 235-237.
3. Epstein LH, Caggiula AR, Perkins KA, Bennett SM and Smith J. Context-dependent tolerance to the heart rate effects of smoking Pharmacol Biochem Behav 1991; 39:15-19.
4. Caggiula AR, Epstein LH, Stiller RL. Changing environmental cues reduces tolerance to nicotine-induced anorexia. Psychopharm 1989; 99: 389-392.
5. Caggiula AR, Epstein LH, Antelman SM, Saylor SS, Perkins KA, Knopf S, Stiller RL. Conditioned tolerance to the anorectic and corticosterone-elevating effects of nicotine. Pharmacol Biochem & Behav 1991; 40: 53-59.
6. Poulos CX, Cappell H. Homeostatic theory of drug tolerance: A general model of physiological adaptation. Psychol Rev 1991; 98(3): 390-408.
7. Clarke PBS and Kumar R. Characterization of the locomotor stimulant action of nicotine in tolerance rats. Br J Pharmacol 1983; 80: 587-594.
8. Caggiula AR, Epstein LH, Perkins KA and Saylor S. Different methods of assessing nicotine-induced antinociception may engage somewhat different neural mechanisms. Int Symposium on Nicotine, Montreal, Canada 1994 (Abstract).
9. Baker TB, Tiffany ST. Morphine tolerance as habituation. Psychological Review 1985; 92: 78-108.
10. MacKenzie-Taylor D and Rech RH. Cellular and learned tolerances for ethanol hypothermia. Pharmacol Biochem Behav 1991; 38: 29-36.
11. Pauly JR and Collins AC. An autoradiographic analysis of alterations in nicotinic cholinergic receptors following 1 week of corticosterone supplement. Neuroendocrin 1993; 57: 262-271.

12. Pauly JR, Grun EU, and Collins AC. Tolerance to nicotine following chronic treatment by injections: a potential role for corticosterone. Psychopharmacology 1992; 108: 33-39.
13. Caggiula AR, Epstein LH, Antelman SM, Saylor S, Knopf S, Perkins KA and Stiller RL Acute stress or corticosterone administration reduces responsiveness to nicotine: Implications for a mechanism of conditioned tolerance Psychopharm 1993; 111: 495-507.

Neuronal, trophic and endocrine effects of nicotine

Effects of Nicotine on
Biological Systems II
Advances in Pharmacological Sciences

DEVELOPMENTAL EFFECTS OF NICOTINE

T.A. Slotkin

Department of Pharmacology
Duke University Medical Center
Durham, NC, 27710, USA

Summary: Nicotine, acting at fetal CNS nicotinic cholinergic receptors, produces neurobehavioral teratogenesis. By contrasting acute nicotine injections to pregnant rats with continuous infusions (implanted minipumps), we have been able to separate the contributing variables of hypoxia/ischemia, maternal nutrition and fetal growth retardation from the direct effects of nicotine exerted on developing neurons. Nicotine mimics the natural actions of acetylcholine as a neurotrophic factor, but because exogenous drug administration elicits the trophic effect prematurely, neural cell replication is arrested at the incorrect time, leading to deficiencies in cell number and synaptic function.

Introduction

Cigarette smoking during pregnancy causes growth retardation as well as neurobehavioral abnormalities and increased incidence of Sudden Infant Death (1-4). Nevertheless, it has been difficult to attribute these alterations to nicotine exposure *per se* because of the presence of substantial hypoxic/ischemic insult during cigarette smoking, as well as a number of other confounding variables present in the maternal-fetal unit (Fig. 1). Accordingly, it has been particularly important to design animal models of fetal nicotine exposure that allow for separation of these variables, and this report summarizes the results of such studies to date in the developing rat.

In general, there are two types of nicotine exposure models in use. Injections of nicotine into pregnant rats reproduce some of the features of cigarette smoking because the peak levels of nicotine produce hypoxia/ischemia (5-7). In contrast, continuous infusions of nicotine via implantable osmotic minipumps avoid hypoxia/ischemia while enabling fetal exposures to be conducted at plasma levels of nicotine comparable to those in typical smokers (8,9).

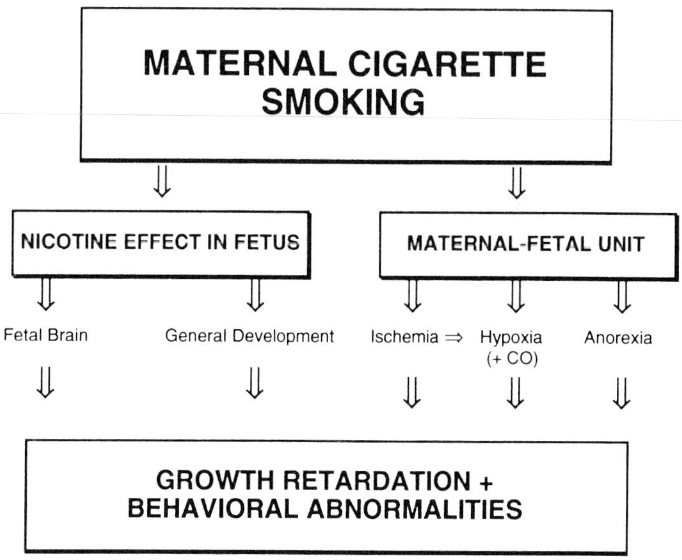

Figure 1. Variables contributing to neurobehavioral abnormalities associated with maternal cigarette smoking.

Nicotine with Fetal Hypoxia/Ischemia: The Nicotine Injection Model

Administration of nicotine subcutaneously to pregnant rats, beginning before the implantation of the embryo in the uterine wall and continued to the end of gestation, produces many of the features common to heavy cigarette smoking, including retarded maternal weight gain, significant fetal loss and intrauterine growth retardation; there are obvious signs of hypoxia/ischemia with each injection (6,10,11). A number of sensitive biochemical tools indicate that this treatment causes cell damage to the fetal brain. First, ornithine decarboxylase (ODC), which is a primary regulator of cell replication/differentiation, is persistently elevated after fetal nicotine exposure by this route (6). Measurement of conventional cell developmental markers, such as nucleic acids and proteins, confirms the interference with cell growth, and the deletion of neurons in favor of glia, a pattern typical of developmental brain damage. At the synaptic level, probes of neuronal activity, such as neurotransmitter biosynthetic enzymes and transmitter turnover, indicate hyperactivity in offspring of animals given nicotine injections during pregnancy, a finding consistent with synaptic hyperproliferation (10,11). Again, this is typical of early neuronal injury (4). But does this reflect direct actions of nicotine on fetal brain, or indirect effects due to hypoxia/ischemia or generalized toxicity in the maternal-fetal unit? A comparison with the effects of hypoxia alone, or even with heavy metals, such as methylmercury, indicates virtually identical spectra of effects (Table 1). In each case, there is growth retardation, postnatal elevation of ODC, increased cell numbers in late-developing regions (increased DNA content, indicative of gliosis), and noradrenergic hyperproliferation/hyperactivity. Thus, the injection model clearly contains components of fetal injury that are not attributable to nicotine acting in the fetal brain, but rather reflect combined actions of the drug on the fetus and on the maternal-fetal unit. This does not obviate the seminal importance of these findings: hypoxia and ischemia are significant components of human cigarette smoking. However, there is clearly a need to evaluate similar markers in the infusion model, in order to separate the confounding variables from direct actions of nicotine in fetal brain. Accordingly, the nicotine infusion model has been developed to address these issues.

Table 1. Fetal Nicotine Exposure via Maternal Injections: Comparisons with Non-Specific Damage

MEASURES	NICOTINE	HYPOXIA	METHYLMERCURY
Cell development	cell damage, gliosis	cell damage, gliosis	cell damage, gliosis
Synaptic development	hyperproliferation	hyperproliferation	hyperproliferation
Transmitter turnover	increase	increase	increase

For detailed references, see Slotkin (4).

Nicotine without Fetal Hypoxia/Ischemia: The Nicotine Infusion Model

Nicotine infusions of 6 mg/kg/day, given to pregnant rats over the same period of gestation as for the injection model, produce plasma levels slightly higher than those in typical smokers (8), and evoke signs of gestational toxicity, including reduced maternal weight gain, fetal resorption and growth retardation (9,12); this is eliminated when the dose level is dropped to 2 mg/kg/day, corresponding to levels lower than those in smokers (13). In either case, brain cell development is still affected adversely (elevated ODC activity), but in this case there is a slowing of cell replication without reactive gliosis (12). By implication, the infusion route is *not* producing the overt neuronal cell death seen with injected nicotine. The effects of nicotine infusions are widespread, affecting virtually all brain regions, instead of targeting late-developing regions as does nicotine injection; this is probably a reflection of the higher effective dose of nicotine delivered to the fetus via continuous infusions as opposed to episodic injections. Similarly, nicotine infusions evoke a different spectrum of effects on synaptic proliferation and activity from those seen with injections. With infusions, synaptic proliferation is relatively unaffected, but synaptic activity is strikingly subnormal, with effects persisting into adulthood (11,14). The neurochemical abnormalities associated with prenatal exposure to nicotine via maternal infusions have been shown to correlate with perturbed behaviors and neuroendocrine parameters linked to dopaminergic and noradrenergic pathways (15,16).

Importantly, all of these effects persist even when the dose of nicotine is reduced to one at which growth of the fetus in unaffected, indicating a targeting of the developing brain by the drug. The reason for this is that developing cholinergic target cells use the onset of cholinergic neurotransmission as a signal to stop dividing and to differentiate into mature neurons. The administration of exogenous nicotine preempts this signal so that the replication/differentiation switch is made prematurely, resulting in permanent deficits (17). Indeed, direct administration of minute amounts of nicotine into the developing brain leads to immediate mitotic arrest, an effect mediated through cholinergic nicotinic receptors (7). The proof that the fetal brain is experiencing prolonged nicotinic stimulation from maternal nicotine administration has been provided by measurements of [^{3}H]nicotine binding sites, which indicate severe desensitization/upregulation even by doses of nicotine that produce plasma levels below those in smokers (9,13).

Conclusions

Animal models of fetal nicotine exposure enable the separation of the multiple variables that produce structural and functional abnormalities in the offspring of women who smoke. First and foremost, we have demonstrated the existence of independent components of hypoxia/ischemia, as compared to the actions of nicotine on the fetus itself. The similarities of the effects of injected nicotine to those of hypoxic brain damage, including cellular death, reactive gliosis and subsequent hyperinnervation and hyperactivity, all indicate that hypoxia is a significant contributor to nicotine's effects in this model. The infusion model, in contrast, does not actually duplicate the maternal-fetal actions associated with smoking, but provides the best means of determining the mechanisms participating in the adverse effects of nicotine on nervous system development. Using infusions, we have shown that nicotine is indeed a neuroteratogen, leading to arrest of cell replication and long-term shortfalls of synaptic activity. Importantly, because nicotine acts on specific neurotransmitter receptors, its actions are exerted at doses below the threshold for growth retardation or other general toxicity markers; this distinguishes nicotine from general teratogens, which typically spare the brain relative to the rest of the fetus. The differences between the two models are summarized in Table 2.

Table 2. Comparative Developmental Effects: Injected vs. Infused Nicotine

Effect	**Injected Nicotine** 6 mg/kg/day	**Infused Nicotine** 2 or 6 mg/kg/day
Hypoxia/Ischemia	Yes	No
Fetal Resorption	Minor	Major (6 mg/kg/day) No (2 mg/kg/day)Ê
Growth Retardation	Yes	Yes (6 mg/kg/day) No (2 mg/kg/day)
Brain Growth Retardation	Yes	Yes (6 mg/kg/day) No (2 mg/kg/day)
Nicotinic Receptor Stimulation in Fetal Brain	Yes	Yes
Neuronal Mitotic Arrest	Yes	Yes
Neuronal Death, Gliosis	Yes	No
Synaptic Hyperproliferation	Yes	No
Eventual Effect on Synaptic Activity	Hyperactivity	Hypoactivity

These findings in animals are important because they can be translated back to the human condition. In the case of smoking and nicotine, if a pregnant woman will not or cannot cease smoking, would nicotine substitution be beneficial? Based on the animal models described here, the answer is a qualified 'yes'. Certainly, the associated hypoxia and ischemia produced by smoking, both as a result of nicotine received in a bolus and the carbon monoxide and cyanide found in cigarette smoke, contribute significantly to alterations in nervous system development. For these epiphenomena of smoking, the variable of intrauterine growth retardation may provide a partial index of the degree of sensitivity of the fetus. However, for the factors related to nicotine itself, the standard indices of safety are inappropriate because the dose-response curve for impaired nervous system development lies to the left of that for growth suppression. Furthermore, if nicotine substitutes are chosen that deliver more nicotine than does smoking, the nicotine-related aspects of fetal nervous system damage may be as bad or worse; this is especially true when typical abuse patterns include smoking in addition to the non-cigarette nicotine source. In light of our findings in rats, the only truly safe course is to discontinue nicotine exposure entirely during pregnancy.

Acknowledgments: *Supported by a grant from the Smokeless Tobacco Research Council.*

References

1. Eriksson M, Larsson G, Zetterstrom R. Abuse of alcohol, drugs and tobacco during pregnancy: consequences for the child. Paediatrician 1979; 8:228-242.
2. Meyer DC, Carr LA. The effects of perinatal exposure to nicotine on plasma LH levels in prepubertal rats. Neurotoxicol Teratol 1987; 9:95-98.
3. Haglund B, Cnattingius S. Cigarette smoking as a risk factor for sudden infant death syndrome: a population-based study. Am J Public Health 1990; 80:29-32.
4. Slotkin TA. In: Maternal Substance Abuse and the Developing Nervous System. Zagon IS, Slotkin TA, Zagon IS, Slotkin TAs, editors. San Diego: Academic Press, 1992: 97-124.
5. Martin JC, Becker RF. The effects of maternal nicotine absorption or hypoxic episodes upon appetitive behavior of rat offspring. Dev Psychobiol 1971; 4:133-147.
6. Slotkin TA, Greer N, Faust J, Cho H, Seidler FJ. Effects of maternal nicotine injections on brain development in the rat: ornithine decarboxylase activity, nucleic acids and proteins in discrete brain regions. Brain Res Bull 1986; 17:41-50.
7. McFarland BJ, Seidler FJ, Slotkin TA. Inhibition of DNA synthesis in neonatal rat brain regions caused by acute nicotine administration. Dev Brain Res 1991; 58:223-229.
8. Lichtensteiger W, Ribary U, Schlumpf M, Odermatt B, Widmer HR. Prenatal adverse effects of nicotine on the developing brain. Prog Brain Res 1988; 73:137-157.
9. Slotkin TA, Orband-Miller L, Queen KL. Development of [^{3}H]nicotine binding sites in brain regions of rats exposed to nicotine prenatally *via* maternal injections or infusions. J Pharmacol Exp Ther 1987; 242:232-237.

10. Slotkin TA, Cho H, Whitmore WL. Effects of prenatal nicotine exposure on neuronal development: selective actions on central and peripheral catecholaminergic pathways. Brain Res Bull 1987; 18:601-611.
11. Navarro HA, Seidler FJ, Whitmore WL, Slotkin TA. Prenatal exposure to nicotine *via* maternal infusions: effects on development of catecholamine systems. J Pharmacol Exp Ther 1988; 244:940-944.
12. Slotkin TA, Orband-Miller L, Queen KL, Whitmore WL, Seidler FJ. Effects of prenatal nicotine exposure on biochemical development of rat brain regions: maternal drug infusions *via* osmotic minipumps. J Pharmacol Exp Ther 1987; 240:602-611.
13. Navarro HA, Seidler FJ, Schwartz RD, Baker FE, Dobbins SS, Slotkin TA. Prenatal exposure to nicotine impairs nervous system development at a dose which does not affect viability or growth. Brain Res Bull 1989; 23:187-192.
14. Seidler FJ, Levin ED, Lappi SE, Slotkin TA. Fetal nicotine exposure ablates the ability of postnatal nicotine challenge to release norepinephrine from rat brain regions. Dev Brain Res 1992; 69:288-291.
15. Lichtensteiger W, Schlumpf M. Prenatal nicotine affects fetal testosterone and sexual dimorphism of saccharin preference. Pharmacol Biochem Behav 1985; 23:439-444.
16. Ribary U, Lichtensteiger W. Effects of acute and chronic prenatal nicotine treatment on central catecholamine systems of male and female rat fetuses and offspring. J Pharmacol Exp Ther 1989; 248:786-792.
17. Navarro HA, Seidler FJ, Eylers JP, Baker FE, Dobbins SS, Lappi SE, Slotkin TA. Effects of prenatal nicotine exposure on development of central and peripheral cholinergic neurotransmitter systems. Evidence for cholinergic trophic influences in developing brain. J Pharmacol Exp Ther 1989; 251:894-900.

Effects of Nicotine on
Biological Systems II
Advances in Pharmacological Sciences

PRESYNAPTIC NICOTINE-GATED CHANNELS POTENTIATE TRANSMISSION AT ACH AND GLUTAMATE-MEDIATED SYNAPSES

D. McGehee and L.W. Role

Columbia University, College of Physicians and Surgeons
Dept. of Cell Biology & Anatomy and The Center for Neurobiology and Behavior
722 W 168th St., New York, N.Y. 10032 USA

Summary: Examination of specific cholinergic and glutamatergic synapses reveals that nanomolar concentrations of nicotine potentiate both spontaneous and evoked transmission. Analysis of miniature excitatory synaptic currents (mepsc) before and after synaptic application of nicotine indicates that presynaptic changes underlie the synaptic facilitation, since nicotine increases mepsc frequency by more than 10 fold without altering the amplitude of the unit mode of mepsc histograms. Image analysis of changes in presynaptic $[Ca]_{int}$, as well as pharmacological and antisense-mediated nicotinic acetylcholine receptor (nAChR) subunit deletion experiments suggest that synaptic transmission is potentiated by direct activation of high affinity presynaptic nAChRs that include the $\alpha 7$ subunit.

Introduction

Despite the diverse actions of nicotine in the CNS - ranging from enhancement of short term memory to alleviating feelings of pain and anxiety - there are strikingly few examples of CNS transmission mediated via postsynaptic nicotinic acetylcholine receptors (nAChRs). In contrast, there is growing evidence for an important role of **pre**synaptic nAChRs in regulating neuronal activity in several regions of the CNS. Earlier studies, that revealed both high affinity nicotine and α-bungarotoxin (α-BTX) binding within the terminal fields of the nigrostriatal and mesolimbic dopamine systems, (e.g. 1) have been extended by recent demonstrations of nicotine-induced changes in transmitter release (for review see 2). However, tests of nicotine-enhanced release *in situ* are limited to indirect measurements of release such as determination of transmitter overflow from synapses treated with pharmacological blockers of transmitter reuptake. Alternatively, investigators have exploited synaptosomal preparations to study nicotine action on synaptic terminals, with relatively few electrophysiological assays of the effects of nicotine on synaptic transmission (but see 3, 4). The goals of the current study are to determine whether CNS nAChRs are generally targeted to presynaptic sites to modify synaptic transmission and then to examine the functional properties, subunit composition and physiological impact of presynaptic nAChRs at identified synapses. To this end we have monitored the effects of nicotine on both evoked and spontaneous transmission at sites of projection of two classes of CNS neurons: those of the medial habenula nucleus (MHN, which synapse upon neurons of the interpeduncular nucleus; IPN) and projections of the visceral motoneurons of the nucleus of Terni (VMT, which synapse upon the neurons of the lumbar sympathetic ganglia; LSG). Our findings indicate that nicotine potentiates both spontaneous and evoked transmission at both cholinergic and glutamatergic CNS synapses by direct activation of high affinity presynaptic nAChRs.

Methods

VMT- LSG co-cultures: Dispersed LSG neurons from embryonic day 10 chickens are prepared and maintained *in vitro* as described in references (5) and (6). Under these conditions the neurons are devoid of non-neuronal cells and are both adrenergic and cholinoceptive (5). Innervation of sympathetic neurons by VMT microexplants is also done according to previously described techniques (5,6). *MHN - IPN co-cultures:* In chick, the bilateral habenula nuclei, are located at the dorsomedial border of the thalamus, extend ≈1 mm from the rostral border of the posterior commissure and are bounded laterally by the stria medularis (7). The IPN is a ventral midline structure caudal to the point where CNIII emerges from the ventral side of the midbrain. Innervated cultures are prepared by preplating habenula explants for 24h followed by IPN neurons dispersed by papain digestion (7).

Patch clamp recording: Recording of macroscopic and synaptic currents employs the whole-cell tight seal recording configuration of the patch clamp technique (8). This technique provides low noise recordings that allow for resolution of elemental synaptic currents. Fabrication of patch electrodes, pipette and bath solutions are all as previously described (9). Currents are recorded with a AXOPATCH 200A patch clamp amplifier and stored on videotape with a PCM digitizer (Instrutech VR-10B) for subsequent off line analysis (see below).

Synaptic current data acquisition and analysis: Continuously recorded (spontaneous) and evoked synaptic currents are stored on videotape and analyzed off-line with software written in Axobasic by D. Madison with an 80486 DX2-66MHz computer equipped with the Axobasic system. The acquisition program samples all events that conform to amplitude and rise-time criteria, both set by the user. Each captured trace includes 20 msec of pre-event baseline data and reliably captures events up to 20 Hz. The analysis software provides amplitude, frequency, rise- and decay-time constant information for each current recorded. Subsequent generation of histograms, cumulative plots, fitting, and statistical analyses are performed with Microsoft Excel 3.0, Sigmaplot 4.1 (Jandel Scientific), and Systat. Statistical analyses of differences between control and treatment groups are evaluated by a two-tailed test. Synaptic current amplitude data are compared by plotting cumulative histograms. These plots are also utilized as estimated cumulative probability distributions for the determination of statistically significant differences between treatment groups, using the Komolgorov-Smirnov test.

Drug Application: Test agents are applied by microperfusion via a large barreled delivery tube with continuous macroperfusion at 1 ml/min. This approach optimizes speed of application (<30 msec), speed of removal and the ease of changing test solutions applied by the same device.

Ca Imaging: Measurement of $[Ca]_{int}$ utilizes fluorescent derivatives of divalent chelating molecules modified for application in the bathing media as a cell permeant, acetoxymethyl ester (-AM), form. Fura 2 was used to measure presynaptic [Ca] because the signal is both a fast and sensitive measure of Ca transients. Emitted light is barrier filtered, transmitted to a CCD camera (Hamamatsu) and the images are background-subtracted and averaged. Fluorescence ratios are calculated in full 8-bit resolution using a digital image processor (ETM Systems). Images provide a representation of overall changes in $[Ca]_{int}$ as well as subcellular localization of changes.

Antisense oligonucleotide design and experimental protocols: Antisense oligonucleotides are targeted to a 15 base sequence spanning the initiation site of each nAChR subunit mRNA. Control oligos include missense sequences of identical composition or oligos mutated at 3 of the 15 bases (same G/C ratio). Our previous studies of the uptake, metabolism, hybridization and block of subunit expression by oligonucleotides determined optimal conditions for specific nAChR subunit block (10,11): under these standard conditions neurons are pretreated with DTT, an irreversible nAChR ligand (bromo-acetylcholine bromide) and DTNB and then incubated for 6-48 hrs with 10 μM oligo in heat inactivated medium. D-oligos are taken up and intact 15mer is maximal within 6 hrs and detected for up to 48 hrs inside the cells (10, 11). In experiments using antisense oligonucleotides to delete nAChR subunits from MHN and VTM microexplants, the deletion is equally efficacious without prior DTT/ BAC/ DTNB treatment *if* the microexplants are incubated with oligonucleotide prior to initiation of neurite outgrowth (i.e. within ≈ 6 hrs of plating).

Results

We examined the role of presynaptic nAChRs in transmission at developing synapses of the MHN to IPN and the VMT to LSG, in an *in vitro* setting because: (a) one can identify and selectively extirpate both the pre and post synaptic partners early on, prior to the establishment of synaptic transmission *in vivo* (b) we developed conditions in which these neurons establish reliable synaptic transmission *in vitro* which, in turn, permits (c) detailed biophysical analyses of synaptic transmission with focal application of nAChR agonists and antagonists, and (d) determination of presynaptic nAChR subunit composition since the expression of specific nAChR subunit genes can be manipulated with antisense oligonucleotides and the effects assayed in detail as in (c). Finally since previous results indicated that both the MHN and VMT neurons express nAChRs and revealed binding sites for nicotine and/or α-BTX within their terminal fields, both classes of neurons appeared to be good candidates for presynaptic nAChR targeting.

1. Nicotine potentiates synaptic transmission at VMT-LSG synapses: Synaptic transmission at VMT-LSG synapses is cholinergic: both spontaneous and evoked synaptic currents recorded in post-synaptic LSG neurons voltage clamped at -60 mV are blocked by curare, hexamethonium, neuronal BTX or mecamylamine (5,12). Within 48 hours of VMT-LSG co-culture both spontaneous and evoked synaptic activity is detected in the majority of neurons within ≈300 μm of VMT explants. Incubation in tetrodotoxin (TTX; 3 μM) blocks evoked release, permitting direct study of the spontaneous synaptic currents. (or "mini" epsc's). Under these conditions, application of nicotine (10-500 nM) in the vicinity of the synapse induces a consistent and dramatic facilitation of transmission characterized by 5-10 fold increase in mepsc frequency. This increase occurs without concomitant alteration in mepsc amplitude indicating that nicotine facilitates transmission selectively by altering presynaptic (rather than postsynaptic) function (12).

2. Nicotine-induced synaptic facilitation at VMT-LSG synapses is mediated by nAChRs distinct from LSG nAChRs in both pharmacology and cellular distribution: Both the agonist and antagonist profiles of nicotine-induced synaptic facilitation differ from those of the nAChRs expressed by the post-synaptic (LSG) neurons. Thus, the concentration of nicotine required for threshold and (approximate) half maximal activation of nAChRs mediating synaptic facilitation are 20 nM and 200 nM respectively, as opposed to 2 μM and 25 μM for postsynaptic nAChR gating (12). In addition, whereas α-BTX has no effect on the amplitude of mepsc's, it blocks nicotine-induced synaptic potentiation (IC_{50} =70 nM). Note, however, that these data do not necessarily indicate a presynaptic locus for the nAChRs mediating synaptic facilitation: pharmacologically distinct nAChRs could localize to non-synaptic areas of LSG neurons and induce presynaptic facilitation via a retrograde messenger mechanism. Results from two preliminary experiments are counter to this view: (a) measurement of $[Ca]_{int}$ in TTX blocked presynaptic (VMT) neurites in the absence of post-synaptic LSG neurons reveals increases from ≈80 nM to ≈1 μM $[Ca]_{int}$ in response to applied nicotine (b) nicotine-induced facilitation of VMT-LSG synapses is unaffected by internal perfusion of LSG neurons with Ca buffering agents (e.g. BAPTA); the latter test derives from observations that retrograde messenger production is largely Ca dependent.

3. Presynaptic nAChRs in VMT neurons include the α7 subunit: In view of the α-BTX sensitivity of presynaptic nAChRs as well as the expression of α7 and α8 subunit encoding genes in the VMT (12), we tested whether the nAChRs mediating facilitation at VMT-LSG synapses include the α7 and/or α8 subunits. Antisense oligonucleotides targeted against each sequence were tested for their ability to block the α-BTX sensitive nicotine-induced synaptic facilitation. Preliminary findings indicate that antisense deletion of α7 blocks α-BTX-sensitive facilitation; the role of α8 is less clear from experiments conducted to date.

4. Nicotine potentiates synaptic transmission at MHN-IPN synapses via high apparent affinity nAChRs: Our studies indicate that synaptic transmission at MHN-IPN synapses is mediated by activation of both NMDA and non NMDA-type glutamate receptors, despite the presence of cholinergic projections from MHN to IPN and the expression of nAChRs on IPN neurons. This finding confirms and extends earlier work by Brown and colleagues (13, 14). Thus, both spontaneous and evoked synaptic currents are blocked by APV plus CNQX and are unaffected by mecamylamine (12). Incubation with TTX and subsequent application of nicotine (10-500 nM) in the vicinity of MHN-IPN synapses facilitates the release of glutamate without changing the magnitude of the spontaneous postsynaptic currents in a manner similar to that at VMT-LSG synapses. That is, nicotine increases mepsc frequency by 4-7 fold without concomitant alteration in mepsc amplitude. Furthermore, direct stimulation of neurites emanating from the MHN microexplant (without TTX), in the presence and absence of nanomolar concentrations of nicotine strongly potentiates evoked responses: epsc amplitudes are increased by more than 2 fold and the number of failures in response to a constant stimulus is decreased by nearly half.

5. Micromolar nicotine depresses synaptic transmission at MHN-IPN: Preliminary studies of higher nicotine concentrations reveal depressive as well as facilitatory presynaptic effects at MHN-IPN synapses. In these studies MHN-IPN co-cultures were incubated with TTX and the synaptic area subsequently exposed to nicotine at a concentration of 10 μM. At these higher agonist concentrations, nicotine profoundly decreases mepsc frequency such that few miniature synaptic currents are recorded (D.M. & L.R., unpublished observations).

Discussion

Examination of ACh-mediated transmission at VMT-LSG synapses reveals pronounced synaptic facilitation by nicotine at concentrations two orders of magnitude lower than those required to activate postsynaptic nAChRs. Nicotine increases calcium in VMT neurites and this response, like the enhanced transmission is blocked by α-BTX. Furthermore, deletion of the α7 subunit gene from the presynaptic neurons by antisense oligonucleotides removes all α-BTX sensitive synaptic facilitation. We also find that nicotine potently regulates spontaneous and evoked synaptic transmission at MHN-IPN synapses, where glutamate appears to act as the primary transmitter. Up to 100 nanomolar nicotine potentiates MHN-IPN synaptic transmission without altering either the holding current or the amplitude of synaptic currents in the innervated IPN neuron. In contrast, higher nicotine concentrations (e.g. 10 μM) profoundly depress transmission. This biphasic effect of nicotine on MHN-IPN transmission may underlie mixed facilitatory and depressive actions of nicotine at other CNS synapses.

Overall, our studies support an important regulatory role for nicotine in the CNS, controlling the release of fast excitatory transmitters such as ACh and glutamate by activation of presynaptic AChRs. In view of the paucity of synapses where post-synaptic nAChRs have been shown to mediate synaptic transmission in the CNS and the mounting evidence for profound effects of nicotinic agonists on transmitter release in brain regions implicated in both behavioral and cognitive functions, we suggest that CNS nACh receptor-ology may require a 180° turn and a closer look at the presynaptic role of nAChRs.

References.

1. Clarke PBS, Pert A. Autoradiographic evidence for nicotine receptors on nigrostriatal and mesolimbic dopamine neurons. Brain Research. 1985; 348:355-358.
2. Wonnacott S, Drasdo A, Sanderson E, Rowell P. Presynaptic nicotinic receptors and the modulation of transmitter release. Ciba-Found-Symp. 1990; 152:87-101.
3. Vidal C, Changeux JP. Pharmacological profile of nicotinic acetylcholine receptors in the rat prefrontal cortex: an electrophysiological study in slice preparation Neurosci. 1989; 29:261-270.
4. Lena C, Changeux J-P, Mulle C. Evidence for "preterminal" nicotinic receptors on GABAergic axons in the rat interpeduncular nucleus. J. Neurosci. 1993; 13: 2680-88.
5. Role LW. Neural regulation of acetylcholine sensitivity of embryonic sympathetic neurons. Proc. Natl. Acad. Sci. USA 1988; 85:2825-2829.
6. Gardette R, Listerud MD, Brussaard AB, Role LW. Developmental changes in transmitter sensitivity and synaptic transmission in embryonic chicken sympathetic neurons innervated *in vitro*. Dev. Biol. 1991; 147:83-95.
7. Brussard AB, Yang X, Doyle JP, Huck S, Role LW. Developmental regulation of multiple nicotinic AChR channel subtypes in embryonic chick habenula neurons: Contributions of both the $\alpha2$ and $\alpha4$ subunit genes. Pflugers Archiv. 1994; in press.
8. Hammil OP, Marty A, Neher E, Sakmann B, Sigworth FJ. Improved patch-clamp techniques for high-resolution current recording from cells and cell-free membrane patches. Pflugers Arch. 1981; 391:85-100.
9. Moss BL, Schuetze SM, Role LW. Functional properties and developmental regulation of nicotinic acetylcholine receptors on embryonic chicken sympathetic neurons. Neuron 1989; 3:597-607.
10. Yu C, Brussaard AB, Yang X, Listerud M, Role LW. Uptake of antisense oligo-nucleotides and functional block of AChR subunit gene expression in primary embryonic neurons. Dev. Genetics. 1993; 14:296-304.
11. Listerud M, Brussaard AB, Devay P, Colman DR, Role LW. Functional contribution of neuronal AChR subunits by antisense oligonucleotides. Science 1991; 254:1518-21.
12. McGehee D, Heath M, Role LW. Potentiation of fast excitatory transmission in the CNS by presynaptic nicotinic AChRs. (1994, submitted to Science.)
13. Brown DA, Docherty RJ, Halliwell JV. Chemical transmission in the rat interpeduncular nucleus in vitro. J. Physiol. 1893; 341:655-670.
14. Brown DA, Docherty RJ, Halliwell JV. The action of cholinomimetric substances on impulse conduction in the habenulo-interpeduncular pathway of the rat *in vitro*. J. Physiol. 1984; 353: 101-109.

Effects of Nicotine on
Biological Systems II
Advances in Pharmacological Sciences

ELECTROPHYSIOLOGY OF NEURONAL NICOTINIC RECEPTORS IN THE CNS

Christophe Mulle, Clément Léna and Jean-Pierre Changeux
Laboratoire de Neurobiologie Moléculaire, URA CNRS D1284,
Institut Pasteur, Paris, France

Summary: We have analyzed the properties of nicotinic acetylcholine receptors (nAChRs) in several regions of rat and mouse central nervous system. We have demonstrated the existence of several functionally distinct subtypes of nAChRs. A direct role of nAChR in synaptic transmission could not be found under our experimental conditions and in these preparations. However, we have shown that nAChRs acting at a preterminal (tetrodotoxin-sensitive) or at a presynaptic (tetrodotoxin-insensitive) level can modulate GABAergic synaptic transmission in several brain regions.

Introduction

Genes encoding subunits of neuronal nicotinic acetylcholine receptors (nAChR) are differentially expressed throughout the CNS. In heterologous expression systems, these subunits can assemble into functionally distinct nAChRs. Several recent studies have begun to unravel the large functional diversity of nAChRs in the vertebrate nervous system, but their role in the physiology of synaptic circuits in the CNS remains elusive.

In order to improve our knowledge of nAChR function in rodent CNS, we have examined the properties of native nAChRs in various diencephalic structures including the medial habenula (MHb), the interpeduncular nucleus (IPN) and the thalamus, using dissociated neurons and slice preparations. These nuclei show a high level of expression for several nAChRs subunits; in addition, both the MHb and the IPN receive a massive cholinergic input. nAChRs present in these various regions share a number of common properties but also show clearly distinct pharmacological profiles. At this stage, we failed to find evidence for a direct role of nAChRs in synaptic transmission in the brain regions studied: however, we show that nAChRs can modulate neurotransmitter release by acting at two different levels on the axon terminal.

Biophysical and pharmacological properties of nAChRs in rodent CNS

Fast application of acetylcholine on neurons from the MHb and the IPN elicits fast inward currents carried by cations (1). These currents exhibit a strong inward rectification, a property shared with most neuronal nAChRs (reviewed in (2)). Similar currents can be activated in these cells by (-)nicotine, cytisine and dimethylphenylpiperazinium (DMPP). Nicotinic responses are reversibly blocked by curare, dihydro-β-erythroidine (DHβE), hexamethonium and mecamylamine, but are insensitive to both α-bungarotoxin and neuronal bungarotoxin. Clear differences in the order of potency of agonists as well as in the IC50 for antagonists between nAChRs in the MHb and the IPN indicate the existence of distinct subtypes of nAChRs in the CNS (1). Differences in the pharmacological profile of these receptors is accompanied with a significant difference in single-channel conductance and kinetics of nAChRs recorded in patches from MHb and IPN neurons.

Thalamic neurons in rodents yet express a different nAChR subtype (C. Léna, P. Vincent, manuscript in preparation). Nicotinic agonists activate an inwardly rectifying current in all neurons recorded from the anteroventral, anterodorsal and laterodorsal nuclei of the thalamus. The order of potency of nicotinic agonists at a concentration of 10 μM is nicotine>DMPP> cytisine. These responses are blocked by micromolar concentrations of hexamethonium, mecamylamine and DHβE. The presence of nAChRs on thalamic neurons is anticipated since the thalamus exhibit one of the highest densities of high affinity nicotine binding sites in the brain (3). Some nicotinic responses have already been recorded in a small fraction of cells in the guinea-pig geniculate nuclei and the cat geniculate nuclei (4), and in the rat ventrobasal thalamus (5). Interestingly, the thalamus specific nuclei only receive a limited cholinergic innervation (except the anteroventral and mediodorsal nuclei (6)) raising the question of how these receptors are activated in vivo. Lesion studies have shown the presence of nAChRs on thalamic axon terminals in the neocortex (rat: (7, 8) cat: (9, 10)).

α3, α4, α7, β2, β3, β4 transcripts are detected in the MHb, α2, α4, α5, β2, β3 transcripts in the IPN and α4 and β2 transcript in the thalamus (12-15). Pairwise combinations of most of these subunits yield functional nAChRs in oocytes. We find a unique and major population of nAChRs expressed on neuronal soma in the MHb and IPN. The comparison of the phar-

Table 1. Pharmacological profile of native and recombinant nAChRs

	agonist efficacy	subunit expressed	antagonist $IC_{50}(\mu M)$			
			Meca	Hexa	DHßE	Curare
MHb	N > C > A > D [a]	α3, α4, α7, β2, β3, β4	0.01	0.5	30	0.2
IPN	C > A > N >> D [a]	α2, α4, α5, α7, β2, β4	0.1	1	2	2
Thalamus	N > D >> C [b]	α4, β2	1	1	0.1	
α3β4 *	C > N = A =D [b]					
α2β4, α4β4*	C > N > A >> D [b]					
α2β2, α4β2*	N > D >> C [b]					

[a] at saturation [b] at 10 μM * taken from (11)
N = nicotine; C = cytisine; A = acetylcholine; D = DMPP

macological profile of recombinant (11) and native receptors indicates a similarity between MHb nAChRs and α3β4 receptor as well as between IPN nAChR and α2β4 receptor (Table 1). The presence of other subtypes of nAChRs on these neurons is not excluded. For instance, in a small fraction of IPN neurons the pharmacology of nAChRs suggested the presence of the β2 subunit, but the rare occurence of these currents prevented their complete pharmacological analysis. The relative order of potency of agonists for thalamic nAChRs at a concentration of 10 μM matches with the pharmacological profile of a2β2 and α4β2 nAChRs expressed in oocytes. Since thalamus specific nuclei only express the α4 and β2 subunits, thalamus nAChRs are likely to be comprised of α4 and β2 subunits.

Calcium and nAChRs

Calcium permeability of nAChRs. Activation of nAChRs in a pure $CaCl_2$ (100 mM) external solution activates an inward current carried by Ca^{2+} (16). Ion substitution experiments indicate that Ca^{2+} and Na^{+} have approximately the same permeability. The permeability of nAChRs to Ca^{2+} was further examined using simultaneous patch-clamp recordings and fura-2 fluorescence measurements. In extracellular Ringer solution, application of nicotine causes a rapid increase in the intracellular Ca^{2+} concentration up to the micromolar range which parallels the amplitude of nAChR currents. nAChRs may thus serve a role in regulating cytosolic Ca^{2+} complementary to that of voltage-dependent Ca^{2+} channels and NMDA receptors which only flux Ca^{2+} at potentials above -40 mV (see (17)). Activation of nAChRs primarily results in cell depolarization. However, sustained opening of nAChRs can result in a significant intracellular Ca^{2+} concentration increase, especially if the membrane remains hyperpolarized. This can be the case if an initial Ca^{2+} influx activates ion channels that tend to maintain the membrane at a hyperpolarized potential. Application of nicotinic agonists in voltage-clamped MHb neurons elicits a cationic current (and a Ca^{2+} influx) followed by a Ca^{2+} activated chloride current (16). Similarly, Ca^{2+} influx through nAChRs has been reported to activate Ca^{2+}-dependent potassium channels in cochlear hair cells (18). As a result, membrane potential tends to remain hyperpolarized, thus maximizing Ca^{2+} influx through nAChR channels. Increasing internal Ca^{2+} concentration can in turn regulate a wide variety of cellular functions. For instance, we have observed a sustained Ca^{2+}-dependent reduction in GABAergic currents caused by activation of nAChRs in MHb neurons.

Modulation of nAChRs by extracellular calcium: Increasing the extracellular Ca^{2+} concentration potentiates the response to nicotine of MHb and IPN neurons in a dose-dependent manner. This effect occurs in a physiological range of extracellular Ca^{2+} concentrations (0.2-2 mM), is voltage-independent and is more pronounced at low concentrations of nicotinic agonists. Other divalent cations (barium, strontium) also strongly potentiate nAChR responses, while magnesium hardly exerts a modulatory effect. Modulation of nicotinic receptor function by external calcium is shared by other native and recombinant neuronal nAChRs (rat parasympathetic neurons (19, 20); bovine adrenal chromaffin cells : (21); pairwise combinations of α2, α3, α4, and β2, β4 (22)) but is not a property of muscle nicotinic receptors.

Divalent cations are known to reduce single-channel conductance of cationic ligand-gated channel such as muscle nicotinic receptors (23) or glutamate receptors (24), by acting at a site within the channel. Increasing extracellular Ca^{2+} concentration from 0 mM to 4 mM also results in a 40% decrease of the elementary conductance of MHb nAChRs, in contradiction with a macroscopic potentiation effect. The single channel opening frequency is however greatly increased, while the open time distribution remains unchanged. In a simple allosteric model of the nAChR kinetics, a 15 fold increase in the on-rate of the activation process could account for all the effects observed.

Thus, changes in extracellular Ca^{2+} may modulate cholinergic synapses in the CNS.

Interestingly, there is no apparent effect of Ca^{2+} at saturating concentrations of acetylcholine, indicating that under these conditions the Ca^{2+}- induced increase in nAChRs probability of opening closely compensates for the single channel conductance diminution. Therefore, modulation of nAChR by extracellular Ca^{2+} could ensure a constant nicotinic response to saturating concentrations of ACh (as occurs at the neuromuscular junction) in conditions of depletion of extracellular calcium due either to high synaptic activity or to calcium entry into the cell through the nAChR.

Physiological functions of nAChRs in the CNS

Despite the relatively high abundance of nAChRs in the mammalian CNS, evidence for a direct involvement of nAChRs in synaptic transmission is scarce. The presence of nAChRs on axon terminals also suggests a role for these receptors in the modulation of neurotransmitter release. We have addressed the question of the physiological function of nAChRs located either in the somato-dendritic compartment or on axon terminals.

A postsynaptic role for nAChRs? The IPN and the MHb receive an abundant cholinergic innervation (see refs in (25, 26)) and exhibit a low abundance of muscarinic binding sites. It is therefore anticipated that stimulation of afferent cholinergic fibers would trigger synaptic events in these structures. In spite of many attempts, extracellular recordings (27) and patch-clamp experiments in thin slices ((28), CM, CL and PV) have failed to demonstrate nicotinic synaptic transmission in MHb and IPN. It is unclear whether nicotinic synapses do not exist in these nuclei or whether the cholinergic fiber stimulation is ineffective under our experimental conditions. Other studies have suggested nicotinic synaptic transmission to occur in the mammalian CNS (pedunculopontine-substantia nigra pathway in rat (29); local connections in the striatum (30, 31); laterodorsal tegmentum-thalamus pathway in cat (32)). We were unable to record any spontaneous nicotinic synaptic currents nor currents evoked by local stimulation in target neurons of these three pathways in rodents. In addition to a putative role in synaptic transmission, synaptic or extrasynaptic nAChRs could ensure a Ca^{2+} influx either localized to a small region of the neuron, or under conditions where NMDA receptors or voltage-activated Ca^{2+} channels are not activated.

Function of nAChRs on axon terminals: Transport of nAChRs along axons has been documented in the primary visual afferent system (33), the septo-habenulo-interpeduncular pathway (34), the nigro-striatal pathway (35) and the thalamo-cortical pathway (7-9). In addition, biochemical measurements of neurotransmitter release from synaptosomes and brain slices indicate the presence of nAChRs on cholinergic, dopaminergic and serotoninergic terminals (36). We sought a physiological counterpart to these morphological and biochemical studies, and we found that nAChRs located on axon terminals could modulate GABAergic transmission through both a TTX-dependent (preterminal effect) and a TTX-independent (presynaptic effect) mechanism.

Low concentrations of nicotinic agonists in thin slices of IPN (37) strongly increase the frequency of GABAergic synaptic currents. The distribution of the amplitude of synaptic currents is significantly shifted toward large values and this effect is blocked by TTX, indicating that the action of nicotinic agonists is mediated by an action potential. The nAChRs responsible for this effect are thus probably located either in the somato-dendritic compartment of GABAergic interneurons or on GABAergic axons within the IPN. In a preparation of acutely dissociated neurons which retain synaptic terminals attached to their soma (as shown by immunoreacivity against synaptophysin), nicotinic agonists also increase the frequency of GABAergic discharge in a TTX-dependent manner. This result suggests that the effect is mediated by nAChRs located on GABAergic axon terminals, but that an action potential is needed to trigger neurotransmitter release (preterminal effect). Our observation could be generalized to other regions, since "preterminal"-like effects of nicotine were observed in the neocortex and in several nuclei of the thalamus (CL and PV, unpublished). Most thalamic nuclei of rat and mouse do not contain local interneurons (38), and inhibition of a given thalamic relay neuron arises from neurons of the reticular nucleus located on the border of the thalamus and probably absent from the slice preparation. The TTX-dependent increase in GABAergic synaptic discharge is thus probably not due to an action at nAChRs located on the soma of GABAergic interneurons. This again suggests an action at preterminal nAChRs located on GABAergic afferents.

In the mouse geniculate nucleus, nicotinic agonists also increase GABAergic synaptic discharge in a TTX-independent manner (CL and PV, unpublished). No change in the distribution of postsynaptic currents amplitude is observed, indicating that nicotine increases the frequency of miniature postsynaptic currents. In addition to preterminal effects on GABAergic transmission, nicotinic agonists can exert a presynaptic effect by acting at nAChRs located directly on presynaptic nerve terminals. The presynaptic effect is not observed in all thalamic nuclei. A large GABAergic innervation to the geniculate nucleus arises from the reticular thalamus which expresses mostly the α4 and β2 nAChR subunits. The effect we observed might thus correspond to the function of α4β2 nAChRs. Consistent with this hypothesis, cytisine is a much less potent presynaptic agonist than nicotine. Recently, presynaptic nAChRs were also observed in the chick ventral lateral geniculate nucleus (39). These results suggest that activation of preterminal as well as presynaptic nAChRs promotes inhibition in the CNS by activating the release of GABA. However nicotinic agonists were also shown to facilitate excitatory transmission in the neocortex (40). Axonal nAChRs appear to serve various functions, which will require futher characterization throughout the CNS.

Conclusion

Our studies of the properties of nAChRs in various regions of the CNS underline the diversity of neuronal nAChRs and of the effects of nicotinic agonists. However, we have failed to demonstrate the presence of nicotinic synapses. This might be due to our unability to properly stimulate cholinergic afferents. Alternatively, it is possible that nAChRs are not activated by ACh released at a conventional synapse, but are rather activated by a paracrine mechanism.

References

1. Mulle C, Vidal C, Benoit P, Changeux JP. Existence of different subtypes of nicotinic acetylcholine receptors in the rat habenulo-interpeduncular system. J Neurosci 1991; 11:2588-97.
2. Sargent PB. The diversity ofneuronal nicotinic acetylcholine receptors. Annu Rev Neurosci 1993; 16:403-43.
3. Clarke PBS, Schwartz RD, Paul SM, Pert CB, Pert A. Nicotinic binding in rat brain: Autoradiographic comparison of [^{3}H]-ACh, [^{3}H]-nicotine and [^{125}I]α-bungarotoxin. J Neurosci. 1985; 5:1307-1313.
4. McCormick DA, Prince DA. Actions of acetylcholine in the guinea-pig and cat medial and lateral geniculate nuclei, in vitro. J Physiol 1987; 392:147-165.
5. McLennan H, Hicks TP. Pharmacological characterization of the excitatory cholinergic receptors ofrat central neurones. Neuropharmacology 1978; 17:329-334.
6. Levey AI, Hallanger AE, Wainer BH. Choline acetyltransferase immunoreactivity in the rat thalamus. J Comp Neurol 1987; 257:317-332.
7. Prusky GT, Arbuckle JM, Cynader MS. Transient concordant distributions of nicotinic receptors and acetylcholinesterase activity in infant rat visual cortex. Dev Brain Res 1988; 39:154-159.
8. Sahin M, Bowen WD, Donoghue JP. Location ofnicotinic and muscarinic cholinergic and mu-opiate receptors in rat cerebral neocortex: evidence from thalamic and cortical lesions. Brain Res 1992; 579:135-47.

9. Prusky GT, Shaw C, Cynader MS. Nicotine receptors are located on lateral geniculate nucleus terminals in cat visual cortex. Brain Res 1987; 412:131-138.
10. Parkinson D, Katz KE, Daw NW. Evidence for a nicotinic component to the action of acetylcholine in the cat visual cortex. Exp Brain Res 1988; 73: 553-568.
11. Luetje CW, Patrick J. Both alpha- and beta-subunits contribute to the agonist sensivity of neuronal nicotinic acetylcholine receptors. J Neurosci 1991; 11:837-45.
12. Wada E, Wada K, Boulter J, Deneris E, Heinemann S, Patrick J, Swanson LW. Distribution of Alpha2, Alpha3, Alpha4, and Beta2 neuronal nicotinic subunit mRNAs in the central nervous system: a hybridization histochemical study in rat. J Comp Neurol 1989; 284:314-335.
13. Wada E, McKinnon D, Heinemann S, Patrick J, Swanson LW. The distribution of mRNA encoded by a new member of the neuronal nicotinic acetylcholine receptor gene family (alpha 5) in the rat central nervous system. Brain Res 1990; 526:45-53.
14. Deneris ES, Boulter J, Swanson LW, Patrick J, Heinemann S. Beta 3: a new member of nicotinic acetylcholine receptor gene family is expressed in brain. J Biol Chem 1989; 264:6268-72.
15. Seguela P, Wadiche J, Dineley MK, Dani JA, Patrick JW. Molecular cloning, functional properties, and distribution of rat brain alpha 7: a nicotinic cation channel highly permeable to calcium. J Neurosci 1993; 13:596-604.
16. Mulle C, Choquet D, Korn H, Changeux JP. Calcium influx through nicotinic receptor in rat central neurons: its relevance to cellular regulation. Neuron 1992; 8:135-43.
17. Schneggenburger R, Zhou Z, Konnerth A, Neher E. Fractional contribution of calcium to the cation current through glutamate receptor channels. Neuron 1993; 11:133-143.
18. Fuchs PA, Murrow BW. Cholinergic inhibition of short (outer) hair cells of the chick's cochlea. J Neurosci 1992; 12:800-809.
19. Fieber LA, Adams DJ. Acetylcholine-evoked currents in cultured neurones dissociated from rat parasympathetic cardiac ganglia. J Physiol Lond 1991; 434:215-37.
20. Adams DJ, Nutter TJ. Calcium permeability and modulation of nicotinic acetylcholine receptor-channels in rat parasympathetic neurons. J Physiol. 1992; 86:67-76.
21. Zhou Z, Neher E. Calcium permeability of nicotinic acetylcholine receptor channels in bovine adrenal chromaffin cells. Pfl=FCgers Arch 1993; 425:511-517.
22. Vernino S, Amador M, Luetje CW, Patrick J, Dani JA. Calcium modulation and high calcium permeability of neuronal nicotinic acetylcholine receptors. Neuron 1992; 8:127-134.
23. Decker ER, Dani JA. Calcium permeability of the nicotinic acetylcholine receptor: the single-channel influx is significant. J Neurosci 1990; 10:3413-3420.
24. Ascher P, Nowak L. The role of divalent cations in the N-methyl-D-aspartate responses of mouse central neurones in culture. J Physiol 1988; 399:247-266.
25. McCormick DA, Prince DA. Acetylcholine causes rapid nicotinic excitation in the medial habenular nucleus of guinea-pig, in vitro. J Neurosci 1987; 7:742-752.
26. Eckenrode TC, Murray M, Haun F. Habenula and thalamus cell transplants mediate different specific patterns of innervation in the interpeduncular nucleus. J Neurosci 1992; 12:3272-3281.
27. Brown DA, Docherty RJ, Halliwell JV. Chemical transmission in the rat interpeduncular nucleus in vitro. J Physiol 1983; 341:655-70.
28. Edwards FA, Alasdair JG, Colquhoun D. ATP receptor-mediated synaptic currents in the central nervous system. Nature 1992; 359: 144-147.
29. Clarke PBS, Hommer DW, Pert A, Skirboll LR. Innervation of substantia nigra neurons by cholinergic afferents from pedunculopontine nucleus in the rat: neuroanatomical and electrophysiological evidence. Neuroscience 1987; 23:1011-9.
30. Misgeld U, Weiler MH, Cheong DK. Atropine enhances nicotinic cholinergic EPSPs in rat neostriatal slices. Brain Res 1982; 253:317-20.
31. Cherubini E, Herrling PL, Lanfumey L, Stanzione P. Excitatory amino acids in synaptic excitation of rat striatal neurones in vitro. J Physiol 1988; 400:677-90.
32. Curro-Dossi R, Pare D, Steriade M. Short-lasting nicotinic and long-lasting muscarinic depolarizing responses of thalamocortical neurons to stimulation of mesopontine cholinergic nuclei. J Neurophysiol 1991; 65:393-406.
33. Swanson LW, Simmons DM, Whiting PJ, Lindstrom J. Immunohistochemical localization of neuronal nicotinic receptor in the rodent central nervous system. J. Neurosci. 1987; 7:3334-3342.

34. Clarke PBS, Hamill GS, Nadi NS, Jacobowitz DM, Pert A. ^{3}H-nicotine and [^{125}I]alpha-bungarotoxin-labeled nicotinic receptors in the interpeduncular nucleus of rats. II. Effects of habenular deafferentation. J Comp Neurol 1986; 251:407-413.
35. Clarke PBS, Pert A. Autoradiographic evidence for nicotine receptor on nigrostriatal and mesolimbic dopaminergic neurons. Brain Res 1985; 348:355-358.
36. Wonnacott S, Irons J, Rapier C, Thorne B, Lunt GG. Presynaptic modulation oftransmitter release by nicotinic receptors. Prog Brain Res 1989; 79:157-63.
37. Lena C, Changeux JP, Mulle C. Evidence for ''preterminal'' nicotinic receptors on GABAergic axons in the rat interpeduncular nucleus. J Neurosci 1993; 13:2680-8.
38. Barbaresi P, Spreafico R, Frassoni C, Rustioni A. GABAergic neurons are present in the dorsal column nuclei but not m the ventroposterior complex ofrats. Brain Res 1986; 382:305-326.
39. MacMahon LL, Kong-Woo Y, Chiappinelli VA. Electrophysiological evidence for presynaptic nicotinic receptors in the avian ventral lateral geniculate nucleus. J Neuropysiol. 1994; 71(2):826-829.
40. Vidal C, Changeux JP. Nicotinic and muscarinic modulations of excitatory synaptic transmission in the rat prefrontal cortex in vitro. Neuroscience 1993; 56:23-32.

Effects of Nicotine on
Biological Systems II
Advances in Pharmacological Sciences

FACTORS CONTROLLING NICOTINIC ACETYLCHOLINE RECEPTOR EXPRESSION ON RAT SYMPATHETIC NEURONS

P. De Koninck and E. Cooper

Department of Physiology, McGill University, Montréal, H3G 1Y6, Canada

Summary: We are studying factors that influence the expression of neuronal nicotinic acetylcholine receptors (nAChR) in rat sympathetic neurons. We find that: 1) nAChR genes are differentially regulated during postnatal development; 2) the developmental increase in α_3 mRNA levels correlate with the increase in ACh-current densities; 3) activity induces α_7 subunit gene expression and increases α-bungarotoxin (α-BGT) binding; 4) choline is a ligand for α_7 receptors in Xenopus oocytes. Finally, we speculate on a possible function for the α-BGT receptor (α-BGTR) in sympathetic ganglia.

Introduction

An important mechanism used by the nervous system to modify the strength of synaptic connections involves altering the expression of postsynaptic receptors. However, little is known about the factors that control postsynaptic receptor numbers on neurons, or how these factors operate. We have been studying various extrinsic influences, such as growth factors and satellite cells, that play a role in controlling neurotransmitter receptor expression in rat autonomic and sensory neurons (1-3). In this study, we are investigating the effects of innervation and neuronal activity on the expression of neuronal nAChRs in sympathetic neurons from neonatal rats. These experiments involve: (a) quantifying mRNA levels for different nAChR transcripts to determine changes in gene expression; and (b) measuring ACh-evoked currents to correlate the appearance of functional nAChRs with changes in gene expression for receptor subunits.

Materials and Methods

Neuronal cultures. Superior cervical ganglia (SCG) were dissected from P1 rats, dissociated and grown in culture as described by Hawrot and Patterson (4) with some modifications (3).

RNase Protection Assays. Total cellular RNA was extracted from freshly dissociated neurons, or cultured neurons for 7 days, as described before (1). The RNase protection assays were based on a method described by Krieg and Melton (5), with minor modifications (1). The hybridization signals were quantified with a phosphor imaging system.

Electrophysiology. ACh-evoked currents were measured with whole-cell patch-clamp techniques on freshly plated neurons (2-12 hrs) or neurons cultured for one week. The recordings were done as described before (3) with some modifications: cells were voltage-clamped at -60 mV while 100 μM or 1 mM ACh, dissolved in perfusion medium, was applied to the cells using a double-barrel fast application system. We always had 500 nM tetrodotoxin in perfusion solutions to avoid activating voltage-gated Na^+ currents by the fast agonist application. For Xenopus oocyte recordings, the oocytes were prepared, injected and recorded as described previously (6).

α-BGT binding. Cultured cells were washed twice with 1 ml of L-15 medium containing 10% horse serum. The neurons were preincubated for 90 minutes in this washing medium with, or without, 100 nM cold α-BGT at 37°C. 150 Kcpm of ^{125}I-α-BGT was then added to the cultures for a 90 min incubation at 37°C. Cells were subsequently washed 6 times over a 30-45 min period, extracted with 0.5 ml 0.5 M NaOH, and the radioactivity was measured in a γ-counter.

Results and Discussion

Postnatal changes in nAChR gene expression in vivo and in culture. Neuronal nAChRs have been studied extensively on autonomic neurons (7). These receptors can be classified according to their binding properties to two snake toxins: alpha-bungarotoxin (α-BGT) and neuronal bungarotoxin (NBT) (7). The receptors that bind NBT are highly localized at synapses, and those that bind α-BGT are found predominantly in the extrasynaptic membrane (8-10). Furthermore, NBT blocks synaptic currents, whereas α-BGT does not (11,12). The molecular identity of these receptors, however, has not been fully resolved, but it is known that the α_7 gene codes for an α-BGT receptor (7). To learn which nAChR genes are expressed in rat sympathetic neurons, we are using RNase protection assays to measure the subunit mRNA levels in SCG neurons. At birth, these neurons express five different nAChR transcripts: α_3,

α_5, α_7, β_2 and β_4, and during the first 2 weeks of postnatal development, the mRNA levels for α_3 and α_7 subunits increase approximately 3 fold, whereas the levels of the other 3 subunits remain relatively constant (1). Since SCG neurons receive the majority of their pre-ganglionic innervation over this period (13), innervation may play a role in the changes in gene expression. To test this, we denervated the ganglia at birth and measured mRNA levels for the nAChR subunits 1-2 weeks later. The effects of denervation on these transcript levels are small; the largest effect is a 30-40% decrease in α_7 mRNA levels (1), indicating that innervation, in part, controls α_7 gene expression. To understand better the mechanisms that control nAChR expression, we cultured SCG neurons under various conditions and measured changes in mRNA levels for nAChR subunits. In culture, the neurons continue to express the same nAChR subunits as *in vivo*. Moreover, after a week in culture, the developmental changes in mRNA levels resemble closely the ones seen from P1 to P7 rats (1): α_3 transcripts increase 2.5 fold, while α_5, β_2 and β_4 transcript levels change little (Table 1).

Table 1. Changes in mRNA levels for nAChR subunits during development in culture.

	Day 7 * (5 mM KCl) % (n=12)	Day 5-7 § (40 mM KCl) % (n=10)	Day 5-7 § (40 mM KCl + 5 μM Verapamil) % (n=3)
α_3	220 ±25	270 ±30	242 ±60
α_5	80 ±10	94 ± 7	79 ±14
α_7	37 ± 5	100 ±10	39 ±14
β_2	120 ±10	130 ± 5	114 ± 8
β_4	115 ± 5	114 ± 3	105 ±20

*Mean changes (±SEM) in mRNA levels at day 7 compared to day 0 (dissociated neurons at the time of plating; =100%). §Treatments for 48 hrs at day 5 to 7.

Activity up-regulates α_7 mRNA synthesis. A notable difference in the pattern of nAChR gene expression between *in vivo* and in culture is that for α_7: its mRNA levels decrease 3 fold within the first few days in culture (Table 1). This finding suggests that extrinsic factors regulate α_7 mRNA levels in neonatal SCG neurons. In view of the effects of denervation on α_7 *in vivo*, a likely factor is the reduced electrical activity in these neurons in culture. To test this, we treated sister cultures with high K^+ for 48 hours to chronically depolarize the neurons.

This treatment increases α_7 mRNA levels 2-3 fold but has only a small effect on the other nAChR subunit mRNA levels (Table 1). Several pathways leading to the induction of gene expression by membrane depolarization involve an elevation in intracellular Ca^{++}; this is also true for the effect of high K^+ on α_7. We find that verapamil, an L-type Ca^{++} channel blocker, completely blocks the induction of α_7 by high K^+ (Table 1). Our preliminary results, using non specific kinase inhibitors, indicate that protein kinases may be involved in this α_7 gene induction pathway.

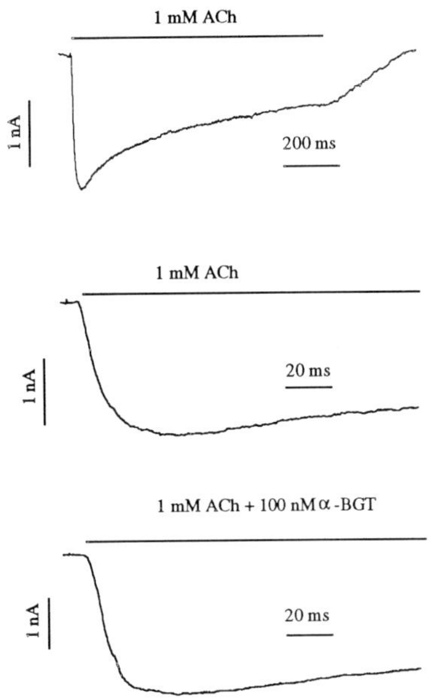

Figure 1. ACh-evoked currents on freshly dissociated neonatal rat SCG neurons. Upper trace: 1 sec rapid application of 1mM ACh; middle trace: same as upper trace but shown on a faster time scale; lower trace: SCG neuron preincubated for 3 hrs in 100 nM α-BGT prior to ACh application.

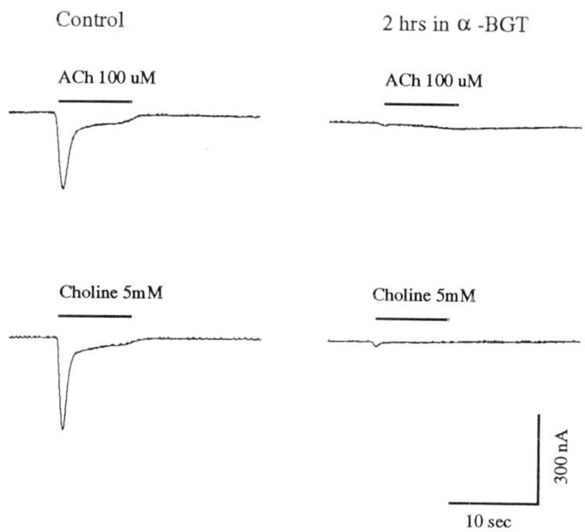

Figure 2. Choline as a ligand for α_7-receptors expressed in Xenopus oocytes. Oocytes were voltage-clamped at -70 mV. Left traces: 100 μM ACh and 5 mM choline; right traces: pre-incubation of the oocytes for 2 hrs in 100 nM α-BGT, prior to agonist application.

Correlative changes in α_3 mRNA levels with ACh-gated current densities. During the first week of development in culture, we observe opposite changes in α_3 and α_7 mRNA levels: α_3 levels increase 2.5 fold while α_7 levels decrease 3 fold. What are the functional consequences of these changes in mRNA levels? To answer this question, we used whole-cell patch clamp techniques to measure ACh-gated current densities on neurons at 2-12 hrs and 7 days in culture (Fig. 1). We find that the ACh-current densities double between 0 and 7 days in culture (Table 2); this increase correlates with the increase in α_3 mRNA levels, suggesting that the amount α_3 subunit may be rate-limiting for the appearance of new functional nAChRs. In contrast, the fact that α_7 mRNA levels decrease over this period suggests that α_7 does not contribute significantly to the ACh-gated currents on SCG neurons. Moreover, several studies have shown that despite the abundance of α-BGT binding sites on autonomic neurons, α-BGT has little effect on ACh-gated currents on these neurons (12) (see also Fig. 1). However, Zhang *et al* (14) showed recently that chick ciliary neurons express a rapidly desensitizing ACh-evoked current that can be blocked by α-BGT. We do not detect comparable currents in SCG neurons (Fig. 1), in spite of the abundant levels of α_7 mRNA in these neurons. This indicates that the α_7 gene product in rat SCG does not form a homomeric receptor with properties similar to that reported for the α_7-receptor in Xenopus oocytes (15).

Correlation of changes in α- mRNA levels with α-BGT binding. To test whether the increase in α_7 mRNA levels after high K^+ treatment translates into more α-BGT binding proteins on the surface, we measured ^{125}I-α-BGT binding on these cultured neurons. Indeed, as the α_7 mRNA levels increase 2-3 fold, the α-BGT binding increases 4-5 fold with high K^+ (Table 2). This finding indicates that the changes in mRNA levels for α_7 reflect changes in the number of surface α-BGTRs and is consistent with the view that the α_7 subunit forms an α-BGT binding protein; however, its role remains unclear.

Table 2. Changes in ACh current densities and α-BGT binding sites in culture.

	ACh[§] current density (pA/pF)	^{125}I-α-BGT binding (cpm/dish)
Day 0	21 ±5(20)	ND
Day 7 (5 mM KCl)	40 ±7(15)	349 ± 53 (3)
Day 5-7 (40 mM KCl)*	49 ±7(10)	1710 ±342 (3)

Values represent mean ±SEM (n). *KCl treatment for 48 hrs at day 5 to 7. [§]100 μM ACh. ND= not determined.

A functional role for the α-BGT receptor. The α-BGTR clearly has properties that differ from the NBT-sensitive synaptic receptors: when expressed in Xenopus oocytes, the rat α_7 gene product assembles into a homomeric functional ACh-gated ion channel that is highly permeable to Ca^{++}, but has a low affinity for ACh (15). In addition, most of the α-BGTRs on autonomic neurons have an extra-synaptic localisation (9, 10), and α-BGT does not block synaptic currents on these neurons (12). Given these properties, it is unlikely that the α-BGTRs have a direct role in synaptic transmission; however, it is possible that they play a modulatory role at synapses. For example, could the α-BGTR be a choline receptor? If so, this receptor may be important at highly active synapses where, as a consequence of acetylcholinesterase, choline would accumulate and diffuse outside the synaptic cleft to where the α-BGTRs are located. The activation of these α-BGTRs would result in a local increase in Ca^{++} concentration post-synaptically thereby providing a mechanism to selectively modulate the efficacy of these active synapses. In addition, if α-BGTRs are located on the pre-synaptic terminals, choline could elevate Ca^{++} concentration in the terminals to affect transmitter release (see Fig. 3). One requirement for this hypothesis however, is that choline act as an agonist on α-BGTRs. Although we did not detect an ACh-evoked current that could be blocked by α-BGT on SCG neurons, if the Ca^{++} current flowing through the α-BGTRs is small and has slow kinetics, it could be difficult to detect because of the large current flowing through the NBT-sensitive synaptic receptors. Interestingly, we find that 5 mM choline is as effective as 100 μM ACh in gating current through α_7 receptors, when expressed in Xenopus oocytes (Fig. 2), and this choline-evoked current is blocked by 100 nM α-BGT. In fact most of the current seen in figure 2 is

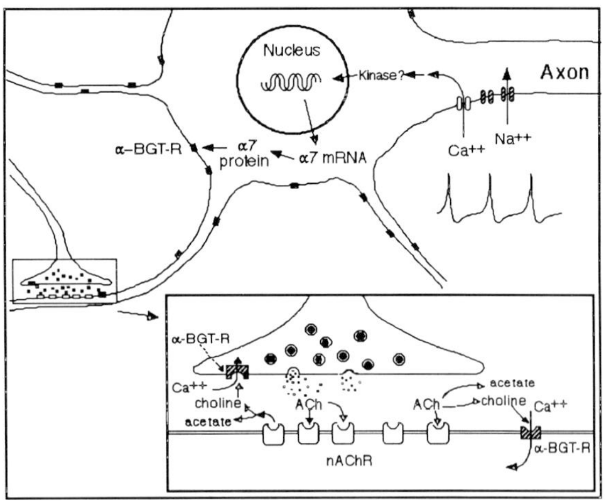

Figure 3. Illustration of the cascade of events leading to the increase in α_7 gene expression and α-BGTRs in rat sympathetic neurons. The inset suggests a role for the α-BGTR as a choline receptor. The activation of this receptor would increase intracellular Ca^{++}, thereby modulating cholinergic synaptic transmission.

due to a Ca^{++}-activated Cl^- conductance (15); the actual inward current flowing through the receptor is small.

Acknowledgements: *We thank Drs ES Deneris and P Séguéla for providing nAChR cDNA clones, Dr. M Quik for her help with the α-BGT binding, A Haghighi for his assistance with Xenopus oocyte experiments, and S Inoue for technical assistance. This work is supported by the MRC of Canada.*

References

1. Mandelzys A, Pié B, Deneris ES, Cooper E. The Developmental increase in ACh current densities on rat sympathetic neurons correlates with changes in nicotinic ACh receptor α-subunit gene expression and occurs independent of innervation. J Neurosci 1994;14:2357-64.
2. Cooper E, Lau M. Factors affecting the expression of acetylcholine receptors on rat sensory neurons in culture. J Physiol (Lond) 1986;377:409-420.
3. Mandelzys A, Cooper E. Effects of ganglionic satellite cells and NGF on the expression of nicotinic acetylcholine currents by rat sensory neurons. J Neurophysiol 1992;67:1213-1221.
4. Hawrot E, Patterson PH. Long term culture of dissociated sympathetic neurons. Meth Enzym 1979;58:574-584.
5. Krieg PA, Melton DA. In vitro RNA synthesis with SP6 polymerase. Meth Enzym 1987;155:397-415.
6. Bertrand D, Cooper E, Valera S, Rungger D, Ballivet M. Electrophysiology of neuronal nicotinic acetylcholine receptors expressed in Xenopus oocytes following nuclear injection of genes or cDNAs. In: Conn PM editor. Methods in Neuronsciences vol 4. San Diego: Academic Press Inc,1991:174-193.
7. Sargent PB. The diversity of neuronal nicotinic acetylcholine receptors. Annu Rev Neurosci 1993;16:403-443.
8. Loring RH, Sah DWY, Landis SC, Zigmond RE. The ultrastructural distribution of putative nicotinic receptors on cultured neurons from rat superior cervical ganglion. Neuroscience 1988;24:1071-1080.
9. Loring RH, Dahm LM, Zigmond RE. Localization of α-bungarotoxin binding sites in the ciliary ganglion of the embryonic chick: autoradiographic study at the light and electron microscopic level. Neuroscience 1985;14:645-660.
10. Jacob MH, Berg DK. The ultrastructural localization of α-bungarotoxin binding sites in relation to synapses on chick ciliary neurons. J Neurosci 1983;3:260-271.
11. Sah DWY, Loring RH, Zigmond RE. Long-term blockade by toxin F of nicotinic synaptic potentials in cultured sympathetic neurons. Neuroscience 1987;20:867-874.
12. Nurse CA, O'Lague PH. Formation of cholinergic synapses between dissociated sympathetic neurons and skeletal myotubes of rat cells in culture. PNAS 1975:72;1955-1959.
13. Rubin E. Development of the rat superior cervical ganglion: initial stages of synapse formation. J Neurosci 1985;5:697-704.
14. Zhang Z-w, Vijayaraghavan S, Berg DK. Neuronal acetylcholine receptors that bind α-bungarotoxin with high affinity function as ligand-gated ion channels. Neuron 1994;12:167-177.
15. Séguéla P, Wadiche J, Dineley-Miller K, Dani J, Patrick JW. Molecular cloning, functional properties and distribution of rat brain α7: a nicotinic cation channel highly permeable to calcium. J Neurosci 1993;13:596-604.

Effects of Nicotine on
Biological Systems II
Advances in Pharmacological Sciences

GROWTH RELATED ROLE FOR THE NICOTINIC α-BUNGAROTOXIN RECEPTOR

M. Quik

Department of Pharmacology, McGill University, 3655 Drummond St., Montreal, Quebec, Canada H3G 1Y6

Summary: Accumulating evidence now suggests that neurotransmitters are not only involved in synaptic transmission but may also exert a trophic or developmental function in the nervous system. This includes acetylcholine which may mediate such a role through an interaction with nicotinic receptors. In neuronal tissue, a nicotinic receptor population which has been implicated in the modulation of neurite outgrowth is the nicotinic α-bungarotoxin receptor. Interestingly, earlier work had shown that activation of nicotinic α-bungarotoxin sites on muscle cells led to muscle cell degeneration. Furthermore, α-bungarotoxin sites present on non-neuronal cells also appear to be involved in a trophic role as exposure of these cells to nicotine elicits an increase in cell number which is blocked by α-bungarotoxin. These combined studies suggest a common role of the α-bungarotoxin receptor in growth related activities.

Introduction

Neurotransmitters such as dopamine, serotonin, GABA, excitatory amino acid and opioid peptides interact with their receptors to modulate cellular growth, in addition to mediating synaptic transmission. Correlative studies have demonstrated the timely presence of different neurotransmitters at very specific stages both pre- and post-natally, findings which may suggest that these agents are used as signals to regulate various aspects of development. A trophic role is also supported from the results of *in vitro* studies which have shown that various neurotransmitters may alter neuronal growth in culture, most likely through a receptor specific interaction (1,2).

Role for acetylcholine in neuronal development/growth

Evidence is also available to support the hypothesis that fluctuations in acetylcholine (Ach) levels control cell growth/development in the nervous system. This response may occur through an interaction with muscarinic and/or nicotinic receptors. Studies indicating an involvement of muscarinic receptors include experiments which show that carbachol stimulates DNA synthesis in brain derived cell lines and that this effect is blocked in the presence of atropine (3). In addition, muscarinic receptors have been identified on nerve growth cones, an observation which may imply that these receptors are involved in neuronal guidance and/or outgrowth (4).

Activation of neuronal nicotinic receptors also appears to influence growth related and/or developmental responses. *In vivo* studies show that pre- and/or postnatal nicotine administration alters a wide variety of CNS functions including choline uptake, choline acetyltransferase activity, tyrosine hydroxylase activity, ornithine decarboxylase activity, DNA and protein levels and various behavioral parameters (5-9). That the nicotine induced response may occur through an interaction at nicotinic receptors is suggested from the results of *in vitro* work. Lipton et al. (2) showed that the nicotinic antagonist d-tubocurarine enhanced neuritic outgrowth in retinal cells in culture, an observation which also implies that the effect of endogenous Ach on these neurons may be of a tonic inhibitory nature. On the other hand, positive trophic responses to Ach, which were inhibited by d-tubocurarine have also been reported by Zheng et al. (10), who demonstrated a turning of neuronal growth cones in response to the presence of an Ach gradient in spinal neurons in culture.

Which receptor subunits are involved in nicotinic receptor mediated trophic responses?

A question which arises is through which nicotinic receptor population such a trophic response may be mediated. Multiple nicotinic receptor ligand binding or α subunit cDNAs have been identified, including $\alpha 1$ to $\alpha 8$. The $\alpha 1$ receptor subunit appears to exist exclusively in muscle, whereas $\alpha 2$ to $\alpha 8$ have been identified in neuronal tissue. In the nervous system, two broad categories of nicotinic Ach receptors may be defined on the basis of their sensitivity to α-bungarotoxin (α-BGT) (11, 12). One of these is the α-BGT insensitive receptors which appear to be composed of $\alpha 2$ to $\alpha 4$ and also ß2 to ß4 subunits; the α-BGT sensitive receptors may consist of $\alpha 7$ and $\alpha 8$, and possibly $\alpha 5$ (11,12). For both the α-BGT sensitive and insensitive subgroups, other as yet unidentified subunits may also be present.

With regard to function, the α-BGT insensitive receptors may primarily be involved in mediating nicotinic transmission, although an involvement in trophic functions is not inconceivable. In contrast, the α-toxin binding sites do not appear to be involved in transmission; instead, evidence is now accumulating which may suggest that these receptors are involved in growth related activities.

Involvement of the α-bungarotoxin receptor in development/growth in the nervous system

Correlative evidence for a growth related role for the α-BGT site stems from the results of ontogenetic studies which show that the α-BGT sites develop prior to other cholinergic markers, findings which may imply that the α-BGT sites exert a trophic effect on incoming neurons (13). The extrasynaptic localization of the α-BGT sites in the nervous system (13), as well as their unique columnar arrangement in the somatosensory cortex also suggests this receptor population may be involved in neuronal guidance (14). More direct evidence that the nicotinic α-BGT receptor modulates neuronal growth is provided by recent observations that the α-toxin prevents nicotine induced alterations in neuritic outgrowth in nerve growth factor treated PC12 cells (15) and ciliary ganglion neurons (16) in culture. Previous work has shown that the nicotinic α-BGT receptor is relatively permeable to calcium levels and, furthermore, that its activation may trigger calcium influx through voltage gated calcium channels (11, 12, 17, 18). These studies led to the suggestion that nicotinic α-BGT mediated effects on neuronal growth may be mediated by alterations in calcium. This idea is supported by recent observations that nicotine cannot modulate neuritic growth in the absence of calcium or the presence of voltage sensitive calcium channel blockers (16). Thus, calcium, which has been shown to be very important in cellular growth processes, may represent a second messenger which mediates the effect of nicotinic α-BGT receptor activation, although α-BGT does not appear to block nicotinic receptor mediated calcium responses in all systems (19).

Findings to date thus suggest that the level of acetylcholine may play a crucial role in regulating neurite length and that it is the nicotinic α-BGT receptor population which regulates this response, possibly through alterations in intracellular calcium.

Growth related role for the α-bungarotoxin site at the neuromuscular junction

Ach has also been reported to influence growth related activities in skeletal muscle. At the neuromuscular junction, acetylcholinesterase inhibitors have been shown to produce a progressive myopathy of various muscle groups. This is most likely due to their ability to increase the concentration of endogenous Ach within the synaptic cleft since treatments which decrease nerve terminal Ach or blocked postsynaptic receptors reversed the degeneration (20). The idea that Ach can modulate muscle growth is further supported by the results of *in vitro* studies which showed that nicotinic Ach receptor mediated muscle cell deterioration occurs after carbachol (21) or nicotine exposure (unpublished observations). Ach also appears to mediate growth promoting effects on motoneurons; several studies have reported that chronic receptor blockade resulted in an increase in survival of incoming neurons (22).

Thus at the neuromuscular junction, Ach may influence muscle growth by acting both pre- and postsynaptically.

Role for the α-bungarotoxin receptor in cell growth in non-neuronal cells

Interestingly, nicotinic α-BGT receptors are not only present in muscle and neuronal tissue but in addition they occur on a variety of tumor cell lines; as might be expected, this includes muscle cell lines or rhabdomyosarcomas, neuroblastomas and thymic epithelial tumors (23,24). In addition, both radiolabelled α-BGT and nicotine binding have been identified on small cell lung carcinoma, adenocarcinoma, epithelioid and epidermoid carcinoma, hepatoma, glioblastoma, as well as others (23,24). The specific nicotinic receptor subunits involved are currently under study. mRNA for α5 is present in numerous tumor cell lines (23), as well as α3 and ß4 in small cell lung carcinoma (25).

The functional significance of nicotinic receptor binding sites in these latter tumor cell lines is not yet known. However, nicotine has been shown to enhance cell number or to reverse opioid induced growth inhibition in neuroendocrine lung tumor cells (26, 27). Furthermore, this effect was blocked by the nicotinic receptor blocker hexamethonium (27). These results suggest that nicotinic receptor activation may mediate tumor cell proliferation. To assess whether the nicotinic α-BGT receptor might be involved in this nicotine induced trophic response, the effect of nicotine was determined on cell growth in the absence and presence of α-BGT. The α-toxin at nM concentrations blocked the nicotine or cytisine induced increase in cell number in small cell lung carcinoma cells (28, 29). This was most likely due to an inhibition of cell proliferation after exposure to nicotine since α-BGT blocks nicotine induced [^{3}H]thymidine incorporation in the same cells (30).

These results suggests that the nicotinic α-BGT site on tumor cells may also be involved in a growth related function.

Conclusion

In neuronal tissue, interaction at the nicotinic α-BGT site results in the modulation of neurite outgrowth. At the neuromuscular junction, enhanced acetylcholine levels result in a decline in muscle cell morphology, which is prevented by the α-toxin. Non-neuronal cells such as small cell lung carcinoma, which are of neuroendocrine origin, also have α-BGT receptors which when stimulated result in an enhancement in cell number. Thus, current evidence suggests that nicotinic α-BGT receptor is involved in a growth related function in neuronal, nonneuronal or muscle tissue.

***Acknowledgements**: Support from the Medical Research Council Canada is gratefully acknowledged.*

References

1. Lauder JM. Neurotransmitters as growth regulatory signals: role of receptors and second messengers. Trends Neurosci. 1993; 16:233-240.
2. Lipton SA, Kater SB. Neurotransmitter regulation of neuronal outgrowth, plasticity and survival. Trends Neurosci. 1989; 12:265-270.
3. Ashkenazi A, Ramachandran J, Capon DJ. Acetylcholine analogue stimulates DNA synthesis in brain derived cells via specific muscarinic receptor subtypes. Nature 1989; 340:146-150.
4. Van Hooff COM, De Graan PNE, Oestreicher AB, Gispen WH. Muscarinic receptor activation stimulates B50/GAP43 phosphorylation in isolated nerve growth cones. J. Neurosci. 1989; 9:3753-3759.
5. Navarro HA, Seidler FJ, Schwartz RD, Baker FE, Dobbins SS, Slotkin TA. Prenatal exposure to nicotine impairs nervous system development at a dose which does not affect viability or growth. Brain Res. Bull. 1989; 23:187-192.
6. Nordberg A, Zhang X, Fredriksson A, Eriksson P. Neonatal nicotine exposure induces permanent changes in brain nicotinic receptors and behaviour in adult mice. Dev. Brain Res. 1991; 63:201-207.
7. Slotkin TA, Greer N, Faust J, Cho H, Seidler FJ. Effects of maternal nicotine injections on brain development in the rat: Ornithine decarboxylase activity, nucleic acids and proteins in discrete brain regions. Brain Res. Bull. 1986; 17:41-50.
8. Slotkin TA, McCook EC, Lappi SE, Seidler FJ. Altered development of basal and forskolin-stimulated adenylate cyclase activity in brain regions of rats exposed to nicotine prenatally. Dev. Brain Res. 1992; 68:233-239.
9. Smith KM, Mitchell SN, Joseph MH. Effects of chronic and subchronic nicotine on tyrosine hydroxylase activity in noradrenergic and dopaminergic neurons in the rat brain. J. Neurochem. 1991; 57:1750-1756.
10. Zheng JQ, Felder M, Connor JA, Poo M. Turning of nerve growth cones induced by neurotransmitters. Nature 1994; 368:140-144.
11. Deneris ES, Connolly J, Rogers SW, Duvoisin R. Pharmacological and functional diversity of neuronal nicotinic acetylcholine receptors. Trends Pharmacol. Sci. 1991; 12:34-40.

12. Luetje CW, Patrick J, Séguéla P. Nicotine receptors in the mammalian brain. FASEB J. 1990; 4:2753-2760.
13. Quik M, Geertsen S. Neuronal nicotinic α-bungarotoxin sites. Can. J. Physiol. Pharmacol. 1988; 66:971-979.
14. Fuchs JL. ^{125}I-α-Bungarotoxin binding marks primary sensory areas of developing rat neocortex. Brain Res. 1989; 501:223-234.
15. Chan J, Quik M. A role for the neuronal nicotinic α-bungarotoxinn receptor in neurite outgrowth in PC12 cells. Neuroscience 1993; 56:441-451.
16. Pugh PC, Berg DK. Neuronal acetylcholine receptors that bind α-bungarotoxin mediate neurite retraction in a calcium dependent manner. J. Neurosci. 1994; 14:889-895.
17. Vijayaraghavan S, Pugh PC, Zhang Z, Rathouz MM, Berg DK. Nicotinic receptors that bind α-bungarotoxin raise intracellular free calcium. Neuron 1992; 8:353-362.
18. Séguéla P, Wadiche J, Dineley-Miller K, Dani J, Patrick J. Molecular cloning, functional properties, and distribution of rat brain α7: a nicotinic cation channel highlty permeable to calcium. J. Neurosci. 1993; 13:595-604.
19. Afar R, Trifaro JM, Quik M. Nicotine induced intracellular calcium changes are not antagonized by α-bungarotoxin in adrenal medullary cells. Brain Res. 1994; 641:127-131.
20. Fenichel GM, Kibler WB, Olson WH, Dettbarn WD. Chronic inhibition of cholinesterase as a cause of myopathy. Neurology 1972; 22:1026-1033.
21. Leonard JP, Salpeter MM. Agonist induced myopathy at the neuromuscular junction is mediated by calcium. J. Cell Biol. 1979; 82:811-819.
22. Brown MC. Sprouting of motor nerves in adult muscles: a recapitulation of ontogeny. Trends Neurosci. 1984; 7:10-14.
23. Chini B, Clementi F, Hukovic N, Sher E. Neuronal-type α-bungarotoxin receptors and the α_5-nicotinic receptor subunit gene are expressed in neuronal and nonneuronal human cell lines. Proc. Natl. Acad. Sci. U.S.A. 1992; 89:1572-1576.
24. Maneckjee R, Minna JD. Opioid and nicotine receptors affect growth regulation of human lung cancer cell lines. Proc. Natl. Acad. Sci. 1990; 87:3294-3298.
25. Tarroni P, Rubboli F, Chini B, Zwart R, Oortgiesen M, Sher E, Clementi F. Neuronal-type nicotinic receptors in human neuroblastoma and small cell lung carcinoma cell lines. FEBS 1992; 312:66-70.
26. Schuller HM. Cell type specific, receptor-mediated modulation of growth kinetics in human lung cancer cell lines by nicotine and tobacco-related nitrosamines. Biochem. Pharmacol. 1989; 38:3439-3442.
27. Schuller HM, Nylen E, Park P, Becker KL. Nicotine, acetylcholine and bombesin are trophic growth factors in neuroendocrine cell lines derived from experimental hamster lung tumors. Life Sciences 1990; 47:571-578.
28. Chan J, Patrick J, Quik M. A role for the nicotinic α-bungarotoxin receptor in cell proliferation in small cell lung carcinoma. Soc. Neurosci. Abstr. 1993; 19:466.
29. Quik M, Chan J, Patrick J. α-Bungarotoxin blocks the nicotinic receptor mediated increase in cell number in a neuroendocrine cell line. Brain Res. 1994; 655:161-167.
30. Codignola A, Tarroni MG, Cattaneo MG, Vincenti L, Clementi F, Sher E. Serotonin release and cell proliferation are under the control of α-bungarotoxin sensitive nicotinic receptors in small cell lung carcinoma cell lines. FEBS letters 1994; 342:286-290.

Effects of Nicotine on
Biological Systems II
Advances in Pharmacological Sciences

MECHANISMS OF NICOTINE STIMULATED CELL PROLIFERATION IN NORMAL AND NEOPLASTIC NEUROENDOCRINE LUNG CELLS

H.M.Schuller

Carcinogenesis and Developmental Therapeutics Program, College of Veterinary Medicine, University of Tennessee, Knoxville TN 37901, USA.

Summary: The secretion of several growth factors by pulmonary neuro-endocrine (PNE) cells is under cholinergic control, and nicotine stimulates the proliferation of fetal PNE cells *in vivo* by binding to the nicotinic acetylcholine receptor (nAChR). Our data show that fetal hamster PNE cells and cell lines derived from human neuroendocrine lung cancers require simultaneous stimulation of the oxygen receptor and the nAChR expressed in these cells in order to proliferate in response to nicotine. The simultaneous stimulation of these two receptor-initiated signal transduction pathways triggers secretion of 5-HT and bombesin, and activates protein kinase C (PKC) and *c-fos* downstream. The importance of this synergism is greatly emphasized by our finding that simultaneous exposure of hamsters to hyperoxia and nicotine causes lung tumors positive for 5-HT and neuron specific enolase (NSE) while nicotine alone or hyperoxia alone cause no tumors.

Introduction

PNE cells are abundant during the perinatal period but sparse in healthy adult mammals. They produce biogenic amines and neuropeptides in response to the relatively hypoxic intrauterine environment and the sudden increase in lung oxygenation concomittant with birth (1). These secretory products have vaso/broncho constrictor/dilator activities, and are thought to help the lungs adjust to changes in lung oxygenation. It was suggested decades ago that the response of PNE cells to changes in pulmonary oxygenation is mediated by an oxygen receptor (2), but this receptor was only identified in 1993 (3).

Human lung cancers with a neuroendocrine phenotype share many morphological and functional features with normal PNE cells. The most malignant variety of this tumor category, small cell lung cancer (SCLC), demonstrates a strong etiological link with cigarette smoking, and is particularly prominent in smokers with a history of chronic obstructive lung disease (4). Among the various neuroendocrine markers expressed by SCLC, neuropeptides of the bombesin family have been most extensively studied. In particular, much recent emphasis has been given to exploring the signal transduction pathways of these autocrine growth factors as targets for chemoprevention and cancer therapy (5). However, the neuropeptides belonging to this family also stimulate the proliferation of other lung cell types including bronchial epithelia and fibroblasts (6). On the other hand, the biogenic amine 5-HT (serotonin) which is secreted by normal PNE cells in response to changes in lung oxygenation (1,7), and is expressed by the vast majority of SCLC (8), has only recently been identified as an autocrine growth factor for normal and neoplastic PNE cells (9-11). Unlike the neuropeptides of the bombesin family, the mitogenic effects of 5-HT are selective for PNE cells while not stimulating other lung cell types. Conditions which stimulate the secretion of 5-HT are thus more likely to provide PNE cells with a selective growth advantage over other lung cell types, and may well contribute to the development of lung tumors with a neuroendocrine phenotype.

A number of physiological, pathological, and experimental conditions which change the pulmonary ventilation and concentration of intrapulmonary gases have been shown to stimulate the secretory activity and hyperplasia of normal PNE cells *in vivo* (1). These responses are mediated by the oxygen receptor expressed in these cells (3). On the other hand, a number of chemicals have been shown to similarly stimulate normal PNE cells *in vivo* (1). Among these are cigarette smoke (12) , nicotine (13), and the carcinogenic nitrosamines N-nitrosodiethylamine (DEN,14,15), and 4-(methylnitrosamino)-1-(3-pyridyl)-1-butanone (NNK,16). Studies with neonatal Syrian golden hamsters have shown that acetylcholine and nicotine both stimulate PNE cells, an effect inhibited by antagonists of the nAChR (13). As DEN has structural similarities with acetylcholine while NNK has similarities with nicotine (from which it is derived), we hypothesized that these nitrosamines also interact with the nAChR. In support of this hypothesis, both nitrosamines bound to the nAChR in radioreceptor assays with cell membrane fractions from hamster lungs in competition experiments with $[^3H]$(-)-nicotine (17). It hence appears that both secretory and proliferative activities of normal PNE cells *in vivo* are regulated by two different signal transduction pathways, one of which is initiated by the oxygen receptor while the other is activated by the nAChR (fig.1).

Neuroendocrine lung cancers demonstrate many of the functional features of normal PNE cells, and in particular express 5-HT and neuropeptides at a high incidence (8). We therefore decided to explore if signal transduction pathways similar to those which govern the biology of normal PNE cells are operative in these tumor cells.

PNE Cell Regulation

Oxygen Receptor Nicotinic Receptor

Secretion of : 5-HT
Neuropeptides

Uptake by: 5-HT-Receptors
Neuropeptide Receptors

Proliferation of PNE Cells

Figure 1. Stimulation of the oxygen receptor and nAChR activates the secretion of 5-HT and neuropeptides by PNE cells. Re-uptake of these growth factors by specific receptors in turn mediates a proliferative response.

The controversy on the effects of nicotine *in vitro*

For our first experiments aimed at studying the role of the nAChR in neoplastic PNE cells, we selected a human cell line derived from a lung carcinoid (NCI-H727) which was very well differentiated morphologically and closely resembled normal PNE cells. The stock material (passage 1) of this cell line was supplied to us by Dr. A. F. Gazdar (formerly NCI-Naval Hospital, now Simmons Cancer Center, Houston TX). This cell line requires an atmosphere of relatively high CO_2 (8-10%), while this environment inhibits the growth of non neuroendocrine lung cancer cell lines. Using simple cell proliferation assays with NCI-H727 cells maintained in 10% CO_2, we were the first to demonstrate a strong and concentration-dependent mitogenic effect of nicotine (9,18). This effect was completely inhibited by two antagonists of the nAChR (hexamethonium and pentolinium), suggesting that binding of nicotine to the nAChR initiated the observed mitogenic response. In the same experiments, we also observed a concentration-dependent stimulation of cell proliferation in response to 5-HT (9), thus identifying this biogenic amine as an autocrine growth factor for these neoplastic

PNE cells. Taking advantage of the selective growth stimulating effect of high CO_2 on lung cells with a neuroendocrine phenotype, we subsequently established cell lines derived from experimentally induced hamster neuroendocrine lung tumors (atypical carcinoids). Using these systems, we confirmed our initial observations on the mitogenic effects of nicotine and also showed that acetylcholine had similar effects (19). Again, these cell proliferation assays were conducted in an atmosphere of 10% CO_2. A year later, our reports on the mitogenic effects of nicotine in neoplastic PNE cells were challenged by Maneckje and Minna (20). These investigators demonstrated the expression of the nAChR in a large panel of human SCLC and nonSCLC cell lines by receptor binding assays but failed to observe a mitogenic effect of nicotine per se in any of these systems. Communication with Dr. Minna revealed that the cell proliferation assays had been conducted in an atmosphere of 5% CO_2. Subsequently, we obtained several of the SCLC lines used by these authors, and began to test them systematically and in parallel with our carcinoid cell lines and a newly developed *in vitro* system of normal fetal hamster PNE cells (11,21). We found, that unlike our carcinoid cell lines and fetal PNE cells, the SCLC lines (NCI-H69,NCI-H82) did grow in an atmosphere of 5% CO_2. However, exposure to 10% CO_2 significantly stimulated the proliferation of these cells (22). Exposure of these two cell lines to nicotine in an atmosphere of 10% CO_2 significantly stimulated cell proliferation while identical exposures in an atmosphere of 5% CO_2 had no effect. Exposure to hyperoxia (40%) had a similar stimulating effect on NCI-H727 and the two SCLC lines but only if exposure to hyperoxia was in pulses of 10 min with intervals of 30 min at 5% CO_2/95% air, while continuous hyperoxia was cytotoxic (22). Subsequently, we showed that the mitogenic effect of elevated CO_2 was concentration dependent and saturated at a concentration of 10% CO_2 (11), thus strongly suggesting a receptor-mediated mechanism. Mitogenesis in response to 10% CO_2 was completely inhibited by inhibitors of PKC (11), an effect which was concentration-dependent (11). Accordingly, the mitogenic effect of elevated CO_2 could be mimicked in normal and neoplastic PNE cells maintained at 5% CO_2 by a phorbol ester which is an established activator of PKC (11). These findings suggested, that the "oxygen" receptor expressed in normal and neoplastic PNE cells either signals both increased and decreased oxygen concentrations or that several subtypes of this receptor category exist, some of which respond to increases in oxygen while others respond to CO_2 or the concomitant decrease in O_2. Moreover, these findings indicated for the first time, that activation of PKC is an important downstream event in the O_2 receptor-initiated signal transduction pathway.

In accordance with our earlier findings (9,18) and in support of the report by Manneckje and Minna (20), fetal PNE cells, the carcinoid NCI-H727, and the SCLC line NCI-H69 failed to proliferate in response to nicotine when maintained in an environment of 5% CO_2 (11). On the other hand, nicotine exerted a strong and concentration-dependent mitogenic effect on all three cell systems in an atmosphere of 10% CO_2 (11), thus supporting our initial findings on the effects of nicotine in carcinoid cell lines maintained at 10% CO_2 (9,18,19). Mitogenesis in response to nicotine in an environment of 10% CO_2 was completely inhibited by the nAChR antagonist hexamethonium and PKC inhibitors, while antagonists of receptors for 5-HT and bombesin yielded significant inhibition, the extent of which varied in the three cell systems (11). Mitogenesis in response to nicotine was mimicked in fetal PNE cells maintained at 5% CO_2 by 5-HT in a concentration-dependent manner. Northern blot analysis revealed a time-dependent overexpression of *c-fos* in fetal PNE cells exposed to nicotine and 10% CO_2 (23). While further analysis of the expression of this particular gene is still in progress, preliminary data at this point suggest that the observed overexpression in *c-fos* is primarily attributable to the stimulating effects of elevated CO_2 (23). However, more detailed studies are necessary to dissect the molecular mechanisms contributing to the observed synergistic effects of CO_2 and nicotine.

In summary, our recent studies have resolved the controversy on the *in vitro* effects of nicotine in PNE cells. Our data clearly show that simultaneous stimulation of the oxygen receptor in PNE cells exposed to nicotine is an essential requirement to yield a mitogenic response to nicotine. Our data are supported by recent reports by two other laboratories which have demonstrated a strong mitogenic effect of nicotine in SCLC lines, including NCI-H69, maintained in an atmosphere of high CO_2 (10,24). In light of these data it is particularly intriguing that an atmosphere of 5% CO_2 (the environment in which nicotine is not mitogenic) is the equivalent of the intraalveolar concentration of this gas in the lungs of healthy adult individuals at an altitude of 0 (25), while concentrations of 10% to 15% CO_2 (the environment in which nicotine is strongly mitogenic) are equivalent to the intraalveolar concentration of this gas in individuals with obstructive lung disease and emphysema (25). As mentioned in the introduction, chronic obstructive lung disease is a predisposing factor for the development of SCLC in smokers (4). In light of our data it appears, that this etiological link is attributable to a synergistic effect of PNE cell specific mitogenic signal transdcution pathwyas initiated by agonists of the oxygen receptor and the nAChR.

To test this hypothesis, we conducted a bioassay experiment in male Syrian golden hamsters exposed to either hyperoxia (60%) plus nicotine (1 mg/kg), hyperoxia alone, or nicotine in ambient air (26). The results of this study fully support our concept on a synergism between the oxygen receptor and the nAChR in that a low but significant number (4 out of 20) of the animals eposed to hyperoxia/nicotine developed lung tumors positive for neuron specific enolase and 5-HT, while none of the animals in the two other groups developed tumors. It is however unclear at this time if the observed synergism between hyperoxia and nicotine in this model is caused by a direct interaction of oxygen with the O_2 receptor or by oxygen toxicity resulting in pulmonary fibrosis, emphysema, and a concomitant increase in pulmonary CO_2.

Conclusion

The data summarized in this manuscript as well as results presented by other participants of this meeting strongly suggest that nicotine contributes significantly to the well documented risk of smokers with chronic lung disease to develop SCLC. In addition to us, two independent laboratories have documented a strong stimulation in the proliferation of SCLC lines *in vitro* by nicotine via an nAChR-mediated mechanism when the cells were maintained in an atmosphere of high CO_2 (10,24). As our data show, the simultaneous stimulation of the oxygen receptor expressed in normal and neoplastic PNE cells acts as an essential promoter of mitogenesis in response to nicotine. The involvement of an autocrine 5-HT loop and the downstream activation of PKC and *c-fos* are exciting findings which may be exploited to develop new strategies for the selective prevention and therapy of this cancer category.

Cancer research of the past and present has greatly overemphasized the importance of classic chemical carcinogens such as the nitrosamines contained in tobacco products as a cause of human lung cancer. NNK, the most potent and most abundant nitrosamine contained in cigarette smoke, is present in cigarettes at concentrations of 15,000 to 30,000 times less than nicotine (27). Although we have shown that the affinity of NNK for the nAChR in hamster lung is about 10 times greater than that of nicotine (17), nicotine,- due to its much higher concentration,- will certainly prevent binding of NNK to this receptor in a smoker's lung. Moreover, with respect to the extrememly low concentration of NNK and other nitrosamines in tobacco products, it is impossible for even a chain smoker to reach the concentrations of these carcinogens which are typically required to cause lung cancer in experimental animals. Accordingly, many of the current dogmas on mechanisms of lung carcinogenesis which are largely based on studies using this class of chemical carcinogens, may have little relevance for human small cell lung cancer.

References

1. Schuller HM. Small cell cancer of the lung: The role of unphysiological pulmonary oxygen levels. Biochem Pharmacol 1988;37:1645-1649.
2. Lauweryns J., Cokelaer M., Lerut T., Theuninck P. Cross circulation studies on the role of hypoxia and hypoxemia on neuroepithelial bodies in young rabbits. Cell Tiss Res 1978;36:767-773.
3. Youngson C, Nurse C, Yeger H, Cutz E. Oxygen sensing in airway chemoreceptors. Nature 1993;365:153-155.
4. Weiss W. COPD and lung cancer. In: Obstructive Pulmonary Disease. Cherniak ED, editor. Philadelphia: WB Saunders, 1991:344-347.
5. Mulshine JL, Treston AM, Natale RB, Kasprzyk PG, Avis I, Nakanishi Y, Cuttitta F. Autocrine growth factors as therapeutic targets in lung cancer. Chest 1989;96:31s-34s.
6. Rozengurth E, Sinnett-Smith J. Bombesin stimulation of fibroblast mitogenesis: specific receptors, signal transduction, and early events. Philos Trans R Soc Lond Biol 1990;327:209-221.
7. Cutz E, Gillan JE, Track NS. Pulmonary endocrine cells in the developing human lung and during neonatal adaptation. In: The Endocrine Lung in Health and Disease. Becker KL, Gazdar AF, editors. Philadelphia: WB Saunders, 1984:210-231.
8. Gazdar AF. The biology of endocrine lung tumors. In: The Endocrine Lung in Health and Disease. Becker KL, Gazdar AF, editors. Philadelphia: WB Saunders, 1984:448-459.
9. Schuller HM, Hegedus TJ. Effects of endogeneous and tobacco-related amines and nitrosamines on cell growth and morphology of a cell line derived from a human neuroendocrine lung cancer. Toxicol *In Vitro* 1989;3:37-43.
10. Cattaneao MG, Codignola A, Vicenti LM, Clementi F, Sher E. Nicotine stimulates a serotonergic autocrine loop in human small cell lung carcinoma. Cancer Res 1993;53:5566-5568.
11. Schuller HM. Carbon dioxide potentiates the mitogenic effects of nicotine and its carcinogenic derivative NNK in normal and neoplastic neuroendocrine lung cells via stimulation of autocrine and protein kinase C-dependent mitogenic pathways. Neurotoxicol (in press).
12. Tabassian AR, Nylen ES, Giron AE, Snider RH, Cassidy MM, Becker KL. Evidence for cigarette smoke-induced calcitonin secretion from lungs of man and hamster. Life Sci 1988;42:2323-2329.
13. Nylen ES, Linnoila IR, Becker KL. Prenatal cholinergic stimulation of pulmonary neuroendocrine cells by nicotine. Acta Physiol Scand 1988;132:117-118.
14. Reznik-Schuller HM. Proliferation of endocrine (APUD-type) cells during early DEN-induced lung carcinogenesis. Cancer Lett 1976;1:255-258.
15. Linnoila IR, Becker KL, Silva OC, Snider RH, Moore CF. Calcitonin as a marker for diethylnitrosamine-induced pulmonary endocrine cell hyperplasia in hamsters. Lab Invest 1984;51:39-45.
16. Schuller HM, Witschi H-P, Nylen ES, Joshi PA, Correa E, Becker KL. Pathobiology of NNK-induced lung tumors in hamsters and the modulating effect of hyperoxia. Cancer Res 1990;50:1960-1965.
17. Schuller HM, Castonguay A, Orloff M, Rossignol G. Modulation of the uptake and metabolism of 4-(methylnitroamino)-1-(3-pyridyl)-1-butanone by nicotine in hamster lung. Cancer Res 1990;50:1960-1965.
18. Schuller HM. Cell type specific, receptor-mediated modulation of growth kinetics in human lung cancer cell lines by nicotine and tobacco-related nitrosamines. Biochem Pharmacol 1989;38:3439-3442.

19. Schuller HM, Nylen ES, Park P, Becker KL. Nicotine, acetylcholine, and bombesin are trophic growth factors in neuroendocrine cell lines derived from experimental hamster lung tumors. Life Sci 1990;47:571-578.
20. Maneckje R, Minna JD. Opioid and nicotine affect growth regulation of human lung cancer cell lines. Proc Natl Acad Sci 1990;87:3294-3298.
21. Linnoila IR, Funa K, Becker KL, Schuller HM, Gazdar AF. Longterm selective culture of hamster pulmonary endocrine cells. Clin Res 1985; 33:468A.
22. Schuller HM, Orloff M. Carbon dioxide and oxygen stimulate proliferation of neuroendocrine lung cancer cell lines via protein kinase C activation. Proc Amer Assoc Cancer Res 1993;34:1049.
23. Miller MS, Schuller HM. Nicotine stimulates the proliferation of fetal hamster pulmonary neuroendocrine cells. Proc Amer Assoc Cancer Res 1994;35:746.
24. Chan J, Patrick J, Quik M. A role for the nicotinic α-bungarotoxin receptor in cell proliferation in small cell lung carcinoma. Soc Neurosci 1993;19:195-20.
25. Lamberdson CJ. The atmosphere and gas exchanges with the lungs and blood. In: Medical Physiology. Mountcastle VB, editor. Saint Louis: CV Mosby Co, 1974:1372-1398.
26. Schuller HM, McGavin D, Riechert A. Nicotine induces tumors in hyperoxic hamstes. Proc Amer Assoc Cancer Res 1994;35:763.
27. Richter E, Tricker AR. Nicotine inhibits the metabolic activation of the tobacco-specific nitrosamine 4-(methylnitrosamino)-1-(3-pyridyl)-1-butanone in rats. Carcinogenesis 1994:15:1061-1064.

ACTIVATION OF THE HYPOTHALAMIC-PITUITARY-ADRENAL AXIS BY NICOTINE: NEUROCHEMICAL AND NEUROANATOMICAL SUBSTRATES

Burt Sharp and Shannon Matta

Endocrine-Neuroscience Research Laboratory, Minneapolis Medical Research Foundation and Departments of Medicine, Hennepin County Medical Center and University of Minnesota, Minneapolis, 55404, U.S.A.

Summary: Nicotine rapidly elevates rat plasma adrenocorticotropin (ACTH) levels, and desensitization occurs after one exposure and persists for hours. Neural structures near the fourth ventricle (4V) mediate this response to systemic nicotine. Hypothalamic α-adrenergic receptors, presumably on corticotropin-releasing factor (CRF) neurons in the paraventricular nucleus (PVN), are involved. Delivering nicotine i.p. or into 4V elevates norepinephrine levels in PVN microdialysates; this response partially desensitizes Selective nicotine microinjection into catecholaminergic regions near 4V shows that nucleus tractus solitarius (NTS)-A_2 and C_2 are important mediators of ACTH secretion. The sensitivity and responsiveness of c-fos protein expression to i.v. nicotine in NTS correlates with that in PVN. Thus, nicotine elevates ACTH, at least in part by activating catecholaminergic afferents near their origin, thereby stimulating PVN CRF neurons.

Introduction

Nicotine stimulates the rodent hypothalamic-pituitary adrenal (H-P-A) axis through central actions that result in secretion of pituitary adrenocorticotropin (ACTH). The ACTH response to systemically administered nicotine is rapid, dose-dependent and inhibited by mecamylamine, but not hexamethonium. Nicotine also stimulates prolactin secretion with a similar pharmacological profile. Both hormonal responses demonstrate desensitization after a single injection which persists for a prolonged time interval. *In vivo* studies of the activation by nicotine of both these hormonal systems have provided insight into its central mechanisms of action at the neurochemical and neuroanatomical levels. The mechanisms of activation of the H-P-A axis by nicotine may be a paradigm for its effects on central autonomic pathways.

Central Mechanism of Action

Nicotine stimulates ACTH secretion on postnatal days 1, 4 and 7, although the magnitude of the responses is markedly reduced compared to days 15 and 25 which resemble the adult rat (1). By day 26, the ACTH response to nicotine is inhibited by mecamylamine, but not hexamethonium which does not cross the blood-brain barrier (2). Nicotine 0.01-0.03 mg/kg i.v. maximally stimulates ACTH secretion in adult male Holtzman rats (1). In contrast, cytisine, a potent nicotinic cholinergic agonist that does not readily access the brain (3), is ineffective at 0.035-0.10 mg/kg. These observations indicate that nicotine stimulates the H-P-A axis through central nicotinic cholinergic mechanisms. Evidence corroborating this idea was obtained from *in vitro* studies with dispersed anterior pituitary cells and from the intracerebroventricular (i.c.v.) administration of nicotine (1). It was shown that corticotropes do not respond to nicotine alone, nor does nicotine modulate corticotrope responsiveness to corticotropin-releasing factor (CRF). However, peak plasma ACTH levels are present within 3-7 minutes of administering nicotine into the upper third ventricle and within 3 minutes when instilled into the lower (hypothalamic) third ventricle. Integrated ACTH responses tend to be greater after nicotine is given into the lower third ventricle, suggesting that the site(s) of action of nicotine is adjacent or distal to the hypothalamus.

Within the hypothalamus, the paraventricular nucleus (PVN), which has been shown to be the central site for the regulation of ACTH (4), exhibits the same complementarity in the distribution of binding sites for ^{125}I-α-bungarotoxin vs. ^{3}H-nicotine as that observed throughout the brain (5,6). Autoradiographic studies have shown that ^{3}H-nicotine binding sites are present in relatively greater density within the neuropil surrounding the PVN compared to within the nucleus (5). In contrast, ^{125}I-α-bungarotoxin sites are most abundant within the PVN, co-distributing with neurophysin-containing neurons within the magnocellular region and much less so with CRF-containing neurons. In view of this evidence for nicotinic receptors within the PVN, studies were undertaken to determine whether mecamylamine, administered bilaterally into the PVN, would inhibit the ACTH response to i.v. nicotine; inhibition was not seen (7).

To ascertain whether structures proximate to the fourth ventricle (4V) mediate the ACTH response to nicotine, the drug was administered to chronically cannulated rats. Nicotine 0.25 μg to 2.5 μg dose-dependently stimulates ACTH secretion, reaching peak levels within 3-7 minutes (7). Mecamylamine 0.5-1.0 mg/kg i.v. inhibits the ACTH response to a low dose of nicotine (0.25 μg) given into the 4V. Furthermore, administering 4.0 μg mecamylamine into the 4V largely eliminates the ACTH response to i.v. nicotine 0.03 mg/kg. Therefore, brain regions accessible from 4V mediate, in large part, the ACTH response to peripherally administered nicotine. Similar studies show that the 4V is involved in nicotine-stimulated prolactin secretion (8).

Studies indicating that norepinephrine and possibly epinephrine are involved in the ACTH response to nicotine, directed attention to catecholaminergic regions of the rat brainstem that are accessible from the 4V. Nicotine 0.25 μg-10.0 μg was selectively microinjected (50 nl/30 sec; 9) into regions that give rise to the norepinephrinergic (nucleus tractus solitarius (NTS)-A_2 and-A_1 region) and epinephrinergic (NTS-C_2 and C_1 region) inputs to the PVN (10,11). The locus coeruleus (LC), which also projects norepinephrinergic axons to the hypothalamus (although not to the parvocellular, CRF-containing PVN) also was studied, since its projections to the hippocampal complex may stimulate neurocircuits that act indirectly upon parvocellular PVN (12). Studies using ^{3}H-nicotine show that the radial spread of nicotine microinjected into the brainstem parenchyma averaged between 400-700 μm (9). For the LC, 80% of the ^{3}H-nicotine is within 371±53 μm, and 90% is within 500±70 μm of the injection site. For the NTS-A_2 region, 80% of the ^{3}H-nicotine is within 513±76 μm of the injection site, and 90% is within 650±73 μm. Therefore, spread of nicotine between LC and these other regions is unlikely. Minimal spread between NTS-A_2 and -C_2 is possible, although the following data do not support this.

Injection of nicotine into NTS-A_2 or -C_2 results in a dose-dependent increase in plasma ACTH that is maximal by 3 minutes, with an ED_{50} in NTS-A_2 of approximately 0.25 μg. In contrast, C_1 is unresponsive and A_1 only shows responses to the highest doses (5 or 10 μg). In LC, nicotine 2.5 μg or higher is required to elevate plasma ACTH. This is approximately 10-fold greater than required in NTS-A_2. Mecamylamine (0.25 mg/kg i.v., 2 minutes before nicotine) completely inhibits the ACTH responses to nicotine in NTS-A_2 and -C_2. Thus, microinjection of nicotine selectively activates catecholaminergic brainstem regions, with a rank order of sensitivity to nicotine that is NTS-A_2 > NTS-C_2 > LC >A_1 >C_1 = control. Since peripherally administered nicotine stimulates ACTH secretion by activating sites accessible from the 4V, it is likely that multiple catecholaminergic brainstem regions, particularly NTS-A_2 and -C_2, are involved.

Expression of c-fos mRNA and c-fos Nuclear Protein

Early rapid response genes, such as c-fos, c-jun, jun B and the fos-related antigens, are induced in the central nervous system by a variety of agents including the psychostimulants cocaine and amphetamine (13). The expression of these genes often leads to interactions with nuclear phosphoproteins (14) which regulate the transcription of other genes such as pre-proenkephalin (15,16). In addition to their potential role in brain plasticity, expression of these genes, in particular c-fos, has been used as an index of neuronal activation.

Studies directed at a limited number of brain regions showed that a single injection of i.p. nicotine increases c-fos mRNA content within 30 minutes (17). Significant elevations are induced by 0.5 mg/kg nicotine in the medial habenula and by 1.0 mg/kg in the hippocampus, dentate gyrus, and piriform cortex. Mecamylamine significantly reduces these responses. Additional studies have shown that a single dose of 0.05 mg/kg of i.v. nicotine significantly increases c-fos mRNA in PVN, and further enhancement is observed with 0.075 and 0.10 mg/kg. With this i.v. dosing regimen, hippocampus, dentate gyrus and cingulate gyrus are stimulated at 0.075-0.10 mg/kg.

Pretreatment with a β-adrenergic antagonist, propranolol, but not a serotonergic antagonist (spiperone), inhibits (>90%) the c-fos mRNA response to nicotine in hippocampus, dentate gyrus and cingulate gyrus. However, PVN is not affected. Similarly, *in situ* hybridization showed that 0.0375 mg/kg of nicotine stimulates c-fos expression in PVN, and propranolol is ineffective at blocking this. Thus, induction of c-fos expression in hippocampus, dentate gyrus and cingulate gyrus appears to require higher concentrations of nicotine than in PVN. Whether differences in nicotinic cholinergic receptor subpopulations or in regional c-fos responsiveness underlie this remains to be clarified. These studies also demonstrate that local secretion of catecholamines, acting through β-adrenergic receptors, appears to be required for nicotine-induced c-fos expression in these extra-hypothalamic regions. Further investigation will clarify the role of catecholamines, acting through β-adrenergic receptors, in the regionally-specific stimulation of neurons by nicotine.

Immunocytochemical detection of c-fos nuclear protein has been used to quantify neuronal activation in catecholaminergic brainstem regions and to correlate this with activation within the hypothalamic PVN and supraoptic nucleus (SON), and within limbic regions such as the hippocampal complex, central nucleus of the amygdala (CeA) and cingulate gyrus (CG; 18).

Nicotine 0.05 mg/kg i.v. stimulates c-fos expression in the parvocellular PVN (pcPVN; containing CRF-positive neurons mediating ACTH secretion) and in the oxytocinergic region of SON; this correlates with c-fos expression in the NTS-A_2 and -C_2 which project directly to pcPVN and SON. At this dose c-fos is detectable within CeA and CG; however, the catecholaminergic regions LC, A_1 and C_1 do not respond, nor does the hippocampal complex. At nicotine 0.10 mg/kg i.v., which produces a brief episode of tremor, c-fos is detectable in LC and dentate gyrus. This is associated with a further increase in the number of c-fos-positive cells in the PVN, primarily through recruitment in the magnocellular region, a known projection field of LC. These studies indicate that nicotine is a selective stimulus of both brainstem and hypothalamic regions regulating ACTH and oxytocin secretion. The relative responsiveness of brainstem catecholaminergic regions to i.v. nicotine-induced c-fos expression correlates with the rank order of sensitivity of these regions as mediators of ACTH secretion due to nicotine microinjection.

Catecholamines and ACTH Secretion

Catecholamines have been shown to stimulate the release of CRF from the hypothalamus *in vivo* (19,20). Our laboratory has taken several approaches to explore the role of catecholamines in the ACTH response to nicotine (21). Experiments with the catecholaminergic neurotoxin, 6-hydroxydopamine, showed that the ACTH response to nicotine delivered i.v. or into the 4V is significantly reduced in lesioned rats (21). Selective inhibitors of epinephrine synthesis, SKF 64139 and 2,3-dichloro-α-methylbenzylamine (DCMB), also reduce the ACTH response to i.v. nicotine, without affecting median eminence CRF content. Finally, administration of the α_1 or α_2 adrenergic receptor antagonists, prazocin or yohimbine, respectively, into the hypothalamic third ventricle significantly inhibits the ACTH response to nicotine instilled into the 4V. In contrast, the β-adrenergic antagonist, propranolol, is ineffective. These studies suggest that central epinephrinergic, and probably norepinephrinergic, input from brainstem regions to the hypothalamus are involved in the ACTH response to nicotine, and that both α_1 and α_2 hypothalamic adrenergic receptors mediate these events.

To determine directly whether nicotine induces norepinephrine secretion in the PVN of alert rats, *in vivo* microdialysis was used (22). Nicotine 0.5 mg/kg i.p. induces peak PVN norepinephrine levels within 20 minutes, and mecamylamine abolishes this, while hexamethonium is ineffective. The norepinephrine response is dose-dependent (ED_{50} of approximately 0.25 mg/kg), and nicotine 0.5 mg/kg stimulated maximal secretion. More recently, nicotine or cytisine was administered into the 4V, using this microdialysis paradigm (23). Following either agonist, dose-dependent secretion of PVN norepinephrine is observed; ED_{50}s for nicotine or cytisine are 1 μg or 6 μg, respectively. Norepinephrine secretion is blocked by prior treatment with mecamylamine into the 4V. In contrast, α-bungarotoxin is ineffective. The results of these studies indicate that nicotine administered systemically or into the 4V can stimulate norepinephrine secretion in the PVN, which would lead to the release of ACTH secretagogues.

Desensitization of Nicotine-Stimulated ACTH and Norepinephrine Secretion

The dose of nicotine and the frequency of its administration appear to be essential determinants of its action on multiple systems including the neuroendocrine regulation of the H-P-A axis. Thus, the effects of repeated injections of nicotine on ACTH and PVN norepinephrine secretion have been studied in our laboratory. Desensitization to the acute stimulatory effects of nicotine on both parameters is induced after a single injection. Thus, when nicotine 1.0 mg/kg i.p. is injected 60 minutes after nicotine 0.5 mg/kg, the ACTH response is markedly attenuated (24). Desensitization persists for at least 6 hours, and a similar phenomenon is found with the prolactin response to nicotine (24).

Using *in vivo* microdialysis of the PVN, norepinephrine secretion in response to the second injection of nicotine 0.5 mg/kg i.p., delivered 100 minutes after the first, is reduced by approximately 50% (22). This degree of desensitization persists, in that norepinephrine responses to a third and fourth injection of the same dose of nicotine are each reduced by 50% in comparison to the initial response, and are not different from each other. Similarly, norepinephrine secretion due to a second injection of nicotine into the 4V is reduced by 50%, when administered 100 minutes after the first dose (23). In contrast, desensitization to repetitive 4V injections of cytisine is complete, and this is observed with either submaximally or maximally stimulative doses. However, cross-desensitization of either agonist to the other did not occur. This may reflect heterogeneity in the nicotinic cholinergic receptor subtypes involved. Thus, nicotine elevates ACTH, at least in part by activating catecholaminergic afferents near their origin through multiple receptor subtypes, thereby stimulating PVN CRF neurons.

Acknowledgements: *This work is supported by U.S. Public Health Service Grant DA-03977. We are grateful to Kathy McAllen and Catherine Foster for their exceptional technical support.*

References

1. Matta SG, Beyer HS, McAllen KM, Sharp BM. Nicotine elevates rat plasma ACTH by a central mechanism. J Pharm Exp Ther 1987; 243: 217-226.
2. Ashgar K, Roth LJ. Entry and distribution of hexamethonium in the central nervous system. Biochem Pharmacol 1971; 20: 2787-2795.
3. Romano C, Goldstein A, Jewell NP. Characterization of the receptor mediating the nicotine discriminative stimulus. Psychopharmacology 1981; 74: 310-315.
4. Antoni FA. Hypothalamic control of adrenocorticotropin secretion: Advances since the discovery of 41-residue corticotropin-releasing factor. Endocrine Reviews 1986; 7:351-378.
5. Sharp BM, Nicol S, Cummings S, Seybold V. Distribution of nicotinic binding sites with respect to CRF and neurophysin immunoreactive perikarya within the rat hypothalamus. Brain Research 1987; 422: 361-366.
6. Clarke PBS, Schwartz RM, Paul SM, Pert CB, Pert A. Nicotinic binding in rat brain: Autoradiographic comparison of ^{3}H-acetylcholine, ^{3}H-nicotine and ^{125}I-alpha-bungarotoxin. J Neurosci 1985; 5: 1307-1315.
7. Matta SG, McAllen KM, Sharp BM. Role of the fourth cerebroventricle in mediating rat plasma ACTH responses to intravenous nicotine. J Pharm Exp Ther 1990; 252: 623-630.
8. Matta SG, Sharp BM. The role of the fourth cerebroventricle in nicotine-stimulated prolactin release in the rat: Involvement of catecholamines. J Pharm Exp Ther 1992; 260: 1285-1291.
9. Matta SG, Foster CA, Sharp BM. Selective administration of nicotine into catecholaminergic regions of rat brainstem stimulates adrenocorticotropin secretion. Endocrinology 1993; 133: 2935-2942.
10. Swanson LW, Sawchenko PE, Berol B, Hartman BK, Helle KB, Van Orden DE. An immunohistochemical study of the organization of catecholaminergic cells and terminal fields in the paraventricular and supraoptic nuclei of the hypothalamus. J Comp Neurol 1981; 196: 271-285.
11. Alonso G, Szafarczyk A, Balmefrezol M, Assenmacher I. Immunocytochemical evidence for stimulatory control by the ventral noradrenergic bundle of parvocellular neurons of the paraventricular nucleus secreting corticotropin releasing hormone and vasopressin in rats. Brain Res 1986; 397: 297-307.
12. Sapolsky RM, Armanini MP, Sutton SW, Plotsky PM. Elevation of hypophysial portal concentrations of adrenocorticotropin secretagogues after fornix transection. Endocrinology 1989; 125: 2881-2887.
13. Graybiel AM, Moratalla R, Robertson HA. Amphetamine and cocaine induce drug-specific activation of the c-fos gene in striosome-matrix compartments and limbic subdivisions of the striatum. Proc Natl Acad Sci USA 1990; 87: 6912-6916.
14. Rivera VM, Greenberg ME. Growth factor-induced gene expression: The ups and downs of c-fos regulation. New Biologist 1990; 2: 751-758.
15. Chiu R, Boyle WJ, Meek J, Smeal T, Hunter T, Karin M. C-fos protein interacts with c-jun/AP-1 to stimulate transcription of AP-1 responsive genes. Cell 1988; 54: 541-552.
16. Sonnenberg JL, Rauscher III FJ, Morgan JI, Curran T. Regulation of proenkephalin by fos and jun. Science 1989; 240: 1622-1625.
17. Sharp BM, Beyer HS, McAllen KM, Hart D, Matta SG. Induction and desensitization of the c-Fos mRNA response to nicotine in rat brain. Mol Cell Neurosci 1993; 4: 199-208.
18. Matta SG, Foster CA, Sharp BM. Nicotine stimulates the expression of c-fos protein in the parvocellular paraventricular nucleus and brainstem catecholaminergic regions. Endocrinology 1993; 132: 2149-2156.
19. Guillame V, Conte-Devolx B, Szafarczyk A, Malaval F, Pares-Herbute N, Grino M, Alonso G, Assenmacher I, Oliver C. The corticotropin-releasing factor release in rat hypophysial port blood is mediated by brain catecholamines. Neuroendocrinology 987; 46: 143.
20. Plotsky PM. Facilitation of immunoreactive corticotropin-releasing factor secretion into the hypophysial-portal circulation after activation of catecholaminergic pathways or central norepinephrine injection. Endocrinology 1987; 121: 924.

21. Matta SG, Singh J, Sharp BM. Catecholamines mediate nicotine-induced adreno-corticotropin secretion via α-adrenergic receptors. Endocrinology 1990; 127: 1646-1655.
22. Sharp BM and Matta SG. Detection by *in vivo* microdialysis of nicotine-induced norepinephrine secretion from the hypothalamic paraventricular nucleus of freely moving rats: Dose-dependency and desensitization. Endocrinology 1993; 133: 11-19.
23. Matta SG, McCoy JG, Foster CA, Sharp BM. Nicotinic agonists administered into the fourth ventricle stimulate norepinephrine secretion in the hypothalamic paraventricular nuclus: An *in vivo* microdialysis study. Submitted, 1994.
24. Sharp BM and Beyer HS. Rapid desensitization of the acute stimulatory effects of nicotine on rat plasma adrenocorticotropin and prolactin. J Pharm Exp Ther 1986; 238: 486-491.

Effects of Nicotine on
Biological Systems II
Advances in Pharmacological Sciences

REGULATION OF PROENKEPHALIN GENE EXPRESSION IN BOVINE ADRENAL MEDULLARY CHROMAFFIN CELLS BY NICOTINE

Bruno Bacher [1], Xiaomin Wang [1] and Volker Höllt [2]

[1] Physiologisches Institut, Universität München
[2] Institut für Pharmakologie und Toxikologie, Universität Magdeburg, Germany

Summary: Nicotine (10 μM) induced a marked elevation of the immediate early gene mRNAs c-fos, c-jun, and jun-B. Maximally elevated levels for c-fos mRNA (about 100-fold) were obtained after 20 min. c-jun and jun-B mRNAs were increased 3- to 4-fold 60 min after nicotine addition. In contrast, the mRNA levels coding for proenkephalin showed, after an apparent lag-phase, a delayed (up to 4-fold) induction between 4 and 24 hours. The expression of the proenkephalin reporter gene PENKCAT-153/+53, containing 153 nucleotides of upstream promotor sequences, was increased about twofold by nicotine or by cotransfection of an expression vector for c-fos, which also enhances expression of PENKCAT(CRE2)2, indicating that nicotine may induce proenkephalin via c-Fos which binds at the ENKCRE-2 element.

Introduction

Activation of acetylcholine receptors in adrenal medullary chromaffin cells by nicotine results in the concomitant secretion of catecholamines and a variety of neuropeptides, such as the enkephalins (1). In addition, continuous stimulation of chromaffin cells with nicotine increases the biosynthesis of the enkephalins as indicated by the elevation of the proenkephalin mRNA levels (2-4). In addition, chronic nicotine infusion by minipumps in rats causes an increase in proenkephalin gene expression in the adrenal medulla of the animals (5).

The mechanism of induction of the proenkephalin gene by nicotine in chromaffin cells is not completely clear. A generally accepted hypothesis suggests that the interaction of nicotine with the acetylcholine receptor causes an influx of sodium ions through the associated channel which, in turn, depolarizes the membrane and opens voltage-dependent calcium channels. The influx of calcium ions activates calcium/calmodulin dependent kinases which phosphorylate transcription factors, such as CREB (cAMP responsive element binding protein), which in turn stimulate the transcription of the proenkephalin gene.

Our previous studies in chromaffin cells, however, indicated that the stimulation of proenkephalin gene expression by nicotine requires the de novo synthesis of proteins (4). In the present paper we will provide additional evidence that the induction of immediate early genes, such as c-fos, rather than the phosphorylation of the constitutive transcription factor CREB is a key feature of the mechanism of proenkephalin gene stimulation in chromaffin cells by nicotine.

Materials and Methods

Primary cultures of bovine chromaffin cells were prepared according to Livett (6). The cells were incubated in DMEM:F12 medium supplemented with 10% fetal calf serum. Twenty-four hours prior to the experiment, the serum containing medium was changed to serum-free medium. The cells were harvested at several time intervals after addition of nicotine.

RNA was isolated from the cells using the method of Chomczynski and Sacchi (7). Samples were subjected to Northern blot analysis as described previously. The filters were hybridized with ^{32}P-labelled-RNA probes complementary to nucleotide sequences to mouse v-fos, mouse jun-B, bovine proenkephalin and mouse α-tubulin. The probes were obtained from Dr. Heumann (v-fos), Dr. Bravo (c-jun, jun-B), Dr. Numa (proenkephalin) and Dr. Vanetti (α-tubulin). The hybridized filters were exposed to x-ray films and the autoradiograms were scanned by densitometry.

For the transfection experiments, chromaffin cells were plated in 5 cm dishes. Ten μg PENKCAT reporter DNA (see below) was transfected using the calcium phosphate method. 16 h later the cells were incubated with nicotine for further 24 h. For co-transfections 5 μg of the PENKCAT reporter plasmid and 5 μg of the expression vector pCMV-fos (a gift from Dr. Curran, New Jersey, USA) were transfected and the cells harvested 48 h later. The chloramphenicol acetyl transferase (CAT) activity was measured by adding ^{14}C-chloramphenicol and acetyl coenzyme A to the cell extracts followed by thin-layer chromatography. The spot on the thin layer-sheets corresponding to the acetylated and nonacetylated forms of chloramphenicol were scraped into scintillation vials and measured for radioactivity.

For the construction of the PENKCAT reporter plasmid a fragment of the rat proenkephalin gene (8) containing 153 nucleotides of upstream sequences and 50 nucleotides of exon 1 was cloned into the vector pBLCAT3.

Results

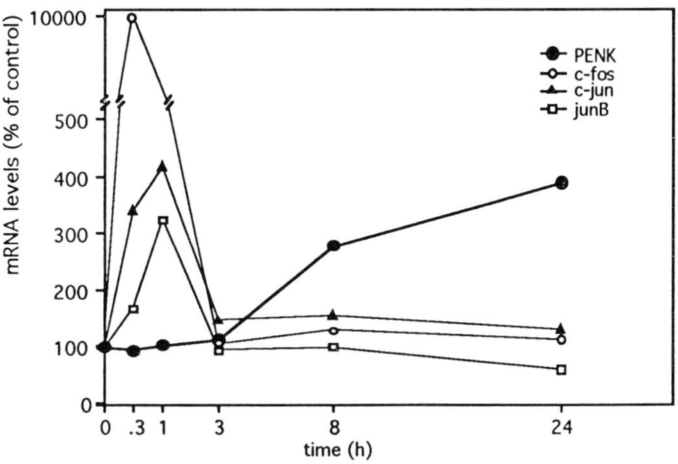

Figure 1. Time-course of the levels of proenkephalin, c-fos, c-jun and jun B mRNAs in bovine chromaffin cells after incubation with 10 μM nicotine. Each point is the mean of three determinations.

Figure 1 shows the time-course of the mRNA levels transcribed by the immediate early genes c-fos, c-jun and jun-B in response to 10 μM nicotine in chromaffin cells. Nicotine causes a rapid induction of c-fos mRNA with highest levels (about 100-fold above controls) at 20 minutes. c-jun and jun-B were maximally elevated at 60 minutes (4-fold and 3-fold) after incubation with nicotine. In contrast, the increase of proenkephalin mRNA showed after an apparent lag phase a delayed (up to 4-fold) induction between 4 and 24 h.

Figure 2 shows that nicotine induces the expression of a PENKCAT reporter gene containing 153 nucleotides of upstream promoter sequences (PENKCAT-153/+50) about twofold. This sequence comprises the enkephalin cAMP responsive elements 1 and 2 (ENKCRE-1 and ENKCRE-2; Fig 2A) which have been shown to be responsible for the induction of the gene by calcium, cAMP and phorbol esters (9,10). Dimers consisting of c-Fos/c-Jun proteins bind to this element in gel retardation experiments (11).

Further evidence that the product of the c-fos gene might be a transcription factor for the proenkephalin gene is illustrated in Figure 3 which shows that cotransfection of the pCMV-fos

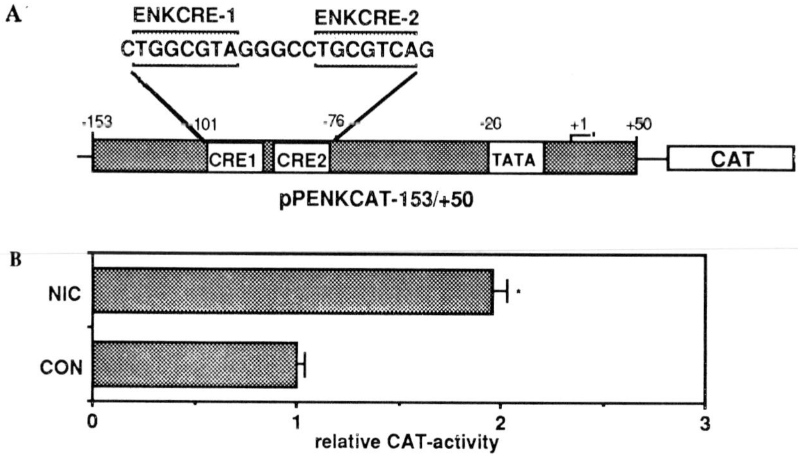

Figure 2. Nicotine induces the expression of proenkephalin CAT reporter genes in chromaffin cells. (A) Structure of the PENKCAT-153/+50 fusion gene. (B) The effect of nicotine (NIC, 10 μM) on the expression of PENKCAT-153/+50. CON represents unstimulated cells transfected with PENKCAT-153/+50 reporter gene. Each bar represents the mean and SEM of four to six determinations. Astericks indicate significant differences ($p < 0.05$).

expression plasmid increases the expression of the PENKCAT-153/+50 reporter plasmid about twofold. In addition, cotransfection of the c-fos expression plasmid also increases the expression of a proenkephalin gene reporter plasmid which contains a dimer of the ENKCRE-2 element in front of a minimal promoter (Fig. 3 A, B).

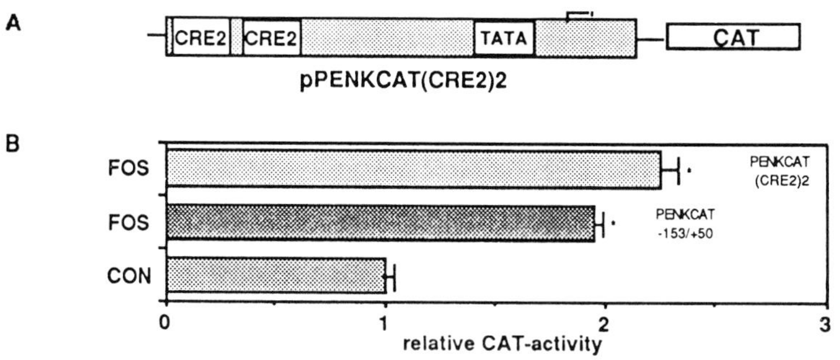

Figure 3. Cotransfection of c-fos increases the expression of proenkephalin CAT reporter genes in chromaffin cells. (A) Structure of the reporter plasmid PENKCAT(CRE2)2 which contains a dimer of the ENKCRE-2 element in front of a minimal promotor. (B) Expression of PENKCAT-153/+50 and PENKCAT(CRE2)2 after cotransfection of the control plasmid (pRc/CMV; CON) and an expression plasmid for c-fos (FOS). Each bar represents the mean and SEM of four to six determinations. Astericks indicate significant differences ($p < 0.05$).

Discussion

AP-1 transcription factors such as heterodimers and/or homodimers of c-Fos, c-Jun and Jun-D proteins have been shown to directly bind to ENKCRE-2 element in band shift experiments in vitro or to transactivate proenkephalin gene expression in undifferentiated F9 cells (11). On the other hand, CREB has also been shown to activate proenkephalin transcription in CV-1 cells (12). Since the c-fos gene itself contains a binding site for CREB (13), it is possible that nicotine activates the proenkephalin gene either directly via CREB or indirectly via an induction of Fos containing AP-1 complexes (e.g. Fos/Jun heterodimers).

The results presented in this study and other data reported by our laboratory support the second hypothesis. Thus, nicotine causes a rapid induction of the mRNA levels coding for the onco-proteins c-Fos, c-Jun and Jun-B whereas the levels of proenkephalin mRNA are increased with a delay of about 3 h. Moreover, the nicotine induced proenkephalin gene expression can be blocked by protein synthesis inhibitors (4) indicating that the proenkephalin gene in chromaffin cells is not stimulated by phosphorylation of a constitutive transcription factor, such as CREB, but via de novo synthesis of AP-1 proteins, such as c-Fos, c-Jun or Jun-B which bind to the ENKCRE-2 element of the proenkephalin gene promoter.

References

1. Höllt V. Regulation of opioid peptide gene expression. In: Herz A, editor. Handbook of experimental pharmacology: opioids I. Berlin, Heidelberg, New York: Springer, 1993:307-346.
2. Eiden LE, Giraud P, Dave JR, Hotchkiss AJ, Affolter HU. Nicotinic receptor stimulation activates enkephalin release and biosynthesis in adrenal chromaffin cells. Nature 1984; 312:661-663.
3. Kley N, Loeffler JP, Pittius CW, Hšllt V. Proenkephalin A gene expression in bovine adrenal chromaffin cells is regulated by changes in electrical activity. EMBO J 1986; 5:967-970.
4. Farin CJ, Kley N, Hšllt V. Mechanisms involved in the transcriptional activation of proenkephalin gene expression in bovine chromaffin cells. J Biol Chem 1990; 265:19116-19121.
5. Höllt V, Horn G. Effect of nicotine on mRNA levels encoding opioid peptides, vasopressin and alpha 3 nicotinic receptor subunit in the rat. Clin Invest 1992; 70:224-231.
6. Livett BG. Adrenal medullary chromaffin cells in vitro. Physiol Rev 1984; 64:1103-1150.
7. Chomczynsky P, Sacchi N. Single-step method of RNA isolation by guanidinium thiocyanate phenol chloroform extraction. Anal Biochem 1987; 162:156-159.
8. Rosen H, Douglass J, Herbert E. Isolation and charakterisation of the rat proenkephalin gene. J Biol Chem 1984; 259:14309-14313.
9. Comb M, Birnberg NC, Seasholtz A, Herbert E, Goodman HM. A cyclic AMP- and phorbolester-inducible DNA element. Nature 1986; 323:353-356.
10. Comb M, Mermod N, Hyman SE, Pearlberg J, Ross ME, Goodman HM. Proteins bound at adjacent DNA elements act synergistically to regulate human proenkephalin cAMP inducible transcription. EMBO J 1988; 7:3793-3805.
11. Sonnenberg JL, Rauscher FJ, Morgan JI, Curran T. Regulation of proenkephalin by Fos and Jun. Science 1989; 246:1622-1625.

12. Huggenvik JI, Collard MW, Stofko RE, Seasholtz AF, Uhler MD. Regulation of the human enkephalin promoter by two isoforms of the catalytic subunit of cyclic adenosine 3',5'-monophosphate-dependent protein kinase. Mol Endocrinol 1991; 5:921-930.
13. Sheng M, Dougan ST, McFadden G, Greenberg ME. Calcium and growth factor pathways of c-fos transcriptional activation reqire distinct upstream regulatory sequences. Mol Cell Biol 1988; 8:2787 2796.

Effects of Nicotine on
Biological Systems II
Advances in Pharmacological Sciences

EFFECTS OF NICOTINE ON THROMBOXANE AND LEUKOTRIENE SYNTHESIS IN CELLULAR SYSTEMS

R.-M. Goerig and S. Koll

Medizinische Klinik 4, University of Erlangen-Nürnberg,
Breslauer Str., 90721 Nürnberg, Germany

Summary: In coculture systems of human macrophage-like cells and human endothelial cells low concentrations of nicotine inhibit thromboxane as well as prostacyclin synthesis and induce a major increase of cell contact-dependent transcellular leukotriene synthesis.

Introduction

Increased bleeding time, chronic inflammatory lung disease, and impaired host defense have been demonstrated in smokers. Animal cells can convert arachidonic acid into prostaglandin, thromboxane, and leukotrienes. These locally produced mediators of inflammatory and hypersensitivity reactions have been implicated in clinically important disease processes. Thromboxane and leukotrienes, major arachidonic acid metabolites of macrophages (1-3), are potent proaggregatory and vasoconstricting molecules or have been related to host defense mechanisms. In contrast, prostacyclin, major arachidonic acid metabolite of endothelial cells, has been shown to be an antiaggregatory and potentially vasodilating substance. The interaction of endothelial cells with mononuclear cells is an essential aspect of acute or chronic vascular inflammation and atherosclerotic transformation of the vessel wall. Monocytes and endothelial cells are able to synergistically cooperate in the synthesis of inflammatory active prostanoids and leukotrienes (eicosanoids) from arachidonic acid (4, 5). The regulation of the transcellular arachidonic acid metabolism may determine spectrum and amount of eicosanoids in inflamed tissues. Nicotine has been shown to be a potent inhibitor of thromboxane formation in myelomonocytic cells (6) and is able to decrease prostacylin formation of endothelial cells (7). In the present investigation we focused attention to the effects of nicotine on trans-

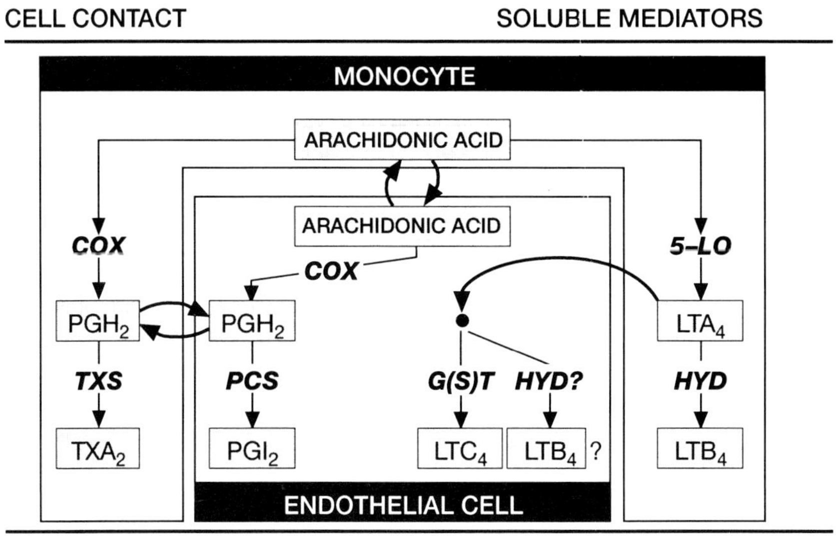

Figure 1. Schematic representation of a monocyte/endothelial cell coculture system with emphasis on the arachidonic acid cascade and the possible pathways of transcellular eicosanoid metabolism.

cellular eicosanoid synthesis in a coculture system of human macrophage-like cells and human endothelial cells (Fig. 1).

Materials and Methods

To investigate mechanisms of the nicotine-mediated effects on eicosanoid synthesis of the vessel wall we used coculture systems of confluent unstimulated human iliac venous or arterial endothelial cells (HIVEC) (the Corriell Institute, New Jersey) and human promyelocytic (HL-60) cells induced to differentiate into macrophages (TPA-treated) (1, 2, 8). To differentiate effects mediated by soluble mediators from those mediated by cell contact we used intercup chambers as described by Saeki et al. (9). Cell culture, determination of thromboxane or leukotriene B_4 or C_4 synthesis, assays of microsomal PGH or thromboxane synthase, PCR of mRNA of cyclooxygenase I or II and thromboxane synthase has been performed as described (1, 2, 10).

Results

1. Our experiments have shown that thromboxane synthesis from exogenously added arachidonic acid is synergistically increased in cocultures of monocytic cells with endothelial cells by a factor of 3.5 - 5.5 when compared with monocultured cells. The synergistic effect depends on a probably cell contact-mediated selective upregulation of the key enzyme cyclooxygenase II (enzyme and mRNA analysis) in endothelial cells. 10 to 100 nM nicotine is an effective inhibitor of the synergistic transcellular thromboxane synthesis in cocultures of endothelial cells and monocytic cells (Fig. 2). Nicotine directly inhibits the activity of thromboxane synthase in cocultered monocytic cells and prevents the upregulation of cyclooxygenase II in endothelial cells (not shown).

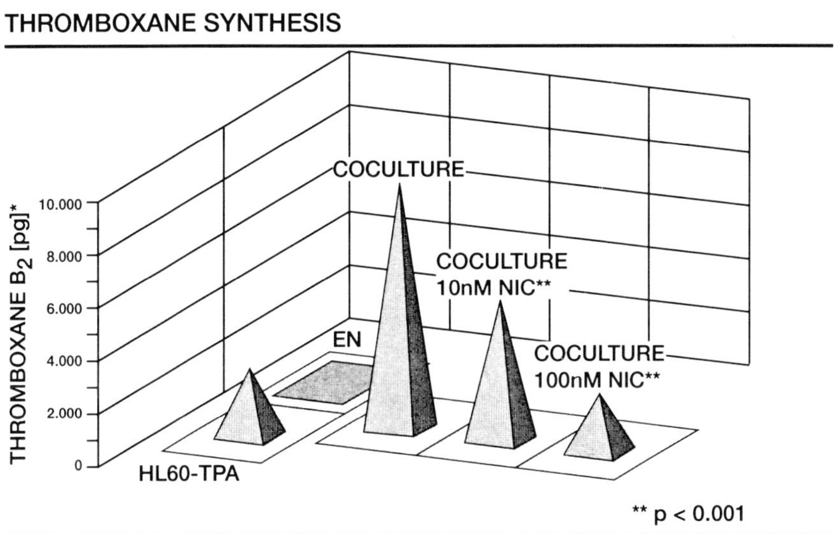

Figure 2. Effects of nicotine on thromboxane synthesis in monocultured and cocultured TPA-treated (100 nM) HL-60 (12 h) cells differentiating into macrophages (1) and unstimulated but confluent human endothelial cells. Both cell types were monocultured or cocultured for 8 hours in the presence or absence of 10 - 100 nM nicotine either with the possibility of cell contact or in intercup chambers. Data of Fig. 2 represent thromboxane synthesis from 100 nM arachidonic acid (60 min.) from cocultures with the possibility of direct cell contact. Data represent mean of the results of three independent experiments.

2. 10 to 100 nM nicotine potentiates leukotriene C_4 (Fig. 3) and leukotriene B_4 (not shown) synthesis in cocultures of endothelial cells with monocytic cells and in cocultures with

granulocytic cells (not shown). This effect is cell contact-dependent (inter-cup chambers as control) and does not appear to be regulated by soluble mediators.

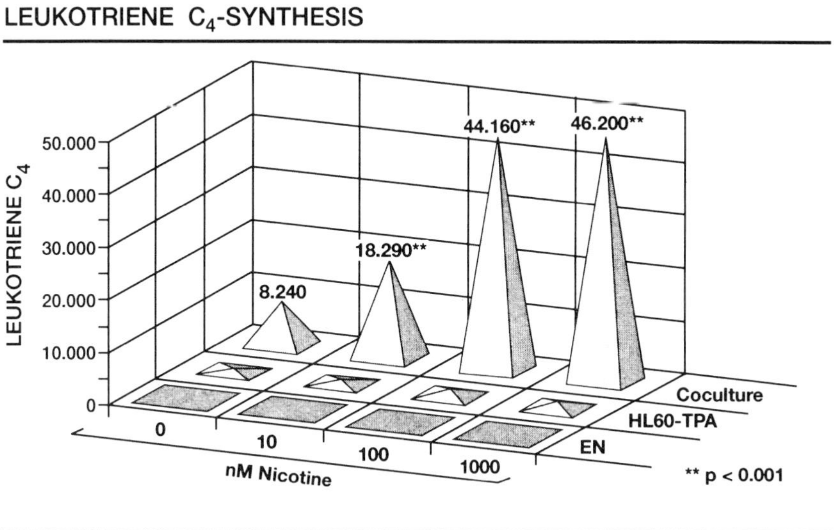

Figure 3. Effects of nicotine on leukotriene C_4 synthesis. Cells were incubated and leukotriene C_4 synthesis from exogenously added arachidonic acid was determined as described in Figure 1 or under Methods. Results represent mean of three independent experiments.

Discussion

Our results indicate that transcellular synthesis of thromboxane is synergistically increased in cocultures of human macrophage-like cells and endothelial cells. This synergistic effect strictly appears to depend on direct cell contact. Preliminary data have shown that this effect may be based on the expression of specific and corresponding adhesion molecules on the surface of the monocytic and the endothelial cells. Nicotine not only directly inhibits thromboxane synthase in monocytic cells but also prevents the induction of cyclooxygenase II in co-cultured endothelial cells. In addition, nicotine dramatically increase the capacity of leukotriene C_4 and leukotriene B_4 synthesis in co-cultured cells. These results indicate that in coculture systems relatively low concentrations of nicotine may dramatically change transcellular synthesis of eicosanoids and may favor leukotriene formation in conditions of vascular inflammation. Our results raise the possibility that the effects of nicotine on eicosanoid synthesis are related to smoking-induced changes in blood coagulation, chronic inflammatory lung disease, and impaired host defense.

References

1. Goerig M, Habenicht A, Heitz R et al. Sn-1,2-diacylglycerols and phorbol diesters stimulate thromboxane synthesis by de novo synthesis of prostaglandin H synthase in human promyelocytic leukemia cells. J Clin Invest 1987; 79:903-911.
2. Goerig M, Habenicht A, Zeh W et al. Evidence for coordinate, selective regulation of eicosanoid synthesis in platelet-derived growth factor-stimulated 3T3 fibroblasts and in HL-60 cells induced to differentiate into macrophages or neutrophils. J Biol Chem 1988; 263:19384-19391.
3. Habenicht A, Goerig M, Rothe D et al. Early erversible induction of leukotriene synthesis in chicken myelomonocytic cells transformed by a temperature-sensitive mutant of avian leukemia virus E26. Proc Natl Acad Sci USA 1989; 86:921-924.
4. Marcus A, Safier L, Ullman H et al. Platelet-neutrophil interactions. J Biol Chem 1988; 263:2223-2231.
5. Feinmark SJ, Cannon PJ. Endothelial cell-neutrophil interactions lead to endothelial cell leukotriene C_4 synthesis. Adv Prostagl Thrombox Leukot Res 1987; 17:120-126 .
6. Goerig M, Ullrich V, Schettler G et al. A new role for nicotine: selective inhibition of thromboxane formation by direct interaction with thromboxane synthase in human promyelocytic leukemia cells differentiating into macrophages. Clin Invest 1992; 70:239-243.
7. Alster P, Wenmalm A. Effect of nicotine on prostacyclin formation in rat aorta. Eur J Pharmacol 1983; 86:441-452
8. Hauser I, Johnson D, Madri JA. Differential expression of VCAM-1 on human iliac venous and arterial endothelial cells and its role in the adhesion of U937 cells. J Immunol 1993; 145:5172-5185.
9. Saeki T, Morioka T, Arakawa M et al. Modulation of mesangial cell proliferation by endothelial cells in coculture. Am J Pathol 1991; 139:949-957.
10. Nüsing R, Goerig M, Habenicht A, Ullrich V. Selective eicosanoid formation during HL-60 macrophage differentiation. FEBS Eur J Biochem 1993; 212:371-376.

Nicotine and smoking: current controversies

Effects of Nicotine on
Biological Systems II
Advances in Pharmacological Sciences

SMOKING-INDUCED ALTERATIONS IN BRAIN ELECTRICAL PROFILES: NORMALIZATION OR ENHANCEMENT ?

V. Knott and A. Harr

Department of Psychiatry, University of Ottawa and Institute of Mental Health Research/Royal Ottawa Hospital, Ottawa, Ontario, Canada, K1Z 7K4.

Summary: Quantitative electroencephalography (EEG) and event-related potentials (ERPs) collected during passive and task activated conditions were employed here to examine putative normalizing/enhancing effects of acute smoking in smokers relative to non-smoking non-smokers. Tasks were specifically selected to tap arousal, attentional and memory processes, which have been previously reported to be affected by smoking and nicotine administration in both normal and neuropsychiatric populations. Evidence indicates that the ability of smoking to affect brain function is response/task dependent and that evidence for both normalization and enhancement are observed with neuroelectric recordings.

Motivational theories have repeatedly argued for the importance of highlighting antecedent trait/state factors which may predispose nonsmokers to initiate an adaptive behavior such as smoking which, once started, persists because of its reinforcing central impact resulting in restoration/normalization or absolute enhancement of emotional, cognitive and behavioral functions (1). Although the process of comparing current habitual smokers with confirmed nonsmokers may be potentially confounded by smoking abstinence effects and/or long term effects of smoking, investigations which compare (relative to nonsmokers) the acute effects of smoking on brain function indices (especially indices which appear to differentiate abstaining smokers from nonsmokers) may provide a unique design opportunity for characterizing the putative normalizing/enhancing qualities frequently attributed to cigarette smoking.

As previous employment of this design strategy has proven useful in isolating significant brain electrical features related to smoking motivation (2), this paper will present recent quantitative electroencephalographic (EEG) and event-related potentials (ERP) findings (from this laboratory) couched in this comparative research paradigm, with smokers after acute sham smoking and cigarette smoking being compared with nonsmokers. For each of the reported study comparisons, smokers (n=10) and nonsmokers (n=10) were matched for age and sex. Smokers were smoking a minimum of 15 cigarettes/day for at least a five year period. Smokers and nonsmokers abstained from caffeine and drugs during the 12 hours prior to testing and smokers further abstained from smoking overnight, up until the test sessions (8:00-11:00 a.m.). Smokers in the active smoking sessions smoked their own cigarette (paced at 1 puff per minute) and in the sham condition they inhaled on an unlit cigarette.

The peak alpha frequency (PAF), i.e. the frequency within the alpha band containing the maximum power, extracted from the resting-state EEG spectra has previously been shown to be slower in smoking deprived (overnight) smokers (DS), relative to nonsmokers (NS). Following the smoking of two cigarettes, PAF in active cigarette smokers (CS) was found to increase to a level comparable to that seen in nonsmokers (2). As PAF is believed to regulate the rate of execution of cognitive/behavioral operations (e.g. faster PAFs are associated with faster reaction times) and as such is of significance for psychological-tool models of smoking motivation, a follow-up study of this apparent normalization effect was carried out which compared, relative to nonsmokers, the effects of day-long smoking (i.e. 8:00 a.m. to 4:00 p.m.; one cigarette per 30 minutes) versus day-long (sham) smoking deprivation on PAF, mood (via analogue scales of subjective alertness) and a rapid visual information processing task (RVIP). As shown in Figure 1, the measurements assessed at the end of the day (4:00 p.m.) confirmed previous findings of slower PAFs in deprived smokers relative to nonsmokers and an acceleration of brain frequency with day-long smoking to a level similar to that seen in nonsmokers. This normalization effect was also extended to mood ratings and behavioral responsivity. Both subjective alertness and response accuracy, which were negatively affected (relative to nonsmokers) by day-long deprivation, were altered by day-long smoking to the degree that smokers and nonsmokers then became similar in these functions. It is of interest to note that the profile effects of day-long smoking mirrored the acute effects observed in the previously reported study, indicating that the central effects of repeated smoking were not cumulative, nor did they habituate with successive smoke exposures.

As previously discussed, the acute power spectrum profile resulting from the smoking of 1-2 cigarettes is characterized by a power shift from slower to faster frequencies with delta, theta, alpha1 bands being decreased and alpha2 and beta bands being increased. Comparison of the resulting power spectra following sham and real smoking (one cigarette) in smokers

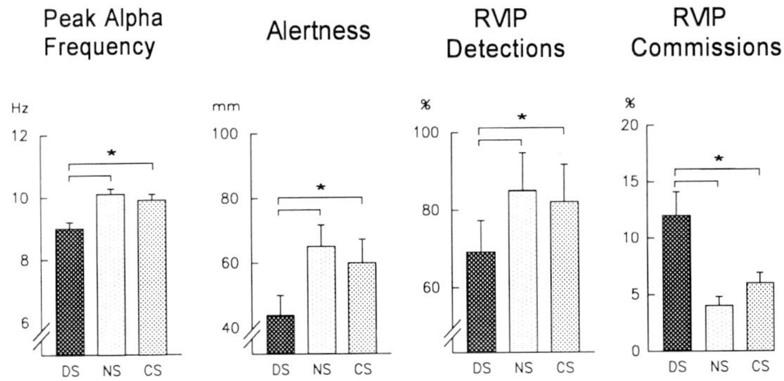

Figure 1. Mean (±SD) peak alpha frequency (O_2-A_1+A_2), subjective alertness ratings, signal detections and commissions in day-long DS, NS, and CS (see text). ★ $p \leq 0.05$

relative to nonsmokers indicated, as shown in Figure 2, that: (a) none of the power band spectra significantly differentiated overnight deprived smokers (after sham smoking) and nonsmokers, but (b) that real smoking appeared to enhance cortical arousal (less alpha1 and more beta) in smokers relative to nonsmokers. The loss of power from slow alpha and the resultant power increase in faster beta frequencies is again typical of psychostimulants and this widespread topographical power increment would indicate that smoking may serve to enhance arousal-associated cognitive/emotional processes to levels beyond that which may be observed in nonsmokers.

The facilitating effects of smoking on mood and behavior offer support for the psychological-tool model of smoking motivation which has frequently adopted an "information processing" approach in studying the effects of smoking on human perception and cognition (3). Although a number of ERP components are believed to reflect various brain transactions or processing stages (e.g. encoding, attention, memory, decision making, etc.), the late parieto-central positivity, which occurs at approximately 300 msec (P_{300}) when a subject detects an

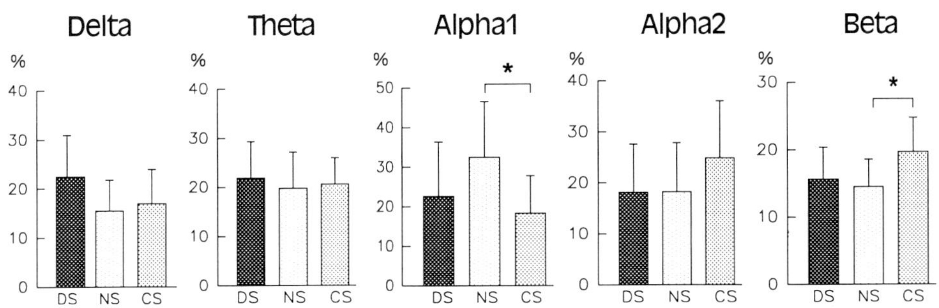

Figure 2. Mean (±SD) relative (%) power (averaged across 21 sites) of delta, theta, alpha1, alpha2, and beta frequency bands in DS, NS, and CS (see text). * $p \leq 0.05$

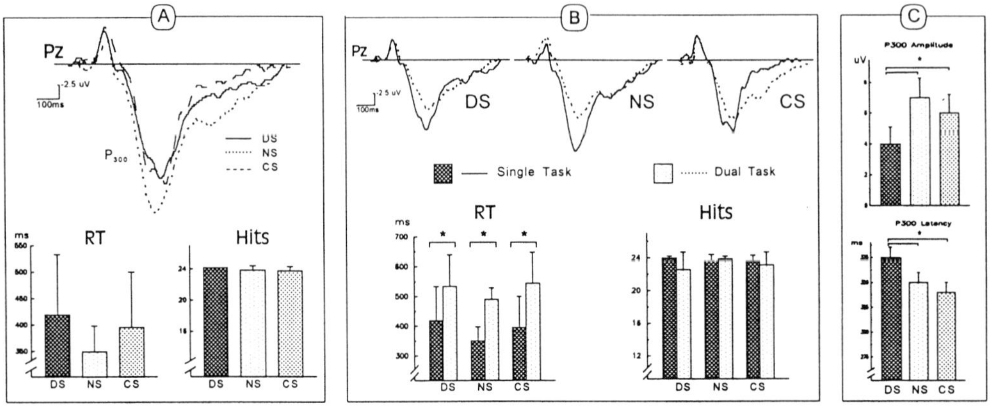

Figure 3. Panel **A**: Group averaged ERP-P_{300} waveforms, RTs and hits in simple auditory oddball task; Panel **B**: Group averaged ERP-P_{300} waveforms, RTs and hits in single and dual task paradigms; Panel **C**: Group averaged ERP-P_{300} amplitudes and latencies (Pz-A_1) during day-long smoking/smoking deprivation. ★ $p \leq 0.05$

informative task relevant stimulus, is the most frequently examined ERP (4). P_{300} indices are seen as a potential marker of attentional load/demand (P_{300} amplitude) and processing speed (P_{300} latency). In a relatively simple, "oddball" task, where subjects are simply required to detect random rare tones (p=.20) among a series of frequent tones, reaction times (RT), hits, P_{300} amplitudes and latencies, as shown in Figure 3 (panel A), are virtually identical in overnight deprived smokers (after sham smoking), non-deprived smokers (after smoking a cigarette) and nonsmokers.

It is quite possible however that putative normalizing/enhancing effects of smoking may not be manifested unless subject and/or task conditions result in excessive loading of the limited capacity central processing system. The effects of smoking in relation to task demands can be examined in dual task paradigms where it has been shown that increases in the cognitive processing requirements of a secondary visual task stimuli result in reduced P_{300} amplitudes evoked by simultaneously presented auditory oddball stimuli (4). In Figure 3 (panel B) however, P_{300} amplitudes elicited by auditory oddball stimuli (which required counting), under dual task conditions (where a secondary visual task involved immediate recognition memory) were significantly but equally attenuated relative to single task conditions, in overnight deprived smokers (after sham smoking), non-deprived smokers (after smoking a cigarette) and nonsmokers. Thus, even when task conditions are taxing and are obviously requiring significant attentional resources, smoking does not appear to have any significant impact on P_{300} related processes. However, in the day-long smoking/smoking deprivation study described earlier, P_{300} amplitudes and latencies assessed at the end of day-long smoking deprivation (4:00 p.m.) (Figure 3, Panel C) were found to be smaller and longer (respectively) than those observed in nonsmokers and day-long smoking resulted in increased amplitudes and faster latencies which were comparable to those observed in nonsmokers. These latter results tentatively suggest that time of day/diurnal rhythms may play a significant role in the appearance of smoker versus nonsmoker differences and in the normalizing/enhancing effects of smoking thereupon.

Late positive potentials have been examined in relation to memory processes and a good number of studies have employed P_{300} measures in memory search paradigms where, in response to probe stimuli, P_{300} latencies (and RT) are found to increase and P_{300} amplitudes are seen to decrease as memory set size increases (5). In such a paradigm where P_{300} latencies (and RT) increased and amplitudes decreased in response to correctly identified in-set and out-of-set probe stimuli (following presentation of 1, 3, and 5 digit memory sets), nonsmokers exhibited similar behavioral and P_{300} responses as overnight deprived smokers (after sham smoking) and smokers allowed to smoke. In general, these findings support the general trend from previous literature which failed to find significant effects of smoking on short term memory processes.

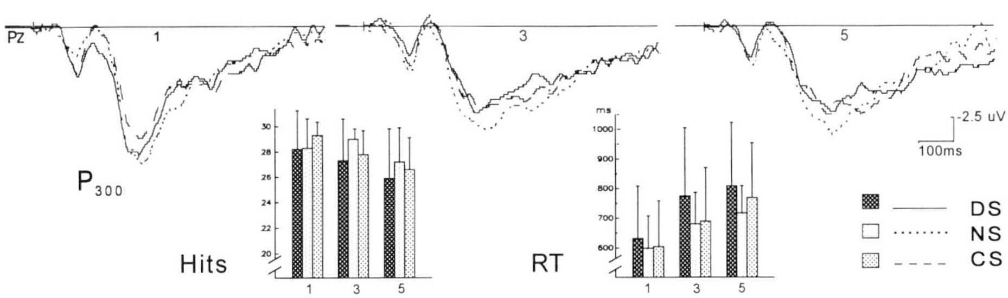

Figure 4. Group averaged ERP-P_{300} waveforms, RTs, and hits during memory search (1, 3, 5 digit sets) for DS, NS, and CS (see text).

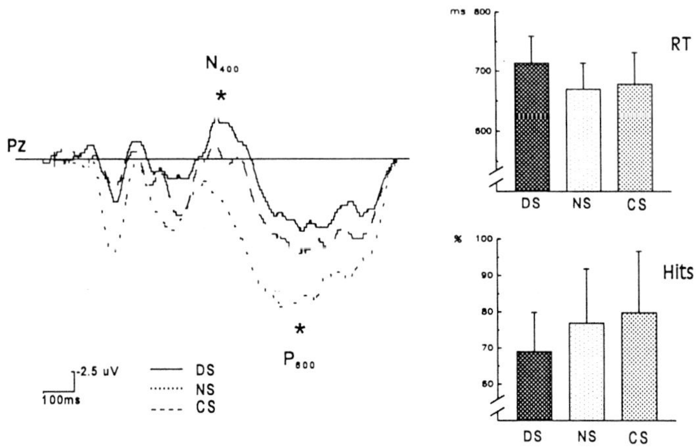

Figure 5. Group averaged ERP-N_{400}/P_{600} waveforms, RTs, and hits during delayed visual word recognition. ★ p≤0.05 .

During continuous recognition memory for visually presented words, ERPs associated with correct identification of previously presented items (i.e. "old words"), in contrast to "new words", exhibit an attenuated "N_{400}" component and a subsequently enhanced "P_{600}" component (6). In a delayed recognition version of this task, correct identification of old words (presented initially 20 minutes earlier), was correlated with faster RTs ($p \leq .06$), reduced N_{400} and larger P_{600} components in nonsmokers (Figure 5) relative to overnight deprived smokers (after sham smoking). Relative to smoking deprivation, smoking resulted in ERP changes, namely N_{400} reductions and P_{600} enlargements, which placed these smoking smokers at an intermediate level between deprived smokers and nonsmokers (i.e. smoking smokers differed significantly from both deprived smokers and nonsmokers). Although groups did not differ in behavioral performance indices, these findings suggest that smoking has a tendency to normalize brain electrical processes related to delayed recognition.

In sentence reading tasks a negative "N_{400}" component is elicited in response to semantically incongruent words and as such it is considered an electrophysiological index of semantic memory processing (7). In a study investigating delayed recognition of true and false sentences based on a passage memorized two to three minutes prior to sentence presentation, N_{400} components elicited by correctly identified false statements (see Figure 6) were found to be of the same amplitude and latency in nonsmokers as smoking deprived (overnight) smokers (after sham smoking). Smoking did not alter N_{400} latencies and amplitudes.

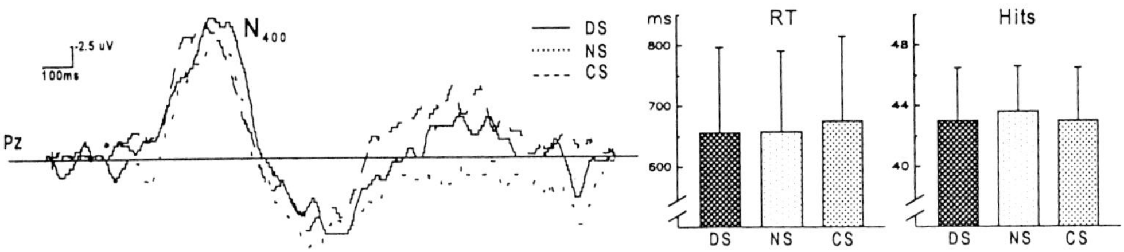

Figure 6. Group averaged ERP-N_{400} waveforms, RTs, and hits during semantic memory task paradigm (see text).

The pharmacological properties of smoking/nicotine are highly complex and although their net effects may be excitatory on basal brain arousal systems, the consequences of the smoking act and /or nicotine administration on higher order perceptual/cognitive processes may well vary as a function of the level of efficiency of these processes prior to smoking/nicotine intake. Nootropics or "cognitive enhancing" compounds appear to exert improvements in information processing in neuropsychiatric subpopulations and as such the normalizing/enhancing effects of smoking and nicotine described in these present studies may be seen to be appreciably augmented when examined in individuals exhibiting neurocognitive deficiencies (8).

References

1. Pomerleau O, Pormerleau C. A behavioral perspective on smoking. In: Smoking and Human Behavior. Ney T, Gayle A, editors. Chichester: John Wiley, 1989: 21-56.
2. Knott V, Venables P. EEG alpha correlates of nonsmokers, smokers, smoking and smoking deprivation. Psychophysiology 1977; 14:150-156.
3. Stepney R. Smoking as a psychological tool. Bull Brit Psychol Soc 1979; 32:341-345.
4. Picton T. The P_{300} wave of the human event-related potential. Jn Clin Neurophysiol 1992; 9:456-479.
5. Strayer D, Wickens C, Braune R. Adult age differences in speed and capacity of information processing: An electrophysiological approach. Psych and Aging 1987; 2:99-110.
6. Rugg M, Doyle M. Event-related potentials and recognition memory for low- and high-frequency words. Jn of Cog Neurosci 1992; 4:69-79.
7. Kutas M, Hillyard S. Reading senseless sentences: Brain potentials reflect semantic incongruity. Science 1980; 207:203-204.
8. Newhouse P. Hughes J. The role of nicotine and nicotine mechanisms in neuropsychiatric disease. Brit Jn Addict 1991; 86:521-526.

Effects of Nicotine on
Biological Systems II
Advances in Pharmacological Sciences

NICOTINE AND COGNITIVE EFFECTS

I. Hindmarch and N. Sherwood

Human Psychopharmacology Research Unit, University of Surrey,
Milford Hospital, Godalming, GU7 1UF, UK.

Summary: Early studies on the cognitive effects of nicotine were limited by inadequacies in test selection and experimental design. Even when these were overcome, the confounding influence of tobacco abstinence still limited the conclusions which could be drawn. However, recent controlled studies among smokers and non-smokers has shown that nicotine has small, but reproducible, specific positive effects on human cognition and psychomotor performance which cannot be simply attributed to a recovery from tobacco abstinence.

Introduction

During tobacco smoking, nicotine is absorbed in sufficient quantities to produce clear pharmacological effects in the central nervous system (1). Functional models of tobacco use have suggested that smokers may use these effects to manipulate their psychological state and benefit from improved cognitive function (2). As such, smoking is seen as a purposeful activity, not just an automatic reaction to genetic make-up, unconscious motives or addiction. Yet despite almost 100 years of research, conclusive evidence of a facilitatory role has only emerged during the last five years. This brief review summarizes research into the cognitive effects of nicotine as three chronological stages.

The Distant Past

The early part of the 20th century saw a surge of interest in the mental effects of smoking which was not repeated until the late 1970's. The general rationale and approach to investigation during this period reflects the era of the "gentleman scientist" and the infancy of the behavioural sciences. In particular the adequacy of some of the tests employed is questionable, as was the understanding and use of control conditions at this time. Discussion of study details is often lacking and statistical comparisons of results are all but absent. Not surprisingly, the results of these experiments (where meaningful comparisons can be made at all) are inconclusive.

Bush (3) compared the performance of 15 smokers on a battery of tests (letter cancellation, chain association, limited association, controlled association, category generation, visual memory, auditory memory, addition and subtraction) before and after smoking a cigar. Scores were then adjusted for variation across the session by those achieved on the same battery by a single non-smoker control subject. A decrease in efficiency after smoking was found in all tests, attributed not to nicotine (which Bush believed would be destroyed in the burning cone of the cigar) but to another alkaloid, pyridine.

A more carefully controlled study by Hull (4) investigated the effects of tobacco on heart-rate, hand steadiness, tapping, muscle fatigue, letter cancellation, oral reaction time, reading speed, mental arithmetic, memory span and rate of rote learning among both smokers and non-smokers who were required to smoke. The study was the first to employ a placebo condition. Hull devised a control pipe which was identical to the tobacco pipe used in the study except that it administered only heated air. Performance on tobacco days showed an increase in pulse rate and loss of hand steadiness, identified today as signs that significant quantities of nicotine had been absorbed. At the same time, the speed and accuracy of letter cancellation and mental arithmetic was improved after tobacco (compared to no tobacco) among smokers, but not non-smokers; conversely, finger tapping rates improved in the non-smokers alone after tobacco. Unfortunately comparisons were presented only as percentage changes with no statistical interpretation.

The last investigation of this period was conducted by Fisher (5). In a first experiment, number cancellation was tested among three non-smokers on several occasions both before and after smoking (or not smoking) a cigarette. An increase in pulse rate on cigarette days seemed to indicate that tobacco had been inhaled, but task performance was not seen to change. A second experiment used a similar experimental design but employed an early continuous attention task (the cancellation of random combinations of lights) and used a further control condition of a de-nicotinised cigarette. Here smoking was seen to improve performance over a continuous 20 minute trial compared either to no smoking or to the de-nicotinised cigarette, leading Fisher to conclude that "In work requiring a sustaining of attention over a considerable period of time, and involving accuracy and promptness of discriminatory response, tobacco smoking increases the efficiency among the few reactors (*subjects*) tested".

In all, the work of this period shows a lack of uniformity of results which cannot be assimilated easily due to the lack of common measures and methods. No investigator attempted to account for results within a psychological or physiological model and little work was undertaken to relate the effects of tobacco to the constituents of tobacco smoke, although the implicit assumption seems to have been that nicotine was the agent responsible for effects. However, these studies did provide some early echoes of findings which have become common in more modern controlled investigations.

The Recent Past

The development of psychology as a scientific discipline was particularly influenced by the demands for data on human performance capabilities arising from of the second world war. The result was the development of models of human behaviour which could be tested empirically through reliable, sensitive measures and controlled investigations. From early beginnings in the work of Broadbent (6), models of Man as an information processing mechanism have proved particularly influential in providing the basis for research into the psychological effects of smoking. This trend continues today, but many of the studies undertaken during this second period are delineated by their use of a particular pharmacological approach to experimental design: Ensure that the system under investigation is free of the compound of interest, administer an effective dose and measure the system response.

This methodology required smokers to abstain from tobacco, usually overnight, to ensure that all but a residual amount of nicotine was flushed from the system. The resulting state was considered the system baseline against which post-treatment effects were then compared. Assessments were made, subjects are asked to smoke, the assessments were repeated and improvements in psychomotor performance were often found. However, the problem remained whether the changes in performance were due to an absolute improvement in psychomotor function after smoking, or merely the recovery of normal, everyday performance levels after a period of tobacco abstinence (7).

Wesnes and Warburton (8) investigated the effects of three cigarettes with different nicotine deliveries (0.28, 0.7 and 1.65 mg respectively) on attention using a rapid visual information processing (RVIP) task. In a repeated measures design, 24 abstinent male smokers performed the task for 10 minutes to establish a baseline, were then given 10 minutes to smoke a cigarette and then performed the task for a further 20 minutes. Compared to baseline, both accuracy of responses (hits) and speed of responses were improved in the first 10 minutes following smoking, with the greatest improvements coming with the highest nicotine delivery cigarette.

However, Keenan, Hatsukami and Anton (9) showed the performance of regular users of smokeless (chewing) tobacco on a similar attention task to be significantly impaired after 24 hours tobacco abstinence compared to a group of users who were allowed to continue to chew tobacco normally. A significant decline in performance was also been found among regular cigarette smokers abstinent for 24 hours compared to a group of smokers who continued to smoke normally (10).

Russell (11) has offered a useful heuristic, that smokers can tolerate a drop in blood nicotine to approximately 2/3 of their average daily nicotine level without major psychological effects, but a smoker with less than 1/3 of this level may exhibit signs of tobacco abstinence. Nicotine has an estimated total half life of 120 minutes (12). On this basis, three or more hours of tobacco abstinence is likely to affect aspects of normal psychomotor function in regular smokers. This conclusion has important implications in that studies which have required smoker subjects to abstain from tobacco for several hours prior to testing can induce a significant level of psychomotor impairment which is then often used as a baseline against which to test the effects of nicotine or smoking. This may well be an appropriate condition against which to test the effects of the infamous "first cigarette of the day", but is frequently employed as being representative of the more common situation later in the day where several cigarettes have been smoked and blood nicotine levels are already raised (13). This is not to question the worth of such studies, but to suggest that their remit must be better understood. Taken alone, the studies of this period provide little compelling evidence of a nicotine induced facilitation of psychomotor performance.

The Present

Where the aim of research has been to investigate the absolute effects of nicotine or smoking on human psychomotor performance apart from the confounding effects of abstinence, there is stronger evidence that nicotine may indeed improve aspects of cognitive function. Improved attentional performance among non-abstinent smokers after smoking has also been found by Hasenfratz, Pfeiffner Pellaud and Bättig (14) using the RVIP task and Pritchard, Robinson and Guy (15) using a simple continuous attention task (CAT) in which subjects were required to spot the digit 0 among a continuous stream of single digits from 0-9. Sherwood Kerr and Hindmarch (16) found that performance of a short-term memory task was improved after 2 mg nicotine gum among a mixed group of non-smokers and non-abstinent smokers. Similarly, West and Hack (17) controlled for any effects of tobacco abstinence by comparing regular smokers with occasional smokers both before and after a period of abstinence using the same memory task. Their results showed that smoking significantly improved the speed of search through the short-term memory set regardless of prior abstinence, suggesting an absolute improvement of memory function.

Work at the Human Psychopharmacology Research Unit has explored other areas of cognitive and psychomotor function (18-20) and has found that nicotine gum or smoking improves the speed of gross motor movement and the accuracy of fine motor manipulation, reaction time responses and complex performance (simulated driving) among non-abstinent smokers. Of interest, the effects of cigarette smoking appear to be somewhat greater than those found using nicotine alone, suggesting that in addition to the presence of nicotine, the total nicotine load delivered across the trial, the control of nicotine delivery or oral-manipulative aspects of smoking may be of importance in mediating the cognitive effects of nicotine.

References

1. Benowitz NL, Porchet H, Jacob P III. Pharmacokinetics, metabolism, and pharmacodynamics of nicotine. In: Wonnacott S, Russell MAH & Stolerman IP (eds) Nicotine Psychopharmacology. Oxford: OUP, 1990: 112-157.
2. Warburton DM. The functions of smoking. In: Martin WR, Van Loon GR, Iwamoto ET & Davis D (eds) Tobacco Smoke and Nicotine: A Neurobiological Approach. New York: Plenum, 1987:51-61.
3. Bush AD. Tobacco smoking and mental efficiency. New York Medical Journal 1914; 94: 519-527.
4. Hull CL. The influence of tobacco smoking on mental and motor efficiency. Psychological Monographs 1924; 3: Nr 23.
5. Fisher VE. An experimental study of the effects of tobacco smoking on certain psychophysical functions. Comparative Psychology Monographs 1927; 4: Nr 19.
6. Broadbent DE. Perception and Communication. Oxford: Pergamon 1958.
7. Hughes JR. Distinguishing withdrawal relief and direct effects of smoking. Psychopharmacology 1991; 104: 409-410.
8. Wesnes K, Warburton DM. Effects of smoking on rapid information processing performance. Neuropsychobiology 1983; 9: 223-229.

9. Keenan RM, Hatsukami DK, Anton DJ. The effects of short-term smokeless tobacco deprivation on performance. Psychopharmacology 1989; 98: 126-130.
10. Hughes JR, Keenan RM, Yellin A. Effect of tobacco withdrawal on sustained attention. Addictive Behaviours 1989; 14: 577-580.
11. Russell MAH. Nicotine intake by smokers: Are rates of absorption or steady-state levels more important? In: Rand MJ, Thurau K (eds) The Pharmacology of Nicotine. Oxford, IRL press, 1987: 375-402.
12. Benowitz NL, Jacob P III, Jones RT, Rosenberg J. Inter-individual variability in metabolism and cardiovascular effects of nicotine in man. Journal of Pharmacology and Experimental Therapeutics 1982; 41: 467-473.
13. Russell MAH, Feyerabend C. Cigarette smoking: A dependence on high nicotine boli. Drug Metabolism Review 1978; 8: 29-57.
14. Hasenfratz M, Pfiffner D, Pellaud K, Bättig K. Post-lunch smoking for pleasure seeking or arousal maintenance? Pharmacology, Biochemistry and Behavior 1989; 34: 631-639.
15. Pritchard WS, Robinson JH, Guy TD. Enhancement of continuous performance task reaction time by smoking in non-deprived smokers. Psychopharmacology 1992; 108: 437-442.
16. Sherwood N, Kerr JS, Hindmarch I. Effects of nicotine gum on short-term memory. In: Adlkofer F (ed) Effects of Nicotine on Biological Systems. Basel: Birkhauser, 1990: 531-535.
17. West R, Hack S. Effect of cigarettes on memory search and subjective ratings. Pharmacology, Biochemistry and Behavior 1991; 38: 281-286.
18. Hindmarch I, Kerr JS, Sherwood N. Effects of nicotine gum on psychomotor performance in smokers and non-smokers. Psychopharmacology 1990; 100: 535-541.
19. Sherwood N, Kerr JS, Hindmarch I. Psychomotor performance in smokers following single and repeated doses of nicotine gum. Psychopharmacology 1992; 108: 432-436.
20. Sherwood N, Fairweather DB, Kerr JS, Hindmarch I. Effects of cigarette smoking on the performance of a simulated driving task (1994; under review).

Effects of Nicotine on
Biological Systems II
Advances in Pharmacological Sciences

EVIDENCE THAT NICOTINE IS ADDICTIVE

I.P. Stolerman[1] and M.J. Jarvis[2]

Section of Behavioural Pharmacology[1] and ICRF Health Behaviour Unit[2],
Institute of Psychiatry, De Crespigny Park, London SE5 8AF, UK

Summary: Views questioning the addictiveness of nicotine continue to be expressed in some quarters. Patterns of use by smokers and the remarkable intractability of smoking point to compulsive use as the norm. Studies in animal and human subjects have shown that nicotine can function as a reinforcer, but under a rather limited range of conditions. There is a resemblance between the discriminative stimulus and other effects of nicotine, amphetamine and cocaine. There is a well-defined nicotine withdrawal syndrome that is alleviated by nicotine replacement. Thus, the evidence identifies nicotine as an addictive agent, comparable to heroin and cocaine.

Introduction

Over the last 20 years opinion among smoking researchers has shifted towards acceptance of nicotine as a powerful drug of addiction and the 1988 report of the US Surgeon General regarded nicotine as addictive in the same sense as drugs such as heroin and cocaine (1). However, views questioning the addictiveness of nicotine continue to be expressed in some quarters (2, 3). Such diverse views need not cause surprise; the same situation has occurred with other drugs. Gawin (4) has commented that as recently as 1980 cocaine was considered incapable of producing dependence, whereas a few years later, it was proclaimed the drug of greatest concern in the USA. Definitions of drug dependence have changed over time and, in response to each drug's unique profile and set of users, have stressed different biological and social effects.

Most recent conceptualisations of drug dependence, recognizing the difficulty of fitting all substances into a rigid framework, and being concerned more with issues of diagnosis of problems in users than with finding a formulation which will permit drugs to be judged as addictive or not, have avoided brief definitions of the nature of dependence, and have instead listed a variety of criteria (5,6). The following formulation, adopted by the World Health Organisation (WHO) as long ago as 1969, has much to recommend it: "A state, psychic and sometimes also physical, resulting from the interaction between a living organism and a drug, characterized by behavioural and other responses that always include a compulsion to take a drug on a continuous or periodic basis in order to experience its psychic effects, and sometimes to avoid the discomfort of its absence. Tolerance may or may not be present" (7). The 1969 WHO definition has at its core the construct of compulsion, and emphasises that the rewarding properties of drugs can be both positive or negative.

This paper summarises some of the evidence relating to the addictiveness of nicotine. We consider patterns of tobacco use by smokers; the dependence potential of nicotine itself; the nicotine withdrawal syndrome and its relief by nicotine replacement. Our conclusion, like those of earlier writers (1, 8-11), is that nicotine is indeed a powerful drug of addiction. Our recent review (12) and its associated commentaries provide a fuller account of this evidence.

Compulsive Use

Whether nicotine should be considered as addictive or merely habit-forming depends to a large extent on the degree of compulsion to take the drug experienced by its users. There are no definitive or quantitative criteria to distinguish where regular recreational and social use ends and compulsion begins, so that this must in the final analysis be a matter of judgement. In the general population, most individual smokers smoke daily at a characteristic consumption level. Male smokers in Britain smokers use an average of 17 cigarettes per day and female smokers average 14 per day (13). Fewer than 1 in 20 smokers (3.7%) smoke less frequently than daily (13). While they do exist (14), stable light smokers of 5 or less cigarettes per day are rare. In a survey of some 14,000 women, 3500 were smokers, but only 60 (2% of the smokers) smoked 5 or less cigarettes per day at both baseline and 6 months later (Hajek, personal communication). Thus the typical smoker is a relatively heavy smoker, and it is unusual to find individuals who smoke only in certain social contexts, on a take it or leave it basis.

There is convincing evidence that points to the stereotyped nature of smoking in many cigarette smokers. Among the general population, 14% of smokers light up within 5 minutes of waking, and over 50% within 30 minutes (15). Among smokers of all ages, 29% had never stopped for as long as a week, and among those who had, in 27% it was longer than 5 years ago. Thus, 48% of smokers had not abstained from smoking for as long as a week in the past 5 years. Smokers self-reports indicate that craving for cigarettes is a common experience. In a national survey of children aged 14-16, 20% of those who smoked regularly reported craving when without cigarettes, and responses to this item were strongly related to nicotine intake as indexed by saliva cotinine (16). Among adult smokers the corresponding percentage was 47% (17). A large proportion of smokers say that they would like to give up. In a recent survey of British adults, 58% of cigarette smokers said that they wanted to quit, of whom only 13% expected to succeed (15). Even among smokers aged 14-15, over half thought that it would be difficult to stop for even a week (16). In 1990, national smoking statistics showed that only 40% of all who had smoked cigarettes regularly had succeeded in stopping (13).

Thus, in the general population of smokers, there are high levels of dependence on cigarette smoking that are not confined to a few extreme cases. The reputation of smoking as notoriously resistant to change is confirmed by studies of individuals suffering from severe smoking-related disease. Forty percent of those who have undergone a laryngectomy try smoking again afterwards (18). Among smokers suffering a heart attack, 40% relapse to smoking while still in hospital, the majority within 48 hours of coming out of intensive care (19). Even after surgery for lung cancer about 50% of smokers who survive resume the habit (20).

Studies of the outcomes of most smoking cessation treatments strengthen the notion of smoking as compulsive behaviour. Rather less than 20% of those who receive intensive treatment succeed in abstaining for as long as a year (21, 22). This is not due to a lack of desire to give up; rather, it implies that dependence places a block between the desire of smokers to give up and their ability to do so. It could be argued that those who come for formal treatment are a tiny and unrepresentative group, and that higher rates of success would be seen in the general population of users. Robins (23) has made this case for heroin. However, studies of tobacco 'self-quitters' indicate that among motivated smokers in general, the outcome of an attempt at giving up is overwhelmingly likely to be unsuccessful. In two studies of smokers attempting to give up on their own, only 3-5% maintained complete abstinence for 6 months (24, 25).

The Nicotine Withdrawal Syndrome and the Effects of Nicotine Replacement

Numerous studies have established the existence of a nicotine withdrawal syndrome in human subjects and a rather smaller number of studies indicate the presence of nicotine withdrawal signs in animals chronically treated with nicotine. Cessation of smoking leads to a variety of signs and symptoms that are predominantly affective in nature (e.g. irritability, difficulty concentrating, anxiety, increased hunger and depressed mood). These findings have been reviewed thoroughly (25, 26). Similar effects are seen after cessation of nicotine chewing gum (27) or of smokeless tobacco (28). That the tobacco withdrawal syndrome is due to loss of nicotine rather than behavioural aspects of use is shown by findings that it is relieved by nicotine replacement but not by placebo, and nicotine replacement by various routes approximately doubles success rates in cessation attempts (12).

Reports also suggest that cessation of the chronic administration of nicotine can produce withdrawal signs in animals, including changes in components of unconditioned behaviour (29), disruptions of learnt behaviour (30), and partial generalization to the discriminative stimulus effects of pentylenetetrazol, an anxiogenic drug (31). Although these withdrawal signs are not shown by gross disturbances of the magnitude of those seen during withdrawal from opioids, barbiturates and alcohol, they are compatible with the signs seen after chronic exposure to psychostimulant drugs such as amphetamine and cocaine.

Positive Reinforcing Effect of Nicotine

The ability of addictive drugs to serve as positive reinforcers is regarded as a core property that supports the development and maintenance of addiction (32). Addiction is characterized by (among other attributes) repetitive self-administration of one or more drugs and the main behavioural basis is the positive reinforcing effect of the drugs, without which it is unlikely that self-administration could develop. In the conditions pertaining in typical laboratory self-administration procedures, many illicit drugs associated with problems of addiction, dependence or abuse have positively reinforcing effects.

Reinforcing effects of drugs are complex and cannot be equated with self-reports of subjective experiences. Rather, any effect of a drug that affects the tendency for it to be self-administered may contribute to or modulate its reinforcing property, and these effects may be neurochemical (e.g. effects on brain reward systems) or behavioural (e.g. facilitation of ongoing behaviour), as well as psychological (e.g. production of euphoria). Thus, the ability of a drug to improve selective attention may, under appropriate circumstances, increase its efficacy as a positive reinforcer while under other circumstances, the effects of the same drug may be aversive. Thus, *drugs are not positive reinforcers*; they can, however, *serve as positive reinforcers* with varying degrees of strength under different conditions.

Solutions of pure nicotine can, in appropriate conditions, serve as a positive reinforcers in at least six species, as reviewed in detail (33). In some experiments, rates of responding maintained by nicotine approximate to those maintained by cocaine under similar conditions (34) whereas in other studies, it had low efficacy as a positive reinforcer (33). In the early days of self-administration research, rates of behaviour maintained by drug reinforcers were in many cases lower than those maintained by conventional reinforcers; as late as 1971, even morphine and cocaine were, by some criteria, weak reinforcers (35). Research on nicotine self-administration may be at a stage like that on self-administration of the classical drugs a decade or two ago.

Brain reward systems

It is thought that many addictive drugs may serve as reinforcers and support compulsive drug-seeking behaviour because they tap directly into brain mechanisms of reward. Nicotine, like classical addictive drugs, can reduce the threshold of current needed to support intracranial self-stimulation behaviour (36). Furthermore, a major neurochemical mechanism for the positive reinforcing effects of many addictive drugs is an enhancement of the synaptic availability of dopamine in the mesolimbic dopamine pathway, an effect shared with nicotine (37); lesions of the mesolimbic dopamine system impair the self-administration of nicotine (38). These commonalities of process suggest that nicotine can produce the neurochemical effect that is regarded as one of the fundamental mechanisms involved in drug-seeking behaviour.

Conclusions

For centuries tobacco smoking was regarded as neither harmful nor addictive. Table 1 illustrates how perceptions of tobacco use have changed. The trend defines the shifting stance of defenders of tobacco use and illustrates the historical inevitability of their ultimate recognition that it is harmful, pharmacologically driven, and addictive. All major lines of evidence on whether nicotine use could be addictive were available by the early 1980's and in 1988, the Surgeon-General delivered an authoritative verdict (1).

The recent attempts (2,3) to reverse this verdict have failed. They have used an outdated definition of addiction that attempted to distinguish between habituating and addicting drugs. As documented above, there is extensive evidence of the compulsive nature of human tobacco use.

1950 →	1960 →	1970 → 1980 →	1990
Not harmful Not drug use Not an addiction	Harmful Not drug use Not an addiction	Harmful Drug use Not an addiction	Harmful Drug use Addiction

Table 1. Changing perceptions of tobacco use.

Unlike some authors (2), we believe that the findings of positive reinforcing effects of nicotine are convincing and that they establish a fundamental psychopharmacological mechanism of tobacco use that has strong similarities to the mechanisms underlying the abuse of the classical addictive drugs. There is no incompatibility between the ability to enhance attention or particular types of memory processes and addictiveness, and the ability to produce a disabling intoxication is not a defining feature of addiction. The nicotine withdrawal syndrome is *prima facie* evidence of a type of dependence that may very well contribute to the continuing use of a substance. In summary, we believe that the time has arrived for the notion that nicotine is habituating but not addictive to be consigned to the archives of history.

References

1. U.S. Department of Health and Human Services. The Health Consequences of Smoking: Nicotine Addiction. A Report of the Surgeon General. Office on Smoking and Health, Maryland, 1988.
2. Robinson JH, Pritchard WS. The role of nicotine in tobacco use. Psychopharmacology 1992; 108: 397-407.
3. Warburton DM. Heroin, cocaine and now nicotine. In: Warburton DM (ed) Addiction Controversies. Harwood Academic Publishers, Switzerland, 1990: 21-35.
4. Gawin FH. Cocaine addiction: psychology and neurophysiology. Science 1991; 251: 1580-1586.

5. American Psychiatric Association. Diagnostic and Statistical Manual of Mental Disorders. 3rd edition. Washington D.C., 1987.
6. World Health Organization. International Classification of Diseases. 9th edition. WHO, Geneva, 1978.
7. World Health Organization. World Health Organization Technical Report Series No. 407. WHO, Geneva, 1969.
8. Jaffe JH. Tobacco smoking and nicotine dependence. In: Wonnacott S, Russell MAH, Stolerman IP (eds) Nicotine psychopharmacology: molecular, cellular and behavioural aspects. Oxford Science Publications, Oxford, 1990: 1-37.
9. Jones RT. What we have learned from nicotine, cocaine and marijuana about addiction. Res Publ Assoc Res Nerv Ment Dis Vol 70. Raven, New York, 1992: 109-122.
10. West RJ. Nicotine addiction: a re-analysis of the arguments. Psychopharmacology 1992a; 108: 408-410.
11. Hughes JR. Smoking is a drug dependence: a reply to Robinson and Pritchard. Psychopharmacology 1993; 113: 282-283.
12. Stolerman IP, Jarvis MJ. The scientific case that nicotine is addictive. Psychopharmacology 1994, in press.
13. Smyth, M, Browne, F. General Household Survey 1990. HMSO, London, 1992.
14. Shiffman S. Tobacco chippers - individual differences in tobacco dependence. Psychopharmacology 1989; 97: 539-547.
15. NOP Omnibus Services. Smoking habits 1991: a report prepared for the Department of Health. Department of Health, London, 1992.
16. Goddard, E. Why Children Start Smoking. HMSO, London, 1990.
17. Russell MAH. Conceptual Framework for nicotine substitution. In: Ockene JK (ed) The Pharmacologic Treatment of Tobacco Dependence. Cambridge, Massuchusetts, Harvard University Institute for the Study of Smoking Behavior and Policy, 1986: 90-107.
18. Himbury S, West R. Smoking habits after laryngectomy. Br Med J 1985; 291: 514-515.
19. Bigelow G, Rand CS, Gross J, Barling TA, Gotlieb SH. Smoking cessation and relapse among cardiac patients. In: Tims FM, Leubefeld CG (eds) Relapse and Recovery in Drug Abuse. NIDA Research Monograph 72. Department of Health and Human Services, Rockville, Maryland, 1963.
20. Davison G, Duffy M. Smoking habits of long-term survivors of surgery for lung cancer. Thorax 1982; 37: 331-333.
21. Hunt WA, Bespalec DA. An evaluation of current methods of modifying smoking behaviour. J Clin Psychol 1974; 30: 431-438.
22. Schwartz, JL. Review and evaluation of smoking cessation methods: the United States and Canada 1978-1985. U.S. Department of Health and Human Services, Washington DC, 1987.
23. Robins LN. Vietnam veterans' rapid recovery from heroin addiction: a fluke or normal expectation. Addiction 1983; 88: 1041-1054.
24. Cohen S, Lichtenstein E, Prochaska JO et al. Debunking myths about self-quitting. Amer Psychol 1989; 44: 1355-1365
25. Hughes JR. Tobacco withdrawal in self-quitters. J Cons Clin Psychol 1992; 60: 689-697.
26. Hughes JR, Hatsukami D. Signs and symptoms of tobacco withdrawal. Arch Gen Psychiatry 1986; 43: 289-294.
27. West RJ, Russell MAH. Effects of withdrawal from long-term nicotine gum use. Psychol Med 1985; 15: 891-893.
28. Hatsukami DK, Gust SW, Keenan RM. Physiologic and subjective changes from smokeless tobacco withdrawal. Clin Pharm Ther 1987; 41: 103-107.
29. Malin DH, Lake JR, Newlin-Maultsby P, Roberts LK, Lanier JG, Carter VA, Cunningham JS, Wilson OB. Rodent model of nicotine abstinence syndrome. Pharmac Biochem Behav 1992; 43: 779-784.
30. Corrigall WA, Herling S, Coen KM. Evidence for a behavioral deficit during withdrawal from chronic nicotine treatment. Pharmac Biochem Behav 1989; 33: 559-562.
31. Harris CM, Emmett-Oglesby MW, Robinson NG, Lal H. Withdrawal from chronic nicotine substitutes partially for the interoceptive stimulus produced by pentylenetetrazol (PTZ). Psychopharmacology 1986; 90: 85-89.
32. Young AM, Herling S. Drugs as reinforcers: studies in laboratory animals. In: Goldberg SR, Stolerman IP (eds) Behavioral analysis of drug dependence. Academic Press, Orlando, Florida, 1986: 9-67.

33. Swedberg MDB, Henningfield JE, Goldberg SR. Nicotine dependency: animal studies. In: Wonnacott S, Russell MAH, Stolerman IP (eds) Nicotine psychopharmacology: molecular, cellular and behavioural aspects. Oxford Science Publications, Oxford, 1990: 38-76.
34. Goldberg SR, Spealman RD, Goldberg DM. Persistent behavior at high rates maintained by intravenous self-administration of nicotine. Science 1981; 214: 573-575.
35. Goldberg SR. Comparable behavior maintained under fixed-ratio and second-order schedules of food presentation, cocaine injection or d-amphetamine injection in the squirrel monkey. J Pharmac Exp Ther 1973; 186: 18-30.
36. Huston-Lyons D, Kornetsky C. Effects of nicotine on the threshold for rewarding brain stimulation in rats. Pharmac Biochem Behav 1992; 41: 755-759.
37. Benwell MEM, Balfour DJK. The effects of acute and repeated nicotine treatment on nucleus accumbens dopamine and locomotor activity. Br J Pharmac 1992; 105: 849-856.
38. Corrigall WA, Franklin KBJ, Coen KM, Clarke PBS. The mesolimbic dopaminergic system is implicated in the reinforcing effects of nicotine. Psychopharmacology 1992; 107: 285-289.

Effects of Nicotine on
Biological Systems II
Advances in Pharmacological Sciences

SELF-ADMINISTERED NICOTINE ACTS THROUGH THE VENTRAL TEGMENTAL AREA: IMPLICATIONS FOR DRUG REINFORCEMENT MECHANISMS

W.A. Corrigall

Addiction Research Foundation and Department of Physiology,
University of Toronto, Toronto, Ontario, Canada M5S 2S1

Summary: Nicotine self-administration is decreased by systemic administration of selective dopamine antagonists, and by neurochemically specific lesions of the dopaminergic projection from the ventral tegmental area (VTA) of the midbrain to the nucleus accumbens. The nicotinic antagonist dihydro-ß-erythroidine reduces nicotine reinforcement when administered focally into the ventral tegmental area, but is without effect in the nucleus accumbens. Collectively these data show that self-administered nicotine targets the mesolimbic system through receptors in the vicinity of the dopamine cells in the VTA. These findings have implications for understanding the broader issue of the cholinergic control of dopamine cells in drug reinforcement.

Introduction

The objectives of this paper are two-fold: (i) to review recent observations that we have made about the nature of nicotine reinforcement mechanisms, and (ii) to discuss the implications of these data for understanding how cholinergic mechanisms might control mesolimbic dopamine cell function and the reinforcement processes which depend on it.

Exploration of the role of dopaminergic mechanisms in the self-administration of nicotine is an obvious experimental question for several reasons. The reinforcing effects of other drugs of abuse are believed to depend on the enhancement of mesolimbic dopamine neurotransmission (1). Neuroanatomical research has demonstrated that there are nicotinic receptors on mesolimbic neurons both at the cell body/dendritic level in the VTA, as well as at their terminal fields in the nucleus accumbens (2). Nicotine can activate mesolimbic dopamine cells (3-5). It is reasonable therefore to suppose that these neurons might be one site at which nicotine engages processes which maintain its voluntary self-administration.

Nicotine Self-Administration

The research described here is of a behavioral neuroscience nature, and relies extensively on operant paradigms to measure reinforcement. While animal self-administration models with nicotine have been difficult for some to establish, particularly with rodents, we find that rats will self-administer this drug reliably (6), and our approach has been recently confirmed by others (7). The techniques of drug self-administration that we use have been reviewed in detail (8). In brief, animals are trained to press a lever for food reinforcement, then prepared with chronic intravenous catheters, and trained to self-administer nicotine (or in some control experiments, cocaine) over a 3-week acquisition period. At this time, subjects either begin treatment, or undergo further surgery to produce lesions or to implant micro-cannulae which allow focal application of drugs in the CNS site of interest. Experimental details are provided in the accompanying references.

Role of Mesolimbic Dopamine Cells in Nicotine Reinforcement

We established several years ago that subtype-selective dopamine antagonists reduce the extent of voluntary nicotine self-administration when administered prior to operant sessions (9). In addition, lesions of the ascending mesolimbic projection from the VTA region, done with focal infusion of the neurotoxin 6-hydroxydopamine (6-OHDA) into the nucleus accumbens, also attenuate nicotine reinforcement (10). In neither of these experiments did the pattern of responding suggest that motor impairment had occurred[1]. These findings therefore provide evidence that the reinforcement processes which control nicotine self-administration depend

---

[1] With other self-administered drugs such as cocaine, a decrease in the reinforcing value, achieved for example by reducing the dose of drug available to the subjects or treating them with a dopamine antagonist, is typically associated with an increase in self-administration behavior. This compensatory responding has often been assumed to be a general characteristic of the self-administration of all drugs. However, even with cocaine, decreases in self-administration can occur after treatment with a dopamine antagonist, depending upon the dose of cocaine examined (11). For nicotine, it is clear that compensatory increases in responding typically do not occur (9,10,12). We believe that this is due to a fundamental difference in how nicotine intake is regulated, one aspect of which may be the way the drug activates dopamine cells.

at least in part on the mesolimbic dopamine projection. Nonetheless, dopamine manipulations need to be interpreted cautiously in view of possible effects on a range of other functions. Moreover, these data do not speak to the site of action of self-administered nicotine, which may be synaptically remote from the mesolimbic dopamine system; the dependence of nicotine self-administration on the mesolimbic dopamine system could alternatively signify the configuration of the latter as a final common pathway in reinforced behavior.

To resolve these issues we have examined the role of the mesolimbic system by a means other than direct manipulation of dopamine. The approach we have chosen is to use the nicotinic antagonist dihydro-ß-erthyroidine (DHßE) to block the effects of self-administered nicotine locally in particular areas of the brain (13). This ability of this compound to act as a nicotine antagonist has been well-established (e.g., 14); doses of DHßE used in these experiments were selected from pilot studies using nicotine-produced locomotor activity.

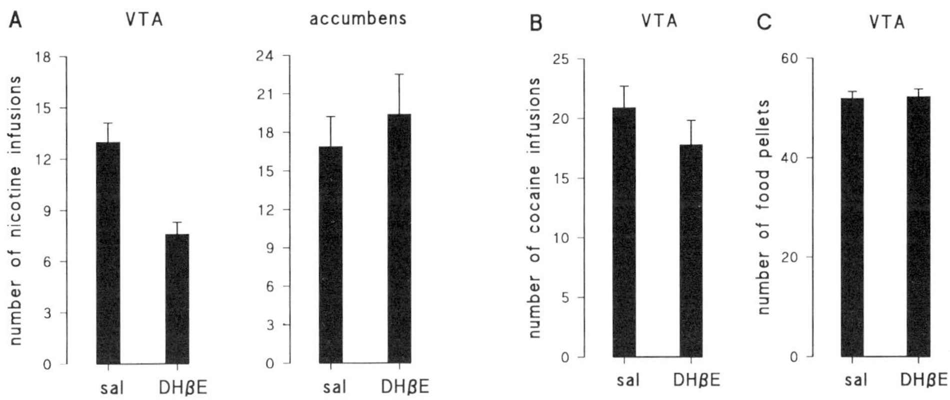

Figure 1. (A) Number of nicotine infusions obtained in self-administration sessions by animals following microinfusions of DHßE into the VTA or nucleus accumbens 10 minutes prior to the start of operant sessions (n=8 at each site). Nicotine self-administration occurred at a unit dose of 30 μg/kg/infusion. Shown here are the results of the highest dose of DHßE tested (30 μg/tissue site). Bars show 1 standard error of the mean. In separate groups of animals, the same treatment protocol and dose of DHßE is without effect on cocaine self-administration **(B)** or responding maintained by food pellets **(C)**. Figure modified from ref. 13.

The results of microinjections of DHßE into each of the VTA and the nucleus accumbens in animals self-administering nicotine are shown in Fig. 1(A). After infusion into the VTA, DHßE produced dose-related and statistically significant decreases in nicotine self-administration. In contrast, there were no significant effects of dose when the antagonist was infused into the nucleus accumbens. It is important to point out that the effects of DHßE within the VTA appear to be due to a selective antagonism of exogenously self-administered nicotine; we make this conclusion for several reasons. First, prior to the start of self-administration sessions, there were no obvious behavioral effects produced by infusion of DHßE into either tissue site. Secondly, the highest dose of DHßE, when infused into the VTA, had no effect on intravenous cocaine self-administration (Fig 1B), or on operant responding maintained by food (Fig. 1C). And thirdly, spontaneous locomotor activity was not altered by intra-VTA infusions of DHßE (data not shown; see 13).

These data confirm that nicotine reinforcement does depend on the mesolimbic system, and show that the drug targets this system directly through receptors in the VTA region and not in the synaptic field in the nucleus accumbens. An alternate interpretation of our data, that DHßE is acting within the VTA to block a tonically active nicotinic cholinergic input to dopamine cells which is critical to reinforcement processes irrespective of nicotine, appears to be untenable given the absence of an effect of DHßE on reinforcement of behavior by either food or cocaine. It would appear therefore that nicotine action in the VTA, probably on dopamine neurons themselves, is a critical component of reinforcement.

A consistent picture showing the VTA to be the site at which nicotine produces effects on the mesolimbic system is beginning to develop. In addition to our studies, recent work has also shown that the VTA is the site at which systemically administered nicotine evokes dopamine release in the nucleus accumbens (4), and is the site at which locomotor activity is produced (15). Using DHßE, we have also confirmed this latter observation (13).

Implications for Cholinergic Control of Mesolimbic Dopamine Cells

In view of these data, one must consider that cholinergic inputs to the VTA are potentially important elements of reinforcement circuitry, since they could modify mesolimbic dopamine cells just as self-administered nicotine does. Anatomically, the only known cholinergic projections to the dopamine cells of the substantia nigra and VTA regions arise from the cholinergic cells of the pedunculopontine tegmental (PPT) and laterodorsal tegmental nuclei (16; see also overview in 13). Both nicotinic and muscarinic effects have been observed at the level of midbrain dopamine neurons following stimulation of neurons within the PPT region (e.g. 18, 19) but in terms of reinforcement little is known about whether or how these neurons regulate dopamine cell activity. Our work shows that nicotinic mechanisms in the VTA are important in regulating dopamine cells. On the other hand, Yeomans and his colleagues have recently shown that muscarinic cholinergic mechanisms in both the PPT and VTA can alter reinforcement of a very different kind, electrical brain self-stimulation (17). Therefore, we have begun to examine whether cholinergic mechanisms of the muscarinic or nicotinic type in each of the VTA and the PPT influence drug reinforcement, with a particular focus on nicotine reinforcement.

Given the mixed nicotinic/muscarinic nature of the PPT projection to the VTA, an obvious question is whether nicotine action in the PPT, in addition to its action in the VTA, contributes to reinforcement processes. To address this, we have carried out a similar experiment to that described above with DHßE, except that in this case the antagonist was infused into the PPT. As it did in the VTA, DHßE attenuated nicotine self-administration in the PPT. Parallel studies using locomotor activity showed that DHßE also reduced nicotine-produced locomotor activity when applied focally in the vicinity of the cholinergic cells of the PPT, but was without effect on spontaneous locomotor activity. Nicotine can therefore also alter the function of mesolimbic dopamine cells, and associated reinforcement processes through action in the PPT region.

Does self-administered nicotine, through this action in the PPT, initiate both nicotinic and muscarinic neurotransmission within the VTA? To test this we have made microinfusions of the muscarinic antagonist atropine into the VTA prior to nicotine self-administration sessions, and we have found that doses of atropine which raise the threshold for electric brain stimulation are without effect on nicotine self-administration (13). Apparently muscarinic mechanisms are not part of a common reward system within the VTA, i.e., activation of muscarinic transmission by each of nicotine or electrical brain stimulation acting through the PPT.

It is clear from this research that nicotine activates the ascending mesolimbic dopamine system directly in the VTA region, and also synaptically more remotely in the PPT nucleus. The different way that nicotine interacts with dopamine cells, compared to cocaine, may be one aspect of the very different regulation of nicotine self-administration. In addition these data suggest that investigation of brain stem cholinergic mechanisms in drug reinforcement might be a fruitful avenue to explore both for a better understanding of basic reinforcement processes, and also because such research may permit a more selective way to develop treatment interventions aimed at regulating dopamine cells.

Acknowledgements: *The Merck Frosst Centre for Therapeutic Research, Dorval, Quebec kindly provided the dihydro-ß-erythroidine used in these experiments. Kathy Coen and Laurie Adamson provided expert technical assistance. Supported by the Addiction Research Foundation of Ontario.*

References

1. Robinson TE, Berridge KC. The neural basis of drug craving: an incentive-sensitization theory of addiction. Brain Res Rev 1993;18:247-291.
2. Clarke PBS, Pert A. Autoradiographic evidence for nicotine receptors on nigrostriatal and mesolimbic dopaminergic neurons. Brain Res 1985;348:355-358.
3. Calabresi P, Lacey MG, North RA. Nicotinic excitation of rat ventral tegmental neurones in vitro studied by intracellular recording. Br J Pharmacol 1989;98:135-140.
4. Imperato A, Mulus A, Di Chiara G. Nicotine preferentially stimulates dopamine release in the limbic system of freely moving rats. Eur J Pharmacol 1986:132;337-338.
5. Nisell M, Nomikos GG, Svensson TH. Systemic nicotine-induced dopamine release in the rat nucleus accumbens is regulated by nicotinic receptors in the ventral tegmental area. Synapse 1994;16:36-44.
6. Corrigall WA, Coen KM. Nicotine maintains robust self-administration in rats on a limited-access schedule. Psychopharmacology 1989;99:473-478.
7. Tessari M, Chiamulera C., Valerio E, Beardsley PM. Nicotine IV self-administration in rat. Eur Behav Pharmacol Soc, Pistoia, Italy, Sept 8-10, 1993.
8. Corrigall WA. A rodent model for nicotine self-administration. In: Animal Models of Drug Addiction Neuromethods, vol. 24. Boulton AA, Baker GB, Wu PH, editors. Clifton, NJ: Humana Press, 1992: 315-344.
9. Corrigall WA, Coen KM. Selective dopamine antagonists reduce nicotine self-administration. Psychopharmacology 1991;104:171-176.
10. Corrigall WA, Franklin KBJ, Coen KM, Clarke PBS. The mesolimbic dopamine system is implicated in the reinforcing effects of nicotine. Psychopharmacology 1992;107:285-289.
11. Corrigall WA, Coen KM. Cocaine self-administration is increased by both D1 and D2 dopamine antagonists. Pharmacol Biochem Behav 1991;39:799-802.
12. Corrigall WA. Regulation of intravenous nicotine self-administration -- dopamine mechanisms. In: Effects of Nicotine on Biological Systems. Adlkofer F, Thurau K, editors. Basel: Birkhauser Verlag, 1991: 423-432.
13. Corrigall WA, Coen KM, Adamson KL. Self-administered nicotine activates the mesolimbic dopamine system through the ventral tegmental area. Brain Res 1994;653:278-284.
14. Egan TM, North RA. Actions of acetylcholine and nicotine on rat locus coeruleus neurons *in vitro*. Neuroscience 1986;19:565-571.
15. Reavill C, Stolerman IP. Locomotor activity in rats after administration of nicotine agonists locally. Br J Pharmacol 1990;99:273-278.
16. Woolf, NJ. Cholinergic systems in mammalian brain and spinal cord. Progr Neurobiol 1991;37:475-524.
17. Yeomans JS, Mathur A, Tampakeras M. Rewarding brain stimulation: role of tegmental cholinergic neurons that activate dopamine neurons. Brain Res 1993;107:1077-1087.

18. Clarke PBS, Hommer DW, Pert A, Skirboll LR. Innervation of substantia nigra neurons by cholinergic afferents from pedunculopontine nucleus in the rat: neuroanatomical and electrophysiological evidence. Neuroscience 1987;23:1011-1019.
19. Niijima K, Yoshida M. Activation of mesencephalic dopamine neurons by chemical stimulation of the nucleus tegmenti pedunculopontinus pars compacta. Brain Res 1988; 451:163-171.

Effects of Nicotine on
Biological Systems II
Advances in Pharmacological Sciences

DESENSITISATION OF THE STIMULATORY EFFECTS OF NICOTINE ON DOPAMINE SECRETION IN THE MESOLIMBIC SYSTEM OF THE RAT

D.J.K. Balfour and M.E.M. Benwell

Department of Pharmacology and Clinical Pharmacology, University Medical School, Ninewells Hospital, Dundee DD1 9SY. Scotland

Summary: It has been suggested that the rewarding properties of nicotine which reinforce its self-administration are related to its ability to stimulate mesoaccumbens dopamine (DA) neurones. This study has shown that the constant infusion of nicotine, at doses which result in plasma levels similar to those found in the plasma of habitual smokers, attenuate or abolish the effects of nicotine injections on (DA) overflow in the nucleus accumbens. The data imply that nicotine administration does not invariably result in increased DA release in the nucleus accumbens and that hypotheses for nicotine addiction which assume that this is the case should be treated with caution.

Introduction

It is now widely accepted that a majority of the people who are habitual tobacco smokers have become addicted to the nicotine present in the smoke and that, as a result, they often find it difficult to quit the habit (1). The neural mechanisms underlying the development of the addiction remain to be established with certainty although there is growing evidence to suggest that the mesoaccumbens dopamine (DA) system forms an important component of the "reward" pathways of the brain and that addictive properties of most of drugs of dependence, particularly psychostimulants, are related to their ability to enhance neurotransmission at mesoaccumbens DA synapses (12). There is convincing evidence that nicotine also stimulates mesoaccumbens neurones (3,7,10) and that the ability of the compound to serve as a reinforcer in a self-administration paradigm depends upon its ability to stimulate mesoaccumbens DA neurones (8). Studies in our laboratory have shown that the mesolimbic DA response to nicotine, measured using *in vivo* microdialysis, is enhanced if the animals are pretreated with 5 daily injections of the drug prior to the test day (3). These data, when taken together, suggest that the mechanisms underlying the addictive properties of nicotine are the same as those thought to mediate the addictive properties of other psychostimulant drugs of dependence. It appears likely, however, that the nicotinic cholinoceptors, thought to mediate many of the effects of nicotine in the brain, are readily desensitised by prolonged or repetitive exposure to nicotine and that this explains the increases in nicotinic receptor density which have been observed in animals when they are chronically exposed to the drug (13). Since similar changes are also observed in human brain tissue taken at postmortem from subjects who were habitual smokers it seems likely that, for many smokers, tobacco smoking also results in prolonged periods when many of the nicotinic receptors in the brain are desensitised (4). If this is true for the receptors which mediate the effects of nicotine on mesoaccumbens neurones then it has important consequences for our understanding of the putative role of this system in maintaining the tobacco smoking habit. The primary objective of the study reported in this paper was to establish the conditions under which the mesoaccumbens DA response to nicotine becomes desensitised.

Materials and methods

In the principal series of experiments, groups of male Sprague-Dawley rats were constantly infused with saline or nicotine (0.25, 1 or 4 mg/kg/day) from subcutaneously-implanted osmotic minipumps (Alzet 2ML2) for 14 days. On days 9 to 13 they also received a subcutaneous (sc) injection of saline or nicotine (0.4 mg/kg) daily. Microdialysis probes were located in the nucleus accumbens (NAc) 3 hours after the last sc injection on day 13. On day 14 the animals were transferred to an activity box, the dialysis probes connected to infusion pumps and, following a period (1 h) for equilibration, 3 x 20 minute dialysate samples collected to establish the baseline levels of DA and its metabolites. All the animals were then given an injection of saline followed, 60 minutes later, by an injection of nicotine (0.4 mg/kg sc). Dialysate samples were collected for a further 140 minutes after the injection of nicotine. The data are expressed as percentages of the mean baseline dialysate concentrations and are presented as means ± sem. Additional details of the procedure can be found in Benwell et al (5). In order to investigate the effects of withdrawing nicotine following a period of infusion, nicotine (4 mg/kg/day) was infused for 14 days. The pumps were then removed and the responses to nicotine (0.4 mg/kg sc) and saline measured 1, 2 and 7 days after nicotine withdrawal, different groups of rats being used for each withdrawal point.

Results

The constant infusion of nicotine at rates of 0.25, 1 and 4 mg/kg/day yielded plasma nicotine levels of 8±4, 24±5 and 87±12 ng/ml respectively. These infusions had no significant effects on the basal levels of DA in the NAc dialysates although the DA concentration did tend to be lower in the samples taken from the animals infused with nicotine. In agreement with our previous studies, in the animals constantly infused with saline, pretreatment with 5 daily injections of nicotine resulted in a significant ($p<0.05$) sensitisation of its effects on DA overflow in the NAc and spontaneous activity (Fig. 1). The responses were significantly attenuated ($p<0.05$) when nicotine was constantly infused at doses of 1 or 4 mg/kg/day. The infusion of the lowest dose tested (0.25 mg/kg/day) also attenuated ($p<0.05$) the locomotor response to subchronic nicotine whereas its apparent effect on DA overflow did not reach significance. Indeed, at this dose the infusions alone caused enhanced responses ($p<0.05$) to nicotine both the locomotor stimulant effects of the drug and its stimulatory effects on DA overflow in the NAc. Since, in non-sensitised rats the increase in DA overflow evoked by an acute injection of nicotine is often small (3) an additional series of experiments was performed in which nomifensine (5 μM) was added to the Ringer solution used to perfuse the probe. Nomifensine is a selective inhibitor of the catecholamine transporters located on nerve terminal membranes and its inclusion in the perfusion causes a substantial increase in DA overflow into NAc dialysates (3). Measurements made in the presence of the inhibitor confirmed that pretreatment with nicotine enhanced the increase in DA overflow in the NAc evoked by an injection of nicotine and that both the enhanced and acute DA responses to a nicotine injection were attenuated by the constant infusion of nicotine at dose of 1 mg/kg (Fig. 2). These data provide clear support for the conclusion that the receptors which mediate the effects of nicotine on mesoaccumbens DA neurones are desensitised by the infusion of the drug at a dose which maintains plasma levels of at approximately 25 ng/ml.

In another series of experiments, the effects of nicotine-withdrawal on the responses to nicotine injections were examined. In these experiments responses were measured for up to 7 days after removal of the pumps which, when in place, had delivered a constant infusion of nicotine at a rate of 4 mg/kg/day for 14 days. Twenty four hours after removal of the pumps, a significantly enhanced ($p<0.01$) locomotor response to a nicotine injection (0.4 mg/kg) was observed whereas the increase in DA overflow in the NAc approached statistical significance. At 48 hours both responses to nicotine were significant (Fig. 3). Sensitised responses were also observed in rats tested with nicotine 7 days after removal of the pumps.

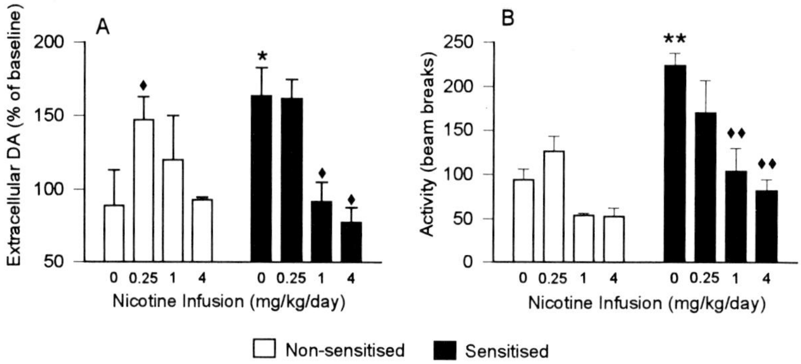

Figure 1. The effects of nicotine infusions on (A) mesoaccumbens dopamine overflow and (B) locomotor responses to nicotine. Groups of rats (N=4-6) were constantly infused with saline or nicotine for 14 days. On days 9 to 13 they were also given daily sc injections of saline (open columns) or nicotine (0.4 mg/kg; filled columns). On day 14, all the rats were challenged with a injection of nicotine (0.4 mg/kg). The results are expressed as means ± sem. Significantly different from rats given saline injections during the pretreatment phase; ★$p<0.05$, ★★$p<0.01$; significantly different from rats infused with saline; ◆ $p<0.05$, ◆ ◆ $p<0.01$

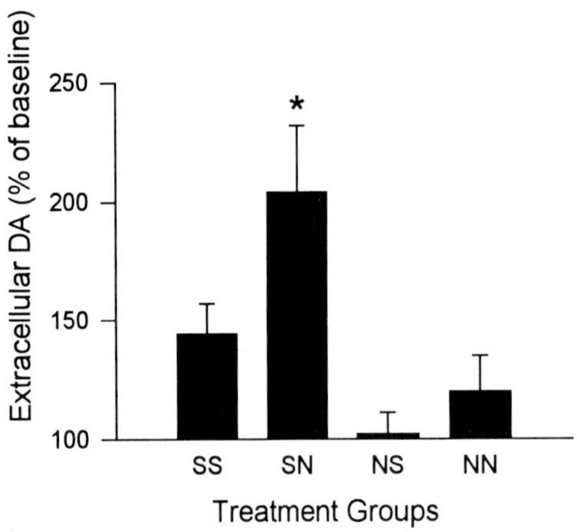

Figure 2. Mesoaccumbens responses to nicotine in the presence of nomifensine. Groups of rats (N=4-6) were constantly infused with saline (SS, SN) or nicotine (NS,NN) for 14 days. On days 9-13 they also received daily injections of saline (SS, NS) or nicotine (0.4 mg/kg; SN, NN). On day 14 all the rats were challenged with an injection of nicotine. The results are means ± sem. Significantly different from the SS group; ★$p<0.05$

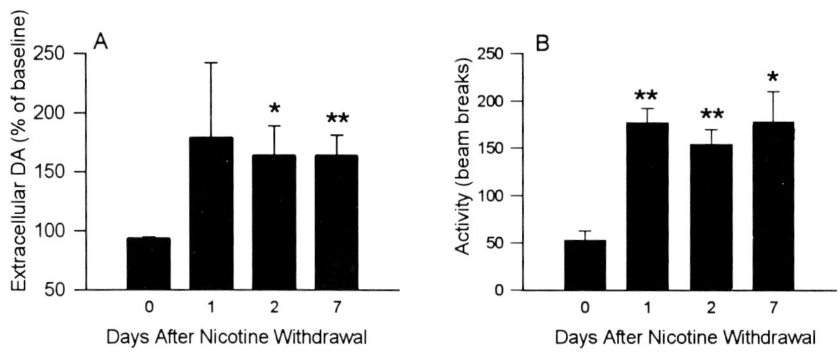

Figure 3. The effects of nicotine withdrawal on (A) mesoaccumbens dopamine overflow and (B) locomotor responses to nicotine. Groups of rats (N=4) were constantly infused with nicotine (4 mg/kg/day) for 14 days. Mesoaccumbens dopamine overflow and locomotor activity were then studied following an injection of nicotine (0.4 mg/kg) in groups of rats with the infusion pumps still in place (0) or 1, 2 and 7 days after their removal. The results are means ± sem. Significantly different from non-withdrawn rats; ⋆$p<0.05$, ⋆$p<0.01$

Discussion

The results of previous studies suggest that nicotine evokes increased DA overflow in the NAc by stimulating nicotinic cholinoceptors located on the somatodendritic membranes of mesoaccumbens DA neurones (6,10,11). The principal results of the present study suggest that the receptors which mediate this response are desensitised when nicotine is infused at a rate which maintains the plasma nicotine level at around 25 ng/ml. This concentration of nicotine is often found in the venous blood of habitual smokers for much of the day (2). The data are, therefore, entirely consistent with the hypothesis that the up-regulation of central nicotinic receptors, observed in brain tissue taken from habitual smokers (4), is caused by prolonged periods when the receptors are desensitised. This conclusion has important consequences for our understanding of the role of nicotine in the tobacco smoking habit since it implies that, for significant parts of the day, the administration of a nicotine bolus in the form of tobacco smoke, may not necessarily result in stimulation of this reward pathway of the brain. It is important to remember, however, that in these experiments, the plasma nicotine level was maintained using a constant infusion of the drug whereas a smoker would inhale the nicotine as a series of boli through the lungs. Nevertheless, the data suggest that hypotheses for nicotine dependence which assume that the drug invariably stimulates mesoaccumbens DA neurones should treated with some caution and that other neural mechanisms may also contribute significantly to the development of nicotine addiction in humans. However, the results presented above imply that the injection of a nicotine bolus, which yields plasma levels in excess of 25 ng/ml, is likely to cause desensitisation of the nicotinic receptors located on mesoaccumbens DA neurones which may persist for a significant period of time following initial stimulation of the receptor complex. This conclusion is supported by results of Hakan and Ksir (9) which showed that the locomotor response to nicotine exhibited acute tolerance to the drug for 60 minutes following a subcutaneous injection of nicotine at a dose of 0.2 mg/kg. The tolerance persisted for a longer period of time if a higher nicotine dose was used. Thus, it is possible that sensitisation to nicotine is primarily a consequence of receptor desensitisation even when the drug is given as a series of repeated bolus injections. However, it is equally possible that the mechanisms underlying the sensitised responses to nicotine which occur following its administration as repeated injections or as infusions are entirely different.

Acknowledgement: *The authors are grateful for the financial support received for these studies from Foundation VERUM and the Wellcome Trust.*

References

1. Balfour DJK The neurochemical mechanisms underlying nicotine tolerance and dependence. In: The Biological Basis of Drug Tolerance and Dependence. Pratt JA, editor. London: Academic Press, 1991: 121-151.
2. Benowitz NL, Porchet H, Jacob P. Pharmacokinetics, metabolism and pharmacodynamics of nicotine. In: Nicotine Psychopharmacology: Molecular, Cellular and Behavioural Aspects. Wonnacott S, Russell MAH, Stolerman IP, editors. Oxford: Oxford University Press, 1990: 112-157.
3. Benwell MEM, Balfour DJK. The effects of acute and repeated nicotine treatment on nucleus accumbens dopamine and locomotor activity. Br J Pharmacol 1992; 105: 849-856.
4. Benwell MEM, Balfour DJK, Anderson JM. Evidence that smoking increases the density of nicotine binding sites in human brain. J Neurochem 1988; 50: 1243-1247.
5. Benwell MEM, Balfour DJK, Khadra LF. Studies on the influence of nicotine infusions on mesolimbic dopamine and locomotr responses to nicotine. Clin Investig 1994; 72: 233-239.
6. Benwell MEM, Balfour DJK, Lucchi HM. The influence of tetrodotoxin and calcium on the stimulation of mesolimbic dopamine activity evoked by systemic nicotine. Psychopharmacology 1993; 112: 467-471.
7. Clarke PBS, Fu DS, Jakubovic A, Fibiger HC. Evidence that mesolimbic dopaminergic activation underlies the locomotr stimulant action of nicotine. J Pharmacol Exp Ther 1988; 246: 701-708.
8. Corrigall WA, Franklin KBJ, Coen KM, Clarke PBS. The mesolimbic dopaminergic system is implicated in the reinforcing effects of nicotine. Psychopharmacology 1992; 107: 285-289.
9. Hakan RL, Ksir C. Acute tolerance to the locomotor stimulant effects of nicotine in the rat. Psychopharmacology 1991; 104: 386-390.
10. Imperato A, Mulas A, Di Chiara G. Nicotine preferentially stimulates dopamine release in the limbic system of freely moving rats. Eur J Pharmacol 1986; 132: 337-338.

11. Nisell M, Nomikos GG, Svensson TH. Systemic nicotine-induced dopamine release in the rat nucleus accumbens is regulated by nicotinic receptors in the ventral tegmental area. Synapse 1994; 16: 36-44.
12. Wise RA, Bozarth MA. A psychomotor stimulant theory of addiction. Psychol Rev 1987; 94: 469-492.
13. Wonnacott S. Characterization of nicotine receptor sites in the brain. In: Nicotine Psychopharmacology: Molecular, Cellular and Behavioural Aspects. Wonnacott S, Russell MAH, Stolerman IP, editors. Oxford: Oxford University Press, 1990: 226-277.

NICOTINE AS A DISCRIMINATIVE STIMULUS: INDIVIDUAL VARIABILITY TO ACUTE TOLERANCE AND THE ROLE OF RECEPTOR DESENSITIZATION

J.A. Rosecrans, L.D. Karan and J.R. James

Department of Pharmacology and Toxicology, Box 890613
Virginia Commonwealth University, Richmond, VA, 23298, USA

Summary: This paper will review experiments designed to study mechanisms of nicotine-induced acute tolerance in two behavioral models, nicotine-induced discriminative stimulus (DS) control of behavior, and nicotine-induced disruption of operant behavior. Acute tolerance to nicotine in both paradigms was explained on the basis that nicotine was able to induce a desensitization of nicotinic cholinergic receptors leading to an attenuation of nicotine's effects.

Introduction

Research conducted in this laboratory has focused on central mechanisms of nicotine action employing a drug discrimination model of nicotine's behavioral effects that parallels the ability of smokers to detect different concentrations of nicotine in tobacco (1,2). This model is based on the ability of nicotine to exert discriminative stimulus (DS) control of behavior. This effect appears to be related to its ability to mimic the effects of endogenous acetylcholine (ACh) at select central nicotinic cholinergic receptors (nAChRs).

To test the hypothesis that nicotine is acting directly or indirectly via an action of ACh at the nAChR, rats trained to discriminate nicotine from vehicle were evaluated as to their ability to generalize to the discriminative stimulus effects elicited by increases in brain ACh (via physostigmine) levels. The results of this study indicated that rats trained to detect nicotine did not perceive physostigmine like nicotine (3). In contrast, rats trained to detect the muscarinic cholinergic receptor (mAChR) agonist, arecoline, did perceive physostigmine as an mAChR agonist which was also antagonized by atropine. This research suggested that nicotine was not acting at a cholinergic receptor, at least a receptor that was sensitive to ACh.

This finding (3) was difficult to interpret, but a hypothesis was put forward suggesting that physostigmine may have induced a desensitization of the nAChR (4) which was interpreted by rats trained to discriminate nicotine as more like vehicle than nicotine. It was also speculated that both physostigmine and nicotine could be shown to antagonize (acute tolerance) the nicotine-elicit DS via a common mechanism, i.e. nAChR desensitization. It should be added that this theoretical model has since been verified (5).

This paper will review experiments designed to test whether acute tolerance (desensitization) can be demonstrated in rats trained to discriminate nicotine. In addition this paper will present data involving acute tolerance to nicotine-induced behavioral disruption of operant behavior that support the concept that desensitization processes may serve an integral role as to how nicotine affects behavior.

Experiment I: Drug discrimination studies

Nicotine-induced acute tolerance in the DS paradigm was evaluated in 52 rats trained to discriminate nicotine (0.4 mg/kg, s.c.) vs. vehicle (3). Once rats learned to discriminate nicotine, a nicotine dose-response study was conducted and each rat was evaluated as to its ability to exhibit acute tolerance to the nicotine DS. For this purpose each rat was administered nicotine (0.8 mg/kg, s.c.) in its home cage and tested for its ability to discriminate nicotine when administered the training dose at 30 min intervals after the first dose (30-360 min) of nicotine. Rats were rated as to their ability to detect nicotine during 2 min test sessions conducted at weekly (Wed.) or bi-weekly intervals (Tues. & Fri.). Normal training was maintained on other days of a 5 day week; the study was repeated 1 month later.

The data obtained indicated that a select group of rats exhibited acute tolerance to nicotine (Fig. 1). Twenty-three rats exhibited at least a 50-70% decrease in their ability to detect nicotine following an initial dose of nicotine (0.8 mg/kg); they were classified as desensitizers. The remaining rats (N=29) were classified as non-desensitizers as they reduced the nicotine DS by less than 30%. Interestingly, the time for maximal desensitization (reduction in % Nicotine discrimination) varied for reach rat (average = 90 min). An evaluation of ED-50 generalization values also revealed that the non-desensitizer was significantly more sensitive to nicotine (0.07 mg/kg) than rats exhibiting desensitization (0.12 mg/kg). This study was replicated one month later with a similar result, i.e. each rat exhibited a similar pattern of acute tolerance indicating that acute tolerance was not state contingent.

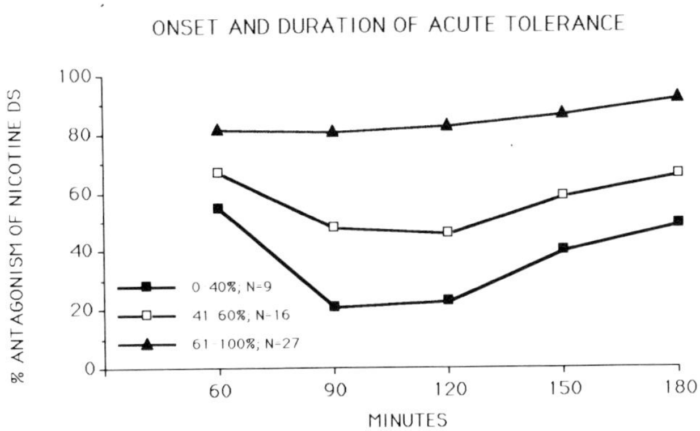

Figure 1. Experiment I: Onset and duration of acute tolerance (% antagonism of the nicotine DS) after a challenge dose of nicotine (0.8 mg/kg, s.c.). followed by the training dose of nicotine (0.4 mg/kg, s.c.). Testing for each time point was done on individual days. Reproduced with permission (5).

Experiment II: Behavioral disruption studies

To further evaluate the pharmacological basis of acute desensitization (acute tolerance), an acute tolerance study was conducted in rats trained to lever-press for food using a Fixed Ratio 30 (FR-30) schedule of reinforcement. Thirty rats were trained to lever press for food and divided into 3 groups of 10 rats each. One group served as a control group and was administered saline at 90 or 180 min (in their home cage) prior to being administered a dose of nicotine (0.4 mg/kg, s.c.). The effects of the 0.4 mg/kg dose were then evaluated in relation to rate of lever-presses during a 30 min behavioral session. The other two experimental groups served as nicotine tolerance-inducing groups. They were administered nicotine (0.8 mg/kg, s.c.) in their home cage 90 or 180 min before the administration a second nicotine dose (0.4 mg/kg, s.c.) and evaluation of lever-pressing was also evaluated for 30 min. Nicotine doses and time parameters were similar to those evaluated in the DS study (Fig. 1).

The results of this experiment indicated that both groups of rats receiving nicotine at 90 or 180 min prior to a second nicotine dose exhibited tolerance to nicotine's disruptive effects on FR-30 behavior when compared to the control group. Initial nicotine doses (0.8 mg/kg) were non-contingent to behavior suggesting that a select pharmacological mechanism was invoked to cause a reduction of nicotine-induced behavioral disruption. This experiment was repeated one week later in which a greater degree of tolerance was observed in all three groups. Behavioral (contingent) tolerance was also evident since groups all groups received nicotine (0.4 mg/kg) during the first behavioral session (Day 1). These data support the concept that nicotine-induced acute tolerance was the result of a pharmacological mechanism at the nAChR, namely desensitization.

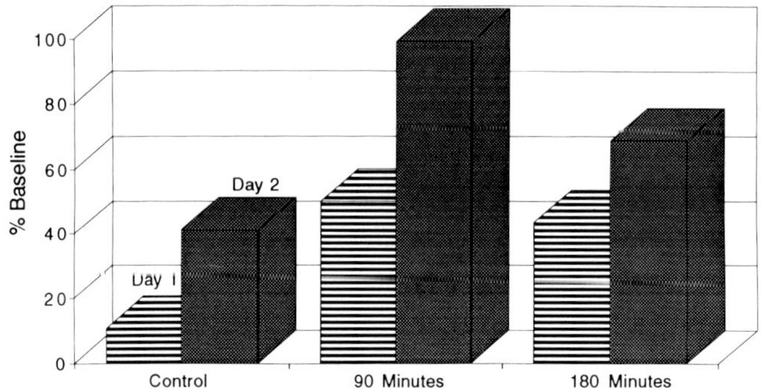

Figure 2. Experiment II: Percent disruption of baseline behavior (FR30) induced by nicotine administration (0.4 mg/kg, s.c.) 90 or 180 min. after vehicle or nicotine administration. The data represents changes in response rates over the first 10 minutes of a 30 min session. Two experiments were conducted 7 days apart, Day 1 and Day 2. (Data were redrawn from ref. 9).

Discussion and Conclusions

Acute tolerance to nicotine's effects on behavior was evaluated in two different behavioral paradigms, nicotine-induced disruption of behavior, and nicotine-elicited discriminative stimulus (DS) control of behavior. Results from studies involving behavioral disruption indicated that a pharmacological mechanism of acute tolerance (i.e nAChR desensitization) was involved when nicotine was administered non-contingently to behavior (Fig. 2; Day 1). Evidence of behavioral tolerance was also observed in the replication of the initial experiment (Day 2) supporting previous work showing that chronic tolerance to nicotine is contingent on both pharmacological and behavioral mechanisms (6).

In contrast to what was observed in the behavioral disruption study, only a select group of rats exhibited acute tolerance to a nicotine-elicited discriminative stimulus (N=23; Fig. 1). Another set of rats (N=29) from the same population exhibited a minimal level, or no tolerance, to acute nicotine injections. The hypothesis formulated suggested that these differential effects were based on the ability of the desensitizer rat to induce a secondary nAChR desensitization following nicotine's initial agonist effect (4,5). A corollary to this hypothesis implies that the non-desensitizer rat was also less likely to exhibit nAChR desensitization under a similar behavioral situation.

At this point it seems plausible to conclude that acute tolerance to the nicotine DS is most likely due to an acute nAChR desensitization. Furthermore, these differential effects may involve a subset of nAChRs not involved in the development of chronic tolerance to nicotine;

chronic tolerance to nicotine may also involve a desensitization of select nAChRs (7,8). It would also seem that rats unable to exhibit acute tolerance differ in relation to nAChR function and possibly molecular structure. These results also have specific ramifications to the molecular genetics of the nAChR. Furthermore, rats not exhibiting acute tolerance may also be unable to develop tolerance to many of nicotine's effects because of a lack of select nAChRs to exhibit desensitization.

This research supports the hypothesis that nicotine is mimicking ACh at its nAChR, directly or via the release of ACh. What also emerges is the fact that nicotine's initial agonist effect is followed by a rapid desensitization of the same nAChR in a select population of rats. Thus, nAChR under these conditions is not able to be activated for a period of time until the receptor is again sensitized. These results may be helpful to better understanding nicotine's effects in the smoker. Rates of nAChR desensitization potentially set the rate of smoking in an individual and may also help us better understand why some individuals initiate and maintain the use of tobacco products. One can develop a hypothesis that contingent upon the development of acute tolerance an individual may or may not find tobacco pleasurable. In fact nicotine's ability to induce a bimodal effect at the nAChR may be important to understanding its effects at a variety of non-cholinergic neurons which may have relevance to nicotine's potential as a therapeutic agent (4).

Acknowledgements: *This research was supported by grants to JAR (German Research Council on Smoking and Health) and LDK (USPHS grant 5K20DA00183-01)*

References

1. Rosecrans JA. Nicotine as a discriminative stimulus: A neurobiological approach to studying central cholinergic mechanisms. J Subs Abuse 1989; 1: 287-300.
2. Rosecrans JA, Karan LD. Neurobehavioral mechanisms of nicotine action: Role in the initiation and maintenance of tobacco dependence. J Subs Abuse Treat 1993; 10: 161-170.
3. Meltzer LT, Rosecrans, JA. Nicotine and arecoline as discriminative stimuli: Involvement of noncholinergic mechanisms for nicotine. Pharmacol Biochem Behav 1988; 29: 587-593.
4. Wonnocott S. The paradox of nicotinic acetylcholine receptor upregulation. Trend Pharmacol Sci 1990; 11: 216-219.
5. James JR, Villanueva HF, Johnson JH Arezo S, Rosecrans JA. Evidence that nicotine can acutely desensitize central nicotinic acetylcholinergic receptors. Psychopharmacology 1994; 114: 456-462.
6. Villanueva HF, Arezo S, James JR, Rosecrans JA. A characterization of nicotine-induced tolerance: Evidence of pharmacological tolerance in the rat. Behav Pharmacol 1992; 3: 255-260.
7. Marks MJ, Burch JB, Collins AC. Effects of chronic nicotine infusion on tolerance development an nicotinic receptors. J Pharmacol Exp Therap 1985; 22: 81-825.
8. Nordberg A, Wahlstrom G, Arnelo U, Larrson C. Effects of long-term nicotine treatment on [^{3}H]nicotine binding sites in the rat brain. Drug Alc Depend 1985; 16: 9-17.
9. Rosecrans JA, Wiley JL, Bass CE, Karan LD. Nicotine-induced acute tolerance: Studies involving schedule-controlled behavior. Brain Res Bull (in press).

Effects of Nicotine on
Biological Systems II
Advances in Pharmacological Sciences

BEHAVIORAL AND BIOCHEMICAL ANALYSIS OF THE DEPENDENCE PROPERTIES OF NICOTINE

Tomoji Yanagita, Kiyoshi Ando, Yoshio Wakasa, and Akira Shimada

Preclinical Research Division, Central Institute for Experimental Animals
1433 Nogawa, Miyamae-ku, Kawasaki 216 Japan

Summary: Our recent studies on the reinforcing properties and psychotoxicity of nicotine in rats and rhesus monkeys are introduced. In drug discrimination experiments in rats, the discriminative effects of subcutaneous nicotine were generalized by the nicotine injected into nucleus accumbens or medical prefrontal cortex. In microdialysis in rats, nicotine increased dopamine (DA) and DOPAC at nucleus accumbens and striatum. In self-administration experiments in monkeys, faster infusion speeds resulted in higher intake rates of nicotine and possible development of physical dependence attenuated monkey's nicotine-seeking behavior. The threshold dose for reinforcing effect of nicotine was found to be 2.5-10 μg/kg.

Methods and Results

1. Nicotine reinforcement and psychotoxicity in rats

1.1 Sites involved in drug discriminative effects of nicotine in the brain. Male Sprague-Dawley rats were trained to discriminate subcutaneous nicotine (0.5 mg/kg) from saline for food reinforcement under a fixed ratio 10 schedule using the 2-lever choice procedure. Guide cannulae were implanted bilaterally into nuclei of the brain. The doses of nicotine tested within the range of 5 to 180 μg/rat. In the generalization tests by intracranial nicotine injection, the nicotine appropriate responses were observed at high rates when nicotine was injected into the lateral ventricle (85.3% at 100 μg/kg), medial prefrontal cortex (88.2% at 40 μg/kg), and nucleus accumbens (77.6% at 100 μg/kg) (1). The response rates were marginal at the ventral tegmental area (60.6% at 60 μg/kg) and dorsal hippocampus (44.1% at 10 μg/kg), and near zero at medial habenular nucleus (3.2% at 10 μg/kg) (Fig 1).

1.2 Microdialysis of DA and DOPAC in drug-sensitized rats. The influences of nicotine, methamphetamine (MAP) and cocaine on extracellular DA and DOPAC were studied in naive

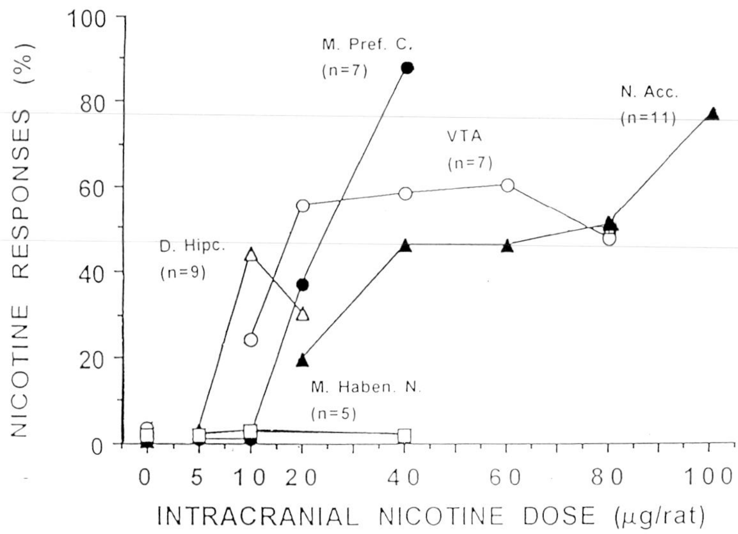

Figure 1. Substitution tests with nicotine administered into various brain areas to rats discriminating subcutaneously administered nicotine at 0.5 mg/kg from saline. Abscissa: Test doses of nicotine in μg per rat (0: vehicle). Ordinate: The mean percent of nicotine-appropriate responses.

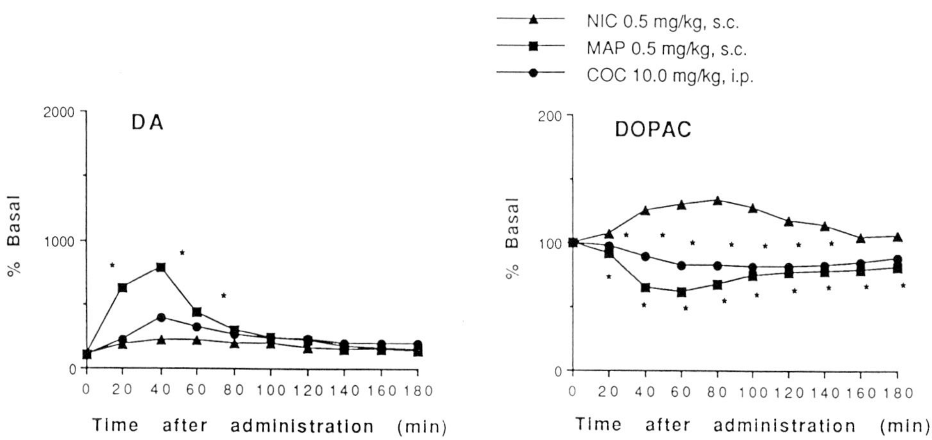

Figure 2. Influence of nicotine (NIC), methamphetamine (MAP), and cocaine (COC) on dopamine (DA) and dihydroxyphenyl acetic acid (DOPAC) levels in the nucleus accumbens in rats sensitized by a test drug. Each value represents the mean of 5 to 6 rats. *: $p<0.05$ vs. nicotine.

and drug-sensitized rats. The dialysates were collected every 20 minutes after drug administration. To sensitize the rats, nicotine, MAP, or cocaine was administered 8 times every 1 or 2 days, and development of sensitization was confirmed by observing spontaneous motor activity. Repeated sc administration of nicotine enhanced the increase of motor activity. As monitored by microdialysis, MAP 0.5 mg/kg sc and cocaine 10 mg/kg ip elevated DA in nucleus accumbens and striatum, particularly with MAP in normal rats in nucleus accumbens However, DOPAC decreased with both drugs. Contrary to MAP and cocaine, nicotine increased both DA and DOPAC (Fig 2, Table 1).

Table 1. Summary of microdialysis in rats.

	DA				DOPAC			
Drug	N. Accumbens		Striatum		N. Accumbens		Striatum	
	Norm	Sens	Norm	Sens	Norm	Sens	Norm	Sens
NIC	→	↑	→	→	↑	↑↑	↑	↑
MAP	↑↑↑	↑↑	↑↑	↑	↓↓	↓↓	↓	↓↓
COC	↑↑	↑↑	-	-	↓	↓	-	-

2. Nicotine self-administration studies in rhesus monkeys

2.1 Influence of infusion speed of nicotine on its reinforcing effect. Self-administration of nicotine was observed around the clock under a fixed-ratio 5 schedule with a 15 min time-out after each intake in one male and 3 female rhesus monkeys. They were allowed to self-administer nicotine at a fixed unit dose of 30 μg/kg delivered at infusion speeds of 0.3, 1.3, and 5.2 μg/sec in 3 separate nicotine administration periods of 8 days each. To determine the blood level of nicotine, 4 other monkeys were used and blood samples were obtained 1 min after nicotine infusion. Prior to studying the different infusion speeds, a control experiment was conducted at a fixed speed of 5.2 μg/sec throughout the 3 experimental periods and stable intake rates during these periods were confirmed. In the experiments with the 3 different speeds, the higher infusion speed resulted in more frequent intake (Fig 3). The plasma nicotine levels were 9.7 ± 0.6 ng/ml at 0.3 μg/sec, 11.3 ± 0.6 at 1.3, and 15.2 ± 1.6 at 5.2.

2.2 Influence of physical dependence on nicotine on its reinforcing effect. The possible influence of physical dependence on self-administration of iv nicotine was then tested under a progressive ratio schedule in 4 rhesus monkeys. Induction of physical dependence on nicotine was attempted by pretreatment with iv nicotine 0.25 mg/kg/h for 4 weeks. In the progressive ratio experiments, the monkeys were trained to obtain a single dose of nicotine 0.25 mg/kg by pressing a lever 100 times. Next, the ratio of lever presses per administration was increased by a factor of the 4th root 2 at each intake. The final ratio achieved is taken to represent the intensity of the reinforcing effect. Each monkey was tested with saline in the first trial, then tested with nicotine following pretreatment with either nicotine or saline in the second trial, and then tested with nicotine following pretreatment with the other substance in the third trial. Saline-pretreated monkeys took nicotine 0.25 mg/kg at final ratios of up to 2,690. Final ratios were lower in nicotine-pretreated monkeys (maximum 1,900; Fig 4). For comparative purposes, some results obtained to date with several dependence-producing drugs

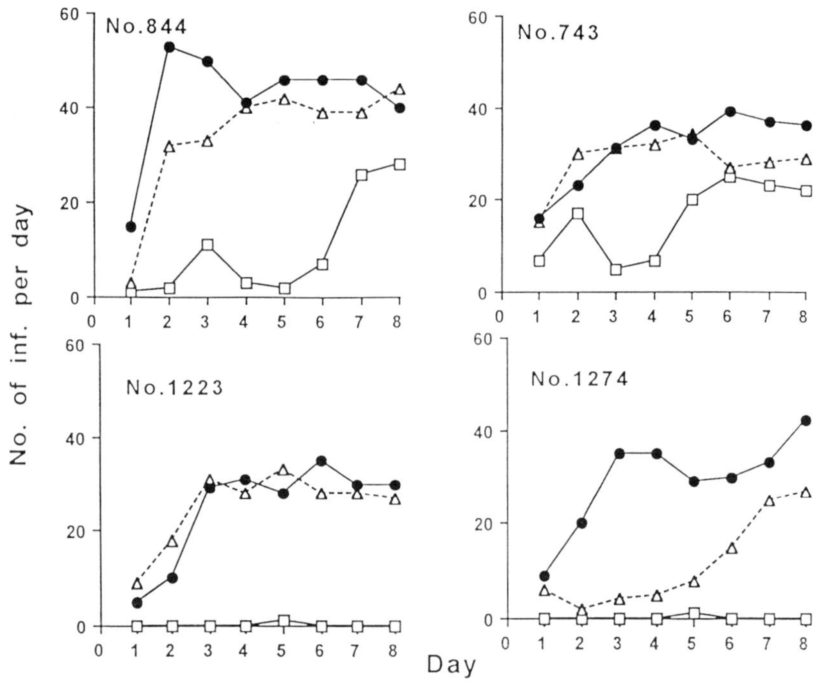

Figure 3. Daily number of intravenous self-administrations of nicotine at different infusion speeds in rhesus monkeys. The monkeys were first allowed to self-administer saline 0.25 ml/kg infusion for 1 week and were then tested with nicotine 30 μg/kg/infusion at infusion speeds of 5.2 (●), 1,3 (△), and 0.3 (□) μg/sec in this order for 8 days each. The nicotine self-administration periods were separated by self-administration of saline for 2 weeks each.

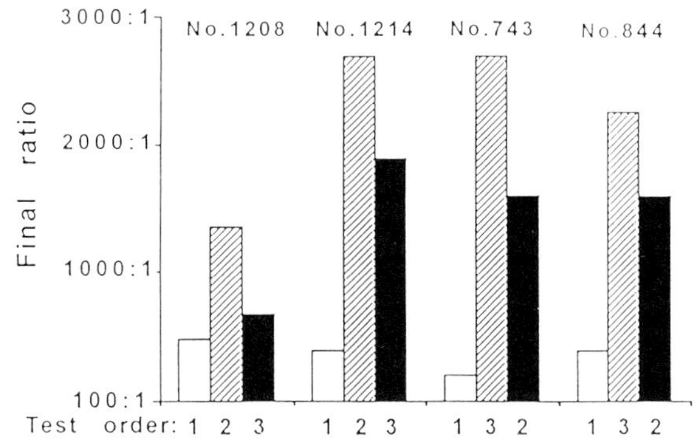

Figure 4. Final ratios obtained in the progressive ratio test on nicotine in monkeys. White bars: saline with no pretreatment; slashed bars: nicotine with pretreatment by saline; black bars: nicotine with pretreatment by nicotine 0.25 mg/kg hourly for 4 weeks.

by the same procedure are shown in Table 2 (2-6). Enhancement of the reinforcing effect by the pretreatment is marked with dihydrocodeine and morphine, and slight with pentazocine and alcohol.

2.3 Threshold doses for reinforcing effect of nicotine. The relationship of dose to reinforcing effect of nicotine in intravenous self-administration was studied in 4 rhesus monkeys. In order to exclude exploratory intakes from active intakes, the self-administration experiment was conducted under a schedule in which a self-administration period of 5 hours a day was divided into 5 sessions, and in each session a time-out period of 1 hour began when either 5 doses were taken or a time interval of 5 minutes from the preceding intake had lapsed and the self-administrations that occurred only once in each session were excluded from the number of intakes. The experiment was continued for each unit dose for 2 to 4 weeks until the daily intake became stable. The unit doses tested were 0.6, 2.5, 10, and 40 μg/kg. These doses were tested in each monkey in either ascending or descending order, and then tested in the other order. The results are shown in Fig 5. The threshold doses were found to be 10 μg/kg in the ascending test but 2.5 μg/kg in the descending test.

Discussion

The results of Experiment 1.1 agree with previous findings suggesting that the nucleus accumbens and medial prefrontal cortex play a key role in mediating the rewarding and reinforcing effects of drugs (7). Nicotine is unique among known dependence-producing drugs, in producing reinforcing but not psychotoxic effects. The animals that are sensitized by repeated administration of amphetamines are regarded to be a good model for amphetamine psychosis, and the sensitization is reported to be closely related to hyperfunction of the DA

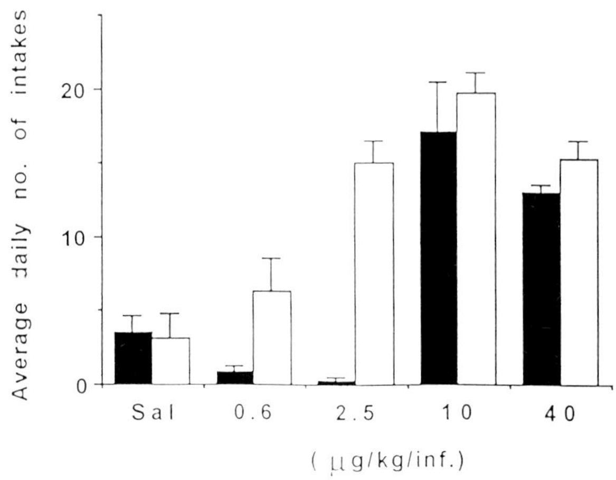

Figure 5. Results of determination of threshold dose for reinforcing effect of nicotine in rhesus monkeys. Saline and test doses of nicotine were made available in either ascending or descending order first, followed by the other order, with 2 to 4 weeks per dose under a block time-out schedule. (see test). Black bars: ascending order; white bars: descending order; vertical lines above bars; standard error.

Table 2. Summary of result of progressive ratio rest in monkeys.

Test Drug	Unit Dose (mg/kg/inj)	Final Ratio for Test Drug after: Saline Pretreatment a)	Test Drug Pretreatment a)
Nicotine	0.25	1,350~ 2,690	670~ 1,900
Morphine	0.25	1,350~ 1,600	2,260~ 6,400
Dihydrocodeine	1.0	950~ 1,900	4,530~10,760
Pentazocine	1.0	1,350~ 3,810	2,260~ 3,810
Alcohol b)	800	1,600~ 6,400	3,200~ 6,400
Cocaine b)	0.11	1,600~ 6,400	800~ 3,200
		No pretreatment	
Cocaine	0.25	2,690~ 9,050	
	1.0	3,810~12,800	
d-Amphetamine	0.06	240~ 670	
	0.25	670~ 4,530	
l-Cathinone	0.25	1,600~ 5,380	
	1.0	1,350~ 7,610	

a) Pretreated by programmed intravenous injections at the unit dose indicated in the table every 0.5 to 2 hrs for 2 to 4 weeks.

b) Experiments were conducted under an old progressive ratio schedule, in which the ratio was doubled at every 4 or 16 injections.

innervation to nucleus accumbens (8, 9). In our experiment, nicotine also sensitized the rats' motor activity as did MAP and cocaine, and moderately elevated DA in the nucleus accumbens although only in the sensitized rats. However, only nicotine elevated DOPAC levels. It is not clear whether this difference is relevant to the lack of psychotoxicity with nicotine.

In self-administration studies (Expt 2.1), faster infusion speeds resulted in higher rates of intake. It is well known that a stronger reinforcing effect occurs when a drug is taken by a route providing more rapid delivery to the brain. Furthermore, the intensity of the reinforcing effect is a function of the unit dose within a certain dose range (9). Although it is not known how infusion speed correlates to drug delivery speed, in our experiment, faster infusion resulted in higher peak plasma concentration for the same dose. This may indicate that a fast infusion provides an efficient way to use a drug at a relatively low dose, thus possibly avoiding its toxic effects. Gross behavioral manifestation of withdrawal signs attributable to the development of physical dependence on nicotine is known to be difficult to observe in laboratory animals even though the animals are treated under the maximally tolerable dose regimen. It was not known whether development of physical dependence on nicotine enhances its reinforcing effect. In our experiments it has become quite evident that possible development of physical dependence will not enhance but rather attenuate the reinforcing effect. These findings indicate that being physically dependent on nicotine may have no practical significance in terms of its hazardous effects. The threshold doses for the reinforcing effect of nicotine were different between the ascending and descending tests (10 and 2.5 μg/kg, respectively). Using this result, the dose to reinforce human smokers was simulated, using the assumption that all factors correlate linearly. Since the plasma level of nicotine was 7.8 ng/ml at a dose of 10 μg/kg, the level at 2.5 μg/kg should be approximately 2 ng/ml. In human smokers, when a cigarette containing nicotine 0.9 mg is smoked, the plasma level is reported to rise by approximately 10 ng/ml. Therefore the dose that is necessary to raise the plasma level by 2 ng/ml will be one-fifth of 0.9 mg, that is 0.18 mg.

Acknowledgements: *These studies were supported by the Smoking Research Foundation Grants for Biomedical Research in Japan.*

References

1. Ando K, Miyata H, Hironaka N, Tsuda T, Yanagita T. The discriminative effects of nicotine and their central sites in rats. Jpn J Psychopharmacol 1993; 13:129-136.

2. Yanagita T. Some methodological problems in assessing dependence-producing properties of drugs in animals. Pharmacol Rev 1976; 27:503-509 .
3. Takada K, Wakasa Y, Yanagita T. Enhancement of reinforcing efficacy by development of physical dependence. Clin Neuropharmacol 1990; 13:450-451.
4. Yanagita T. In: Methods of assessing the reinforcing properties of abused drugs. Bozarth MA editor. Verlag New York 1987: 189-198.
5. Yanagita T. Intravenous self-administration of l-cathinone and aminodimethoxymethyl phenylpropane in rhesus monkeys. Drug Alcoh Depend 1986; 17:135-141.
6. Yanagita T, Wakasa Y, Oinuma N. The dependence potential of nomifensine tested in rhesus monkeys (in Jap with Eng summary and table). Precln Rep Cent Inst Exp Animals 1981; 7:101-113.
7. Di Chiara G, Imperato A. Drugs abused by humans preferentially increase synaptic dopamine concentrations in the mesolimbic system of freely moving rats. Proc Natl Acad Sci USA 1988; 85:5274-5278.
8. Robinson TE, Dianne M, Camp M. Does amphetamine preferentially increase the extracellular concentration of dopamine in the mesolimbic system of freely moving rats? Neuropsychopharmacol 1990:163-173.
9. Robinson TA, Jurson PA, Nennett JA, Bentgen KM. Persistent sensitization of dopamine neurotransmission in ventral striatum (nucleus accumbens) produced by prior experience with (+)-amphetamine: a microdialysis study in freely moving rats. Brain Research 1988; 462:211-222

Effects of Nicotine on
Biological Systems II
Advances in Pharmacological Sciences

NICOTINE BOOST PER CIGARETTE AS THE CONTROLLING FACTOR OF INTAKE REGULATION BY SMOKERS

Michael A.H. Russell, John A. Stapleton, Colin Feyerabend

Health Behaviour Unit, Institute of Psychiatry and Maudsley Hospital
London, SE5 8AF, UK

Summary: Whether nicotine intake is regulated by smokers is not in doubt, but two recent studies raised questions about the precision of the regulation. On reviewing the data it seems clear that smoke intake is regulated to obtain a constant blood nicotine boost, averaging 10 ng/ml per cigarette. A 10 ng/ml boost in 10 min or less in venous blood is characteristic of oral and nasal snuff use and can be obtained from a nicotine nasal spray. This is sufficiently rapid for positive subjective effects in contrast with nicotine gum and patches which provide only negative reinforcement through relief of withdrawal. Positive reinforcement appears particularly important to the most highly dependent smokers.

Introduction

It is well established that nicotine is the major factor underlying the maintenance of smoking. Most smokers (at least 90%) absorb sufficient nicotine to obtain pharmacological effects many of which are potentially rewarding. Although peak blood nicotine levels just after a cigarette vary widely between smokers (< 5 to > 80, around a mean of about 35 ng/ml) the levels of individual smokers are kept remarkably constant from one day to the next (r = 0.8 - 0.9). A variety of approaches have been used to demonstrate that smokers alter their pattern of puffing and inhalation and thereby regulate their nicotine intake independently of other smoke components to maintain their customary blood, and presumably brain, levels (1,2). These include down-regulation of intake from smoking when nicotine is administered from another source (3-6), its upregulation by more intensive puffing (7), and inhalation to give higher carbon monoxide (CO) and blood nicotine levels when the effects are blocked by mecamylamine (8, 9), and its regulation either way to correct for alterations in urinary excretion induced by manipulations of urinary pH (10).

Self-regulation of nicotine intake, or nicotine titration as it is sometimes called, provides critical evidence of the central role of nicotine in the control of smoking behaviour. Such studies, together with a host of less direct studies involving brand switching and measures of puffing patterns, often without blood nicotine measures or use of other biochemical markers, were a major interest throughout the 1970s and 1980s. Up to 1990, data from brand-switching and the more direct studies were consistent with the view that down-regulation to avoid aversive effects of exceeding usual peak blood nicotine levels was very precise, whereas upregulation was only partial with smokers able to tolerate and adjust quite rapidly to a drop to about two-thirds of their usual peak levels (see 2 for review).

This position has been blown apart by two recent reports on the effect of nicotine administration on ad lib smoking, one involving 14-hr nicotine infusions by Benowitz and Jacob (5), the other, by Foulds et al. (6), using transdermal nicotine patches. Both showed that down-regulation of nicotine intake from smoking was far from precise and that subjects tolerated blood nicotine levels substantially higher than their usual levels and those of placebo control conditions.

In this paper we review the Foulds et al. (6) data, together with data from a variety of other kinds of study and find consistency for a nicotine boost hypothesis. Namely, that the blood nicotine boost per cigarette is the controlling factor, and that its maintenance by intake regulation appears essential for maximal nicotine reinforcement.

Nicotine intake is regulated to obtain a constant boost per cigarette

Whether nicotine intake is regulated by smokers is not in doubt. This is well established by studies of the kind outlined above. Where uncertainties arise is with the precision of nicotine intake regulation. To clarify this and explain discrepancies it is necessary to consider whether it is the peak level, the trough level, or indeed the blood nicotine boost (from trough before, to peak just after, a cigarette) that smokers are most motivated to keep constant. One problem is that so few studies include the two venepunctures, before and after a cigarette, that are necessary to measure the nicotine boost. One exception is the Foulds et al. (6) double blind, placebo controlled, crossover study of the effect of transdermal nicotine patches on ad libitum cigarette smoking over the course of one week. Despite significant partial downregulation of both smoke intake (measured by drop in CO) and nicotine intake from smoking, the smokers tolerated substantially higher peak and trough blood nicotine levels to maintain their usual nicotine boost, averaging about 10 ng/ml per cigarette. Figure 1, based on data from the Foulds et al study (6), illustrates the lack of precision of both peak and trough regulation which are "sacrificed" to enable the mean 10 ng/ml nicotine boost per cigarette to be kept constant. Since these subjects were clearly not suffering from nicotine withdrawal, motivation to maintain the boosts was probably based on a desire for positive reinforcement (conditioned and/or pharmacological). Once the primacy of the nicotine boost per cigarette is recognised, apparent discrepancies between studies can be explained.

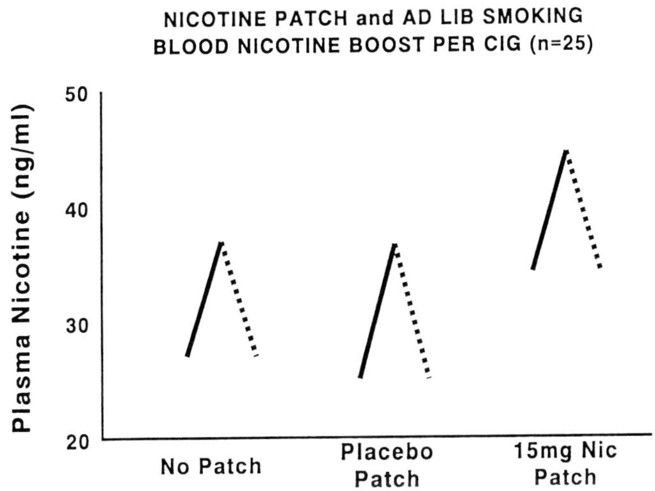

Figure 1. Blood nicotine boost per cigarette (means in 25 subjects) during ad libitum smoking with and without a nicotine skin patch. The peaks as shown are mainly schematic. The only components based on real data are the vertical dimensions of the solid lines.

What is special about a venous blood nicotine boost of 10 ng/ml within 10 min ?

Due to rapid absorption and distribution of nicotine, the blood nicotine profile of the smoker who inhales is characterised by a series of sharp peaks coinciding with each cigarette. The blood nicotine boost per cigarette, during steady-state, averages about 10 ng/ml (11) and reflects the average 'smoking dose' of 1.0 mg/cigarette of systemically available nicotine (12,13). This 1.0 mg dose is taken typically as about ten 0.1 mg puff-by-puff doses absorbed sufficiently rapidly through the lungs to deliver high nicotine boli in arterial blood, which give transient concentrations of 100-150 ng/ml within 10 seconds of each inhaled puff (14,15). Peak levels in venous blood average about 30-40 ng/ml, but it is the size of the boost per cigarette, rather than peak level just after the cigarette, that would correlate best with the concentration in arterial boli. A very small venous blood boost would indicate insignificant arterial blood boli. Since some 50% of smokers smoke at least once every hour and start within 30 min of waking, most of their nicotinic receptors may be desensitised to agonist effects at venous blood levels. In those with prominent boosts and sharp peaks after each cigarette, the boli may activate some non-desensitised receptors, whereas those with less prominent peaks and boli may be motivated more by the need to maintain a desensitised state. Since the size of the boost in venous blood reflects concentrations in arterial boli, it is not surprising that it has priority as the controlling factor of intake regulation.

Further evidence for the reinforcing effects of a 10 ng/ml blood nicotine boost within 10 min comes from the use of other tobacco products. The blood nicotine levels of regular users of oral 'wet' snuff and chewing tobacco are similar to those of regular smokers, and standard doses are capable of raising blood nicotine levels by 10 ng/ml in 10 min (16). Dry nasal snuff, widely used in the 18th Century, is still available in the UK. The rate of nicotine absorption from nasal snuff is very rapid. In 11 regular users, the average blood nicotine boost of 12.6 ng/ml to a peak of 36.1 ng/ml, after their usual dose taken during steady state, is similar to the data from cigarette smokers (11). Although lacking post-inhalation boli, the trough and peak levels and blood nicotine boosts of smokeless tobacco users and smokers are strikingly similar. It is unlikely that such similarities arising from such different products and rituals is merely coincidental, and suggests that the rituals are determined by the nicotine concentrations they produce.

The only example of natural tobacco use without nicotine boosts of 10 ng/ml in 10 min is non-inhaled smoking of a pipe or large cigars. Although so-called 'primary' cigar and pipe smokers, who have never smoked cigarettes regularly, appear to be satisfied by slow buccal absorption of nicotine, without inhalation, there is abundant evidence that cigarette smokers who switch to a pipe or cigars continue to inhale like cigarette smokers, and have blood nicotine and CO levels as high or higher than those of typical cigarette smokers. It seems that once experienced regularly, inhalation nicotine boli are difficult to forego.

The most common acute subjective effect of nicotine is a brief drug-like feeling of slight dizziness, lightheadedness or slight euphoria. However, it is subject to acute tolerance and seldom lasts longer than 5 to 10 min. Regular smokers tend to experience it only during the first cigarette of the day or after a few hours of abstinence, as do smokeless tobacco users after a period of prior abstinence. It occurs after nicotine administration by IV injection or infusion, subcutaneous injection and nasal spray or droplet provided the doses are sufficient to raise blood nicotine levels by 10 ng/ml within 10 min, which standard dose nicotine patches and 2 mg gum are unable to provide.

Nicotine intake by slow absorption

Due to the interaction between acute tolerance and rate of absorption, effects which are subject to acute tolerance will not occur if absorption is so slow that desensitisation or acute tolerance develop before threshold levels for the effect are reached. Although nicotine gum and patches are effective at alleviating the negative effects of cigarette withdrawal, they do not as a rule produce positive subjective effects of satisfaction due probably to the slow rate of nicotine absorption. Indeed, far from being satisfying and reinforcing, when nicotine gum dosage is increased to produce blood nicotine concentrations up to peak smoking levels of 35-40 ng/ml the effects are aversive rather than rewarding, with dose-related decreases in liking (2). This may explain why blood nicotine levels during clinical gum use for cigarette withdrawal average around 10-15 ng/ml, ie roughly one third of peak smoking levels. This appears to be the upper limit for reinforcement by the slow delivery provided by the gum. Indeed, such subjects seem reluctant to take more gum, despite continuing to experience cigarette withdrawal symptoms and being pressed by their therapist to take higher doses. What appears to be lacking are the positively reinforcing effects of more rapidly absorbed nicotine boosts. Despite these limitations, nicotine gum has several significant effects as a negative reinforcer to alleviate withdrawal. It reverses impairment of concentration and performance on some cognitive tasks following cigarette withdrawal, and reverses abstinence induced EEG changes (17).

Conclusions

New evidence has been outlined that smokers regulate their nicotine intake to obtain a constant dose per cigarette. Even in the absence of post-inhalation arterial blood boli, a venous blood nicotine boost of 10 ng/ml within 10 min produces positive effects, possibly through agonist action at non-desensitised receptors. The effect of slower absorption via nicotine gum or patches appears limited to the negative reinforcement of withdrawal relief, possibly through maintenance of desensitisation. Evidence of a qualitative difference in the clinical response to nicotine nasal spray and patches used during treatment for smoking cessation provides support for this view (18, 19).

References

1. Russell MAH, Wilson C, Patel UA, Feyerabend C, Cole PV. Plasma nicotine levels after smoking cigarettes with high, medium and low nicotine yields. Br Med J 1975; 2:414-416.
2. Russell MAH. Nicotine intake and its control over smoking. In: "Nicotine Psychopharmacology: Molecular, Cellular and Behavioural Aspects". (eds) Wonnacott S, Russell MAH, Stolerman IP. Oxford: Oxford University Press, 1990: 374-418.
3. Lucchesi BR, Schuster CR, Emley GS. The role of nicotine as a determinant of cigarette smoking frequency in man with observations of certain cardiovascular effects associated with the tobacco alkaloid. Clin Pharmacol & Therap 1967; 8:789-796.
4. Ebert RV, McNabb ME, Snow SL. Effect of nicotine chewing gum on plasma nicotine levels of cigarette smokers. Clin Pharmacol & Therap 1984; 35:495-598.
5. Benowitz NL, Jacob P. Intravenous nicotine replacement suppresses nicotine intake from cigarette smoking J Pharm Exp Ther 1990; 254:1000-1005.
6. Foulds J, Stapleton J, Feyerabend C, Vesey C, Jarvis M, Russell MAH. Effect of transdermal nicotine patches on cigarette smoking: a double-blind crossover study Psychopharmacology 1992; 106:421-427.
7. Stolerman IP, Fink R, Jarvik ME. Influencing cigarette smoking with nicotine antagonists. Psychopharmacology 1973; 28:247-259.
8. Nemeth-Coslett R, Henningfield JE, O'Keefe MK, Griffiths RR. Effects of mecamylamine on human cigarette smoking and subjective ratings. Psychopharmacology 1986; 88:420-425.
9. Pomerleau CS, Pomerleau OF, Majchrzak MJ. Mecamylamine pretreatment increases subsequent nicotine self-administration as indicated by changes in plasma nicotine level. Psychopharmacology 1987; 91:391-393.10.
10. Benowitz NL, Jacob P. Nicotine renal excretion rate influences nicotine intake during cigarette smoking. J. Pharm & Exper Ther 1985; 234: 153-155.
11. Russell MAH, Jarvis MJ, Devitt G, Feyerabend C. Nicotine intake by snuff users. Brit Med J 1981; 283:814-817.
12. Benowitz NL, Jacob P. Daily intake of nicotine during cigarette smoking. Clin Pharmacol Ther 1984; 35:499-504.
13. Feyerabend C, Ings RMJ, Russell MAH. Nicotine pharmacokinetics and its application to intake from smoking. Br J Clin Pharmac 1985; 19:239-247.
14. Isaac PF, Rand MJ. Blood levels of nicotine and physiological effects after inhalation of tobacco smoke Eur J Pharm 1986; 132:337-338.
15. Rand MJ. Neuropharmacological effects of nicotine in relation to cholinergic mechanisms. In Nordberg A, Fuxe K, Holmstedt B, Sundwall A. (eds) Progress in Brain Research 1989; 79:3-11. Elsevier, Amsterdam.
16. Holm H, Jarvis, MJ, Russell MAH, Feyerabend C. Nicotine intake and dependence in Swedish snuff takers. Psychopharmacology 1992; 108: 512-518.
17. Pickworth WB, Herning RI, Henningfield JE. Electroencephalographic effects of nicotine chewing gum in humans. Pharm Biochem & Behav 1986; 25:879-882.
18. Sutherland G, Stapleton JA, Russell MAH, Jarvis MJ, Hajek P, Belcher M, Feyerabend C. Randomised controlled trial of nasal nicotine spray in smoking cessation. Lancet 1992; 340: 324-329.
19. Stapleton JA, Russell MAH, Feyerabend C et al. Dose effects and predictors of outcome in a randomised trial of transdermal nicotine patches in general practice. Addiction 1995, In Press.

Effects of Nicotine on
Biological Systems II
Advances in Pharmacological Sciences

THERE IS MORE TO SMOKING THAN THE CNS EFFECTS OF NICOTINE

Jed E. Rose and Frederique M. Behm

Duke University and V.A. Medical Center
508 Fulton St.
Durham, N.C. 27705, U.S.A.

Summary: Cigarette smoking presents a constellation of sensory and behavioral cues along with the CNS effects of nicotine. Several studies have shown that craving for cigarettes is difficult to alleviate with alternative nicotine delivery systems that do not provide these cues. Conversely, presentation of the sensory/behavioral cues of smoking using de-nicotinized cigarettes, effectively alleviates craving. The effects of de-nicotinized cigarette smoke and nicotine skin patches were compared in a sample of 20 cigarette smokers, who smoked *ad lib* for 2 hours while wearing nicotine or placebo skin patches that had been applied the previous night. Craving was substantially alleviated by the de-nicotinized cigarettes but only slightly affected by the nicotine patch. These results suggest that smoking cessation treatment strategies should provide replacements for the sensory/behavioral components of smoking.

Introduction

There is considerable evidence that nicotine is a key reinforcer for smoking behavior (1-3). That is, if no nicotine were provided in cigarettes, subjects would probably not smoke, or would smoke to a much lesser extent. Evidence in support of this hypothesis has been summarized in several publications, such as the Surgeon General's Report on Nicotine Addiction (4). The rewarding effects of nicotine have been demonstrated by findings that: a) when allowed to control the nicotine content of each puff, smokers select higher nicotine concentrations in smoke as a function of cigarette deprivation (5) or when administered a centrally acting nicotine antagonist (6). b) Nicotine provides subjective and behavioral effects which are desired, including pleasurable relaxation, anxiety relief, reduced hunger, decreased irritability, reduction in craving for cigarettes, increased alertness and improved cognitive functioning (7, 8). These effects involve to some extent relief of withdrawal symptoms, which can be conceptualized as negative reinforcement by nicotine; however, some effects (e.g. improved working memory) can be demonstrated in nondependent animal subjects (9, 10). c) Under some conditions, smokers, as well as naive animal subjects, will self-administer i.v. nicotine (11, 12).

These observations lead to a tendency on the part of many to equate cigarette smoking with the obtaining of desired CNS effects of nicotine. However, it is clear that cigarette smoking is far more rewarding that alternative methods of delivering that do not include the act of smoking. The majority of smokers who are actively trying to quit eventually relapse even when provided with nicotine gum, patches or nasal spray (13-16). The potent reinforcing effects of cigarettes may only partially be accounted for by the rewarding effects of pure nicotine. While the rapid pharmacokinetics of the lung-to-brain nicotine delivery when smoking a cigarette may be important (3), the fact remains that other rapid forms of nicotine delivery, such as nicotine nasal spray (16-18) or i.v. bolus injections (12, 19) are usually not as desirable to smokers as smoking a cigarette. The pleasurable sensations accompanying smoking, including the act of puffing and inhaling, the taste and aroma, and especially the respiratory tract sensations accompanying each puff of smoke, provide a rich set of conditioned reinforcing cues. These cues, which may have become reinforcing indirectly through their association with the pharmacologic effects of nicotine, are important to many smokers (20-22). An analogous process is seen with eating behavior; the caloric nourishment ultimately received from a meal may be reinforcing, but greater pleasure is obviously associated with the behavior of eating a meal *vs* i.v. feeding. In addition to providing added reinforcement for smoking cigarettes, the sensory/behavioral cues may also provide a chain of stimulus-response associations that facilitate the conditioning of smoking behavior in response to certain situational cues that provoke relapse.

Thus, while nicotine may be necessary in the maintenance of smoking, the CNS pharmacologic effects alone are probably not *sufficient* to satisfy smokers. We have reported several studies over the last ten years in which the sensory and behavioral components of smoking, most notably the respiratory tract sensations elicited by inhaled smoke, have been critical in providing an immediate satisfaction of smokers' craving for cigarettes. For example, local anesthesia of the upper airways significantly attenuated the ability of a controlled doses of nicotine inhaled in cigarette smoke to reduce craving (23). Conversely, providing sensory cues without significant doses of nicotine reduces craving for cigarettes. Previously we have reported that citric acid aerosol (24), a low-nicotine extract of tobacco condensate (25), capsaicin-laden low tar and nicotine cigarettes (26), and inhalation of pepper vapor (27) all reduce smokers' craving for cigarettes. These manipulations provide some of the airway sensory feedback to which smokers are accustomed, but their taste and aroma differ from conventional cigarette smoke. Therefore, we conducted a study to assess the effects of smoking de-nicotinized cigarettes, which have very similar sensory attributes as typical commercial cigarettes. We directly compared the short-term effects of these cigarette with those of nicotine skin patches to help quantify the relative role of different components in smoking reward.

Materials and Methods

Subjects: Twenty smokers (12 males, 8 females), having a mean age of 45.2 (SD=12.93), who smoked an average of 26.8 cigarettes/day (SD=10.52) of nonmenthol brands having a mean nicotine delivery (by FTC method) of 0.98 mg (SD=0.265), participated in the study. Subjects had smoked for an average of 26.8 yr (SD=13.01) and scored an average of 6.9 (SD=1.8) on the Fagerström Test for Nicotine Dependence.

Cigarettes: De-nicotinized cigarettes were obtained from Phillip Morris, Inc. These cigarettes contain tobacco from which the nicotine has been selectively extracted by high-pressure CO_2, and have a taste and tar delivery similar to nicotine-containing brands of cigarette. When smoked by FTC criteria, these cigarettes deliver approximately 9 mg tar and less than 0.1 mg nicotine. CO delivery is approximately 10 mg, comparable to subjects' habitual brands. Hasenfratz et al. (28) measured smoking behavior and nicotine intake after smoking de-nicotinized *vs* nicotine-containing cigarettes and found that plasma nicotine levels increased less than 2 ng/ml after smoking the de-nicotinized cigarette. In studies using these cigarettes, we have verified that they produce no increase in heart rate, a sensitive index of nicotine delivery in deprived smokers.

Procedure. Subjects participated in five sessions, conducted after overnight abstinence from smoking (confirmed by expired air CO measurement). The conditions, presented in counterbalanced order using a digram-balanced Latin square design, were as follows:

1. **Nicotine** skin patch, **No Smoking**
2. **Nicotine** skin patch, **De-nicotinized cigarettes**
3. **Placebo** skin patch, **No smoking**
4. **Placebo** skin patch, **De-nicotinized cigarettes**
5. **Placebo** skin patch, **High nicotine cigarettes**

Condition 5 presented cigarettes of the same nicotine delivery as subjects' habitual brands; cigarettes were masked to keep their appearance similar. Subjects were allowed to smoke *ad lib* on smoking days during the two hour session. Nicotine skin patches (Ciba-Geigy, Inc., Summit NJ) delivering 21 mg/24 hr or placebo patches were applied the night before in order to allow plasma nicotine levels to attain a steady state of 10-15 ng/ml prior to the beginning of the session.

Smoking withdrawal symptom assessment: We used a modified Shiffman-Jarvik questionnaire (29) which we have employed previously in several laboratory studies and clinical trials involving a total of approximately 600 subjects, e.g. (30). It has been sensitive in detecting effects of cigarette deprivation and pharmacologic treatments. The items are rated from 1 (not at all) to 7 (extremely), and comprise six subscales: *craving* (craved a cigarette, would have liked a cigarette, thought of cigarettes, missed a cigarette, had urges to smoke and, scored oppositely, would have refused a cigarette); *negative affect* (tense, irritable, and scored oppositely, calm, content); *arousal* (wide awake, able to concentrate, and, scored oppositely, sleepy); *somatic anxiety* (fluttery feelings in chest, heart beat faster than usual, hands shake); *respiratory symptoms* (cough, sore throat), *gastrointestinal symptoms* (nausea, upset stomach, dizziness), *appetite* (hungrier than usual, craved sweets, craved salty foods); *habit* (missed something to do with the hands, missed having something in the mouth).

Smoking behavior measurement: Subjects' expired air CO concentrations were measured using a handheld CO monitor (Vitalograph, Lenexa, KS). Expired air CO concentration were calculated by subtracting the background (ambient) CO from the peak CO reading. The number of puffs taken from each cigarette and total number of cigarettes were also be counted.

Results

The main variable of interest was reported craving for cigarettes; at the beginning of the session there was a slight, yet statistically significant reduction in craving for cigarettes in the nicotine, as opposed to placebo patch conditions (craving was 4.6, SD=1.48 in the nicotine patch conditions *vs* 5.0, SD=1.32 in the placebo patch conditions, $p<.05$). However, by the end of the session the effect of the nicotine patch was not significant (see Fig. 1). In contrast, smoking the de-nicotinized cigarettes caused a substantial decrease in craving (end of session craving of 4.0, SD=1.49 in the de-nicotinized cigarette conditions *vs* 4.8, SD=1.34 in the no cigarette conditions, $p<.005$), which was equivalent to that produced by smoking the high nicotine cigarettes.

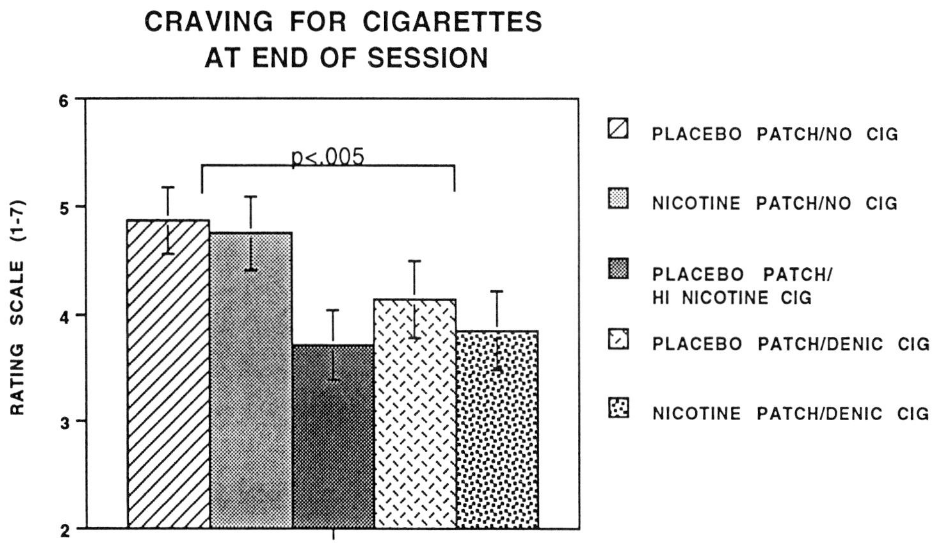

Figure 1. Mean (± s.e.m.) reported craving for cigarettes after two hours in the five experimental conditions.

Interestingly, subjects smoked the de-nicotinized cigarettes more intensively in the placebo patch condition than the nicotine patch condition. The CO boost during the session was 7.5 ppm (SD=9.29) in the nicotine patch condition vs 14.2 ppm (SD=11.55) in the placebo patch condition ($p<.0001$). The CO boost in the high nicotine cigarette condition was 7.4 ppm

(SD=7.66).

Heart rate, a sensitive index of nicotine absorption, was an average of 5.7 bpm higher in the nicotine patch conditions than the placebo patch conditions (71.8 bpm, SD=10.52 *vs* 66.1 bpm, SD=8.95, $p<.0001$) showing significant nicotine delivery and pharmacologic effect. The high nicotine cigarette condition also increased heart rate by the end of the session 4.4 bpm relative to the placebo patch, no smoking condition (change from beginning of the session was -2.5 bpm, SD=5.99 *vs* -6.9 bpm, SD=5.47, $p<.05$). Smoking the de-nicotinized cigarettes had no measurable effect on heart rate, consistent with their low nicotine delivery.

Discussion

Our results showed that craving for cigarettes was relieved by de-nicotinized cigarettes to a similar extent as by cigarettes delivering an average of nearly 1 mg. This supports the view that sensory/behavioral factors are important in the immediate reduction in craving accompanying cigarette smoking. While the de-nicotinized cigarettes have a small amount of nicotine, compensatory smoking and resultant increase in nicotine delivery from these cigarettes cannot account for this effect, because the CO boost and number of cigarettes smoked in the nicotine patch, de-nicotinized smoke condition was comparable to that in the high nicotine cigarette condition. The effect of providing the sensory and behavioral cues of smoking with these cigarettes reduced craving well beyond the small effect of the nicotine patch on craving. The effect of the patch on craving, while detectable at the beginning of the session, was minimal. Previous clinical studies using nicotine skin patches have shown that while craving seems to diminish more quickly over weeks in subjects receiving nicotine patch treatment, the initial effects on craving are slight (30-32). There is evidence that more rapid nicotine administration, as with nicotine nasal sprays, may relieve craving more effectively than slower forms such as nicotine chewing gum and skin patches (16-18). However, these nicotine delivery systems also present significant sensory cues, such as throat irritation which may affect craving (33).

While craving was only slightly affected by the nicotine patch, the fact that subjects smoked the de-nicotinized cigarettes more intensively in the placebo patch condition shows an important effect of systemic nicotine levels in modulating smoking behavior. Thus, the CNS effects of nicotine may potentiate the efficacy of smoking related sensorimotor cues in satisfying craving for cigarettes. However, nicotine alone, in the dose tested here, is not sufficient to alleviate craving in the absence of smoking behavior. Thus, in smoking cessation treatment, it may be wise to develop substitutes for the sensory and behavioral components of cigarette smoking as well as administering nicotine replacement. Not only would this lead to more effective relief of craving early in treatment, but repeated presentation of smoking related cues without nicotine reinforcement may help extinguish the conditioned reinforcing value of these stimuli and discourage relapse (22).

***Acknowledgments**: This work was supported by the Medical Research Service of the Department of Veterans Affairs.*

References

1. Jarvik ME, Gritz ER. Nicotine and tobacco. In: Jarvik ME, ed. Psychopharmacology in the Practice of Medicine. New York: Appleton-Century-Crofts, 1977: 483-495.
2. Russell MAH. Nicotine replacement: The role of blood nicotine levels, their rate of change, and nicotine tolerance. In: Pomerleau OF, Pomerleau CS, editors. Nicotine Replacement: A Critical Evaluation. New York: Alan R. Liss, Inc., 1988: 63-94. vol 261.
3. Russell MAH, Feyerabend C. Cigarette smoking: A dependence on high-nicotine boli. Drug Metabol. Rev. 1978;8:29-57.
4. U.S.D.H.H.S. The Health Consequences of Smoking: Nicotine Addiction.Rockville, MD: Office on Smoking and Health, 1988
5. Rose JE, Jarvik ME, Ananda S. Nicotine preference increases after cigarette deprivation. Pharmacol Biochem Behav 1984;20:55-58.
6. Rose JE, Sampson A, Levin ED, Henningfield JE. Mecamylamine increases nicotine preference and attenuates nicotine discrimination. Pharmacol Biochem Behav 1989;32:933-938.
7. Jarvik ME. Beneficial effects of nicotine. Br J Addict 1991;86: 571-576.
8. Warburton DM. Nicotine and the smoker. Rev. Environ. Health 1985;5:343-390.
9. Levin ED, Briggs SJ, Christopher NC, Rose JE. Chronic nicotinic stimulation and blockade effects on working memory. Behav Pharmacol 1993;4:179-182.
10. Levin ED, Rose JE. Nicotinic and muscarinic interactions and choice accuracy in the radial-arm maze. Br Res Bull 1991;27:125-128.
11. Corrigall WA, Coen KM. Nicotine maintains robust self-administration in rats on a limited-access schedule. Psychopharmacology 1989;99:473-478.
12. Henningfield J, Miyasato K, Jasinski D. Cigarette smokers self-administer intravenous nicotine. Pharmacol Biochem Behav 1983;19:887-890.
13. Fagerström KO. Efficacy of nicotine chewing gum: A review. In: Pomerleau OF, Pomerleau CS, ed. Nicotine Replacement: A Critical Evaluation. New York: Alan R. Liss, Inc., 1988: 109-128.
14. Hughes JR, Glaser M. Transdermal nicotine for smoking cessation. J Health Behav Educ Promotion 1993;17(2):25-31.
15. Palmer KJ, Buckley MM, Faulds D. Transdermal nicotine: A review of its pharmacodynamic and pharmacokinetic properties, and therapeutic efficacy as an aid to smoking cessation. Drugs 1992;44(3):498-529.
16. Sutherland G, Stapleton JA, Russell MAH, et al. Randomised controlled trial of nasal nicotine spray in smoking cessation. Lancet 1992;340:324-329.
17. Perkins KA, Grobe JE, Stiller RL, Fonte C, Goettler JE. Nasal spray nicotine replacement suppresses cigarette smoking desire and behavior. Clin Pharmacol Ther 1992;52:627-634.

18. Pomerleau OF, Flessland KA, Pomerleau CS, Hariharan M. Controlled dosing of nicotine via and *i*ntranasal *a*erosol *d*elivery *d*evice (INADD). Psychopharmacology 1992;108:519-526.
19. Henningfield J, Goldberg S. Control of behavior by intravenous nicotine injections in human subjects. Pharmacol Biochem Behav 1983;19:1021-1026.
20. Rose JE. The role of upper airway stimulation in smoking. In: Pomerleau OF, Pomerleau CS, editors. Nicotine Replacement: A Critical Evaluation. New York: Alan R. Liss, Inc., 1988: 95-106.
21. Rose JE, Behm FM, E.D. L. The role of nicotine dose and sensory cues in the regulation of smoke intake. Pharmacol Biochem Behav 1993;44:891-900.
22. Rose JE, Levin ED. Inter-relationships between conditioned and primary reinforcement in the maintenance of cigarette smoking. Br J Addict 1991;86: 605-610.
23. Rose JE, Tashkin DP, Ertle A, Zinser MC, Lafer R. Sensory blockade of smoking satisfaction. Pharmacol. Biochem. Behav 1985;23:289-293.
24. Rose JE, Hickman CS. Citric acid aerosol as a potential smoking cessation aid. Chest 1987;92:1005-1008.
25. Rose JE, Behm FM. Refined cigarette smoke as a method for reducing nicotine intake. Pharmacol Biochem Behav 1987;28:305-310.
26. Behm FM, Rose JE. Reducing craving for cigarettes while decreasing smoke intake using capsaicin-enhanced low-tar cigarettes. Exp Clin Psychopharmacol; in press.
27. Rose JE, Behm FM. Inhalation of vapor from black pepper extract reduces smoking withdrawal symptoms. Drug Alc Depend 1994;34:225-229.
28. Hasenfratz M, Baldinger B, Battig K. Nicotine or tar titration in cigarette smoking behavior? Psychopharmacology 1993;112:253-258.
29. Shiffman SM, Jarvik ME. Smoking withdrawal symptoms in two weeks of abstinence. Psychopharmacology 1976;50:35-39.
30. Westman EC, Levin ED, Rose JE. The nicotine patch in smoking cessation: A randomized trial with telephone counseling. Arch Int Med 1993;153:1917-1923.
31. Abelin T, Buehler A, Muller P, Vesanen K, Imhof PR. Controlled trial of transdermal nicotine patch in tobacco withdrawal. Lancet 1989;(January 7):7-10.
32. Tonnesen P, Norregaard J, Simonsen K, Sawe U. A double-blind trial of a 16-hour transdermal nicotine patch in smoking cessation. N Engl J Med 1991;325:311-315.
33. Schneider NG. Nicotine nasal spray. Health Values 1994;18(3):10-14.

Effects of Nicotine on
Biological Systems II
Advances in Pharmacological Sciences

PHARMACOLOGICAL DETERMINANTS OF CIGARETTE SMOKING

Jack E. Henningfield, Leslie M. Schuh, and Stephen J. Heishman

Addiction Research Center, National Institute on Drug Abuse and
The Johns Hopkins University School of Medicine, Baltimore, MD , USA

Summary: Tobacco has been used ritualistically to produce intoxication and hallucinations. Most of today's cigarette smokers, however, use tobacco daily, at intervals of less than one hour during waking hours. Since early in the twentieth century, it has been understood that this behavior is driven largely by nicotine's pharmacological actions and has little to do with the taste of nicotine. Nicotine's role in tobacco use parallels that of morphine in the use of opium derivatives. It produces CNS reinforcing effects, tolerance and physiological dependence, pharmacological effects smokers find valuable, and pleasurable effects induced by the central actions of nicotine. Nicotine and other smoke constituents also provide sensory stimuli which become conditioned reinforcers and further strengthen tobacco self-administration.

Introduction

It has been widely understood by pharmacologists and physiologists since the turn of the century that nicotine has effects in the nervous system and that some of these effects are important determinants of why people use tobacco (1,2). Clinicians, too, have recognized for years the powerful hold that cigarettes can exert over behavior as they have seen 50% or more of people suffering a major cigarette-caused illness such as a heart attack revert to smoking within days or weeks after leaving the hospital (2). There may be semantic and legalistic debates on the best term to describe such extreme nicotine seeking behavior, but there can be little debate about the phenomenon of highly driven use of a psychoactive substance known by most users to be harmful to their health.

Drug dependence and abuse are generally driven by multiple factors including pharmacological and nonpharmacological determinants. It is this multilayered convergence that makes certain drugs appeal to a broader range of persons than others and makes abstinence so difficult. This review focuses on the central nervous system (CNS) pharmacology of nicotine, the factor that clearly differentiates use of addictive drugs from other behaviors such as sky diving and candy or saccharine consumption.

Pharmacological determinants of tobacco use include the nicotine dose and dosage form, clearance rate, reinforcing effects, tolerance and physiological dependence, interactions with other drugs, beneficial effects including weight control and mood alteration, and the sensory effects of nicotine itself. A complete understanding of the forces driving people to use tobacco must include an analysis of the many factors involved, in the same way that such an analysis has been helpful in understanding addictions to other drugs such as heroin, cocaine, and alcohol.

Pharmacological Effects of Nicotine

Most drugs, including addictive drugs, have diverse pharmacologic effects which vary as a function of dose, rate of administration, time since the last administration, and characteristics of the individual. Clearly, addictive drugs control behavior through some mechanisms common across drugs as well as other mechanisms specific to certain drug classes. Even within a drug class, the behavior of different people may be controlled to varying degrees by the different pharmacologic actions produced by a single drug. For example, if intoxication is rewarding to an individual, this effect is readily obtained by rapidly consuming high ethanol doses. If keeping alert during protracted and boring tasks is important, amphetamine and nicotine would be more effective and alcohol would be avoided. As this example suggests, intoxication is not necessary for drug dependence (2,3).

It would be naive to attribute the powerful reinforcing effects of cigarettes primarily to their taste or the sensory effects of nicotine in the mouth and nose. Nicotine produces diverse effects throughout the body including modifying CNS functions such as EEG, cerebral metabolism, and neurotransmitter levels and altering cellular structures that lead to increased expression of nicotinic receptors (2). The primary mechanism of nicotine's reinforcing effects appears to be the activation of CNS nicotinic receptors which release dopamine; modulation of other transmitter systems such as catecholamines are also undoubtedly important and may vary within and across individuals.

The Cigarette as an Optimal Dosage Form (To Produce Dependence)

The cigarette is an optimal dosage form to establish dependence. It readily enables the user to self-administer nicotine doses repeatedly and rapidly and incorporates a rich confluence of sensory elements. In essence, the cigarette does for tobacco what crack did for cocaine. It provides a rapid onset of effects, producing an arterial bolus that may be ten times more concentrated than venous blood. The cigarette also provides rich sensory stimuli that Wikler, Goldberg, and many others have found so important in building strong patterns of drug seeking behavior (4,5). Smoke inhalation mimics a rapid intravenous injection and exposes the heart, brain, and fetus to high nicotine concentrations that dissipate within a few minutes (6). The smoking route is preferred by users because it is easy to administer the desired unit dose and it avoids the disease risks of injections. However, smoking carries risks of its own, namely exposure of the lungs to carbon monoxide and many other toxicologically significant pyrolysis products. Rose's work elucidates the role of sensory stimuli in cigarette smoking (7, cf. also Rose, this volume) and is comparable in some respects to that of O'Brien and colleagues with heroin and cocaine (8).

Controlled Behavior

Cigarette smoking is a controlled behavior. Patterns of cigarette use are orderly, regular, and determined by the same kinds of factors determining other drug use. Manipulations such as dose, nicotine preloading, deprivation, and central blockade of nicotine's effects with mecamylamine produce expected changes in tobacco intake. The pioneering work of Lucchesi, Schuster, and Emley (9) demonstrated that even nondiscriminable doses of nicotine infused into smokers could reduce their tobacco intake. Likewise, nicotine preloading and antagonism of the central, but not peripheral, effects of nicotine by mecamylamine decrease tobacco intake (10) and smoking deprivation increases tobacco intake (2). Cigarette smoking may also increase when some psychological need arises, e.g., in stressful or cognitively demanding situations (2).

Nicotine Reinforcement

Unlike food, water and sex, there appear to be no biologically reinforcing or basic needs fulfilled by smoking other than nicotine's ability to trigger brain reward mechanisms. Nor does the taste of cigarettes substitute for intrinsically reinforcing tastes such as sweet or salty. Use is sustained without food deprivation, although food deprivation does appear to enhance nicotine's reinforcing effects as is true with cocaine, alcohol and opioids (4). Self-administration also varies as a function of factors such as dose and the schedule under which nicotine is available (11).

Conditioning processes are maximally effective when the drug effect is discrete, paired with readily discriminable stimuli, and follows a specific behavior within a few seconds (12). The paradigm is optimal for smoking to become powerfully conditioned because each cigarette provides approximately 10 units (puffs) of nicotine reinforcement, each carried by a sensorally sating cloud of smoke and delivered to the brain in seconds. Tolerance and physical dependence potentiate the process by establishing a new motivational state in the individual. Thus, smoking is reinforced both by the direct positively reinforcing actions of nicotine on the brain and by the necessity of continued nicotine administration to prevent withdrawal symptoms. Goldberg and colleagues (13) established high rates of lever pressing for nicotine injections in squirrel monkeys using small intermittently available doses, and pairing them with environmental stimuli. The cigarette quite naturally provides the transportable equivalent of this apparatus and program.

At least three kinds of nicotine effects then can contribute to the dependence development: 1) nicotine delivery produces reinforcing effects mediated by reward systems in the brain; 2) tolerance and physical dependence are produced such that nicotine abstinence is accompanied by adverse effects; and 3) at least those dependent on nicotine may derive useful effects on mood, appetite, and cognition. These effects are not mutually exclusive and they often interact.

Tolerance

Nicotine produces tolerance such that people self administer much more nicotine over time than they could have physiologically tolerated when they began smoking. Nicotine tolerance appears to be substantially acquired during youth as smokers progress from a few to many cigarettes to obtain the same effects (14). Many smokers continue to escalate their daily dose for eight years or more (2). Administering nicotine to a tobacco deprived smoker can substantially increase heart rate and euphoria and decrease knee reflex strength. With repeated doses, heart rate stabilizes at a level intermediate to that produced by the first dose and that occurring when nicotine deprived, subjective effects are minimal, and the knee reflex may appear normal (2). Tolerance to a variety of the behavioral, physiologic, and subjective effects of nicotine have been studied (2). There are several physiologic mechanisms of nicotine tolerance including decreased responsiveness to the drug at the site of drug action and increased nicotinic receptor number and some degree of increased metabolism (2,15). Cigarette smokers lose a substantial degree of tolerance while sleeping each day and regain it upon resumption of smoking. A single nicotine exposure induces short-lived tolerance to its psychoactive, cardiovascular and other effects, and is thus referred to as tachyphylaxis (16).

Physiological Dependence

The cellular and neurological adaptations that produce tolerance also lead to physiological dependence. After at least several weeks of repeated nicotine exposure (3), physiological dependence develops, and, when deprived for more than a few hours, smokers report withdrawal symptoms that are generally opposite to the effects initially produced by nicotine (2). In actuality, compared to nonsmokers, the cigarette smoker has an elevated pulse (5-7 bpm), elevated circulating catecholamines, lower body weight (5-8 lbs), and increased nicotine receptor binding sites (15).

The nicotine withdrawal syndrome has been described in detail (3). Onset begins within a few hours of the last cigarette; symptoms include increased tendency to smoke, craving, anxiety, irritability, appetite and decreased cognitive capabilities and heart rate (17,18). Altered brain electrical potentials and hormonal output are primarily opposite to those produced by acute nicotine administration, and decrements in evoked electrical potentials of the brain indicate impaired information processing (2).

The severity of the syndrome and specific prominent symptoms vary across individuals but it is generally unpleasant and frequently intolerable (2), with most patients relapsing before the syndrome begins to subside (11,18). The time course varies across individuals and responses, but, generally, withdrawal symptoms peak within a few days, then begin to subside in several weeks. Severity is related to prior nicotine intake (2). It is precipitated by an approximately 50-60% reduction in smoking (19). Similarly, symptoms are relieved by readministering nicotine. The degree of relief appears related to the nicotine dose (2) with significant relief of physical withdrawal signs provided by 60% replacement of plasma nicotine and greater relief by higher levels of replacement (17). Laboratory studies of the nicotine withdrawal syndrome (2,17) show the time course of measures of brain and cognitive function are parallel, and symptoms are reversed (in dose related fashion) by nicotine polacrilex gum. The psychological components of withdrawal can be attenuated by sensory stimuli. It is not clear that either pharmacologic or sensory stimuli alone could completely eliminate all components of the withdrawal syndrome.

Beneficial Effects of Nicotine

Even those who primarily study effects sometimes referred to as "beneficial" or "psychological resources" understand that these effects have little if anything to do with nicotine's taste, but much to do with pharmacological actions in the brain and endocrine system. The putative beneficial effects of nicotine could strengthen addiction to tobacco and they are factors in acquisition, dependence, and relapse to smoking. They include mood control, reduction of stress and boredom, weight and appetite control, preventing withdrawal in dependent persons, and therapeutic applications in psychiatric disorders and ulcerative colitis.

Some researchers have concluded that nicotine improves performance over a wide range of human abilities and circumstances (20, 21, cf. also Hindmarch, this volume; Warburton, this volume). However, because the performance effects of any drug are determined by many variables, most notably drug dose, nature of the ability being tested, environmental conditions surrounding the performance, and the individual's degree of tolerance, it is too simplistic to characterize a drug as either enhancing *or* impairing performance. Rather, research should be aimed at delineating the conditions under which drugs may enhance *and* impair behavior.

In a recent literature review on the effects of nicotine and smoking on human performance, Heishman et al. (22) found differential effects of nicotine depending on whether individuals were in a state of nicotine deprivation. Results from studies testing abstinent smokers support the conclusion that nicotine deprivation functions to maintain smoking in nicotine-dependent persons, in part, because nicotine can reverse withdrawal-induced deficits in several areas of performance. However, in nonabstinent smokers and nonsmokers, nicotine was found to enhance only finger tapping and motor responses in tests of attention; cognitive functioning was not reliably enhanced. It is unlikely that these limited performance-enhancing effects of nicotine play an important role in the acquisition phase of cigarette smoking.

Several questions exist about the importance of these beneficial effects. Do they explain the high incidence of graduation to dependence in children aware of risks? Indeed, nearly half of young smokers in America graduate from a few cigarettes per day to dependence, with 70% regretting have started smoking and 50% having tried and failed to quit by age 17 (23). Should "psychological resource" benefits be regulated as "drug effects" or are they trivial? Do the benefits of nicotine help keep nicotine from beating the addiction rap any more than do the beneficial effects of opioids, stimulants or sedatives keep those drugs from being so labeled and controlled? Finally, are the "psychological resource" benefits strong enough to warrant approval by regulatory agencies or to risk 30-40% rate of premature death (24)?

Concluding Comments

In the end, healthy debates will occur over the relative contribution of each factor conspiring to make cigarette smoking one of the deadliest forms of addiction. Some of the debate will be productive, leading to a better understanding of the differences and similarities among tobacco dependent persons and across drug addictions that may help us more effectively treat people dependent upon nicotine and other drugs. For example, if there are subpopulations that suffer cognitive disorders that are restored by the administration of nicotinic agonists it will be important to understand how to identify those people and how best to treat them. If some people do need nicotine as diabetics need insulin, then it would seem absurd to tell them that the source of their medication is as a highly toxic delivery vehicle. Would they not be better served to receive their drug in forms tested and evaluated in the same way as other medications?

Semantic and legalistic issues also exist, such as whether the term "addiction" is an appropriate label for nicotine dependence or, for that matter, for cocaine dependence. This debate is often supported by observations that the pharmacological effects of nicotine are not identical to those of heroin, alcohol, or cocaine. In fact, nicotine is often posed against a hypothetical "hard drug" that produces the intoxication of ethanol, withdrawal of ethanol and heroin, euphoriant effects of cocaine, and strength or reinforcement in animals and naive humans of cocaine-like drugs. One may select any one factor to argue that one of these drugs is more or less addicting than the others. However, this exercise makes it clear that addiction severity and society's level of concern about drug use are best evaluated by assessing several variables. Table 1 presents such an evaluation for five psychoactive drugs. It is evident that tobacco addiction is a serious cause for concern

Table 1. Ranking of Addiction: Factors of Concern (Ratings by authors; see 2,25-28 for supporting data.) [a,b,c]

1. Dependence Among Users:
nicotine > heroin > cocaine > alcohol > caffeine

2. Difficulty Achieving Abstinence:
(alcohol = cocaine = heroin = nicotine) > caffeine

3. Tolerance:
(alcohol = heroin = nicotine) > cocaine > caffeine

4. Physical Withdrawal Severity:
alcohol > heroin > nicotine > cocaine > caffeine

5. Societal Impact:
serious effects due to secondary deaths (nicotine), accidents (alcohol) or crime (heroin, cocaine); no substantial impact for caffeine

6. Deaths:
nicotine > alcohol > (cocaine = heroin) > caffeine

7. Importance in User's Daily Life:
(alcohol = cocaine = heroin = nicotine) > caffeine

8. Intoxication:
alcohol > (cocaine = heroin) > caffeine > nicotine

9. Animal Self-Administration:
cocaine > heroin > (alcohol = nicotine) > caffeine

10. Liking by Non Drug Abusers:
cocaine > (alcohol = caffeine = heroin = nicotine)

11. Prevalence:
caffeine > nicotine > alcohol > (cocaine = heroin)

[a] "=" indicates the authors cannot reliably differentiate between the drugs on this parameter.
[b] Note that drug effects differ depending on factors such as dose and dosage form, and the rankings presented above are meant as general guidelines.
[c] Caffeine is not recognized as a dependence producing drug by the American Psychiatric Association or the World Health Organization and is included in this table to provide contrast with the dependence producing drugs.

Some challenges for future research in nicotine dependence are to delineate the relative contribution of physical dependence, peripheral effects, and those effects that at least some users believe to be beneficial in their daily lives. A better understanding of these variables may lead to treatment strategies that more effectively enable the millions of people who make a choice to quit smoking (cf. West, this volume) to fulfill that goal. There is no longer any serious challenge to the conclusion that cigarette smoking is substantially determined by the pharmacologic actions of nicotine and that nicotine dependence and withdrawal are diagnosable and treatable clinical entities.

References

1. Langley JN. On the reaction of cells and of nerve endings to certain poisons, chiefly as regards the reaction of striated muscle to nicotine and to curari. J Physiol 1905;33:374-413.
2. U.S. Department of Health and Human Services (US DHHS). The health consequences of smoking: Nicotine addiction. A report of the Surgeon General. U.S. Department of Health and Human Services, Public Health Service, Office on Smoking and Health. DHHS Publication No. (CDC) 88-8406, Washington DC: U.S. Government Printing Office, 1988.
3. American Psychiatric Association. Diagnostic and Statistical Manual of Mental Disorders, 4th edition. Washington, D.C.: American Psychiatric Association, 1994.
4. Swedberg MDB, Henningfield JE, Goldberg SR. Nicotine dependency: Animal studies. In: Nicotine Psychopharmacology: Molecular, Cellular, and Behavioral Aspects. Wonnacott S, Russell MAH, Stolerman IP editors. Oxford: Oxford University Press, 1990: 38-70.
5. Henningfield JE, Schuh LM, Jarvik ME. Pathophysiology of tobacco dependence. In: Pychopharmacology: The fourth generation of progress. Bloom FE, Kupfer DJ editors. New York: Raven Press, In press.
6. Henningfield JE, Stapleton JM, Benowitz NL, Grayson RF, London ED. Higher levels of nicotine in arterial than in venous blood after cigarette smoking. Drug Alcohol Depend 1993;33:23-9.
7. Rose JE, Levin ED. Inter-relationships between conditioned and primary reinforcement in the maintenance of cigarette smoking. Br J Addiction 1991; 86:605-9.
8. O'Brien CP, Childress AR, McLellan AT, Ehrman R. A learning model of addiction. In: Addictive States. O'Brien CP, Jaffee JH editors. New York: Raven Press, 1992:157-77.
9. Lucchesi BR, Schuster CR, Emley AB. The role of nicotine as a determinant of cigarette smoking frequency in man with observations of certain cardiovascular effects associated with the tobacco alkaloid. Clin Pharmacol Ther 1967:789-96.
10. Stolerman IP, Goldfarb T, Fink R, Jarvik ME. Influencing cigarette smoking with nicotinic antagonists. Psychopharmacol, 1973;28:247-59.
11. Henningfield JE, Keenan RM. Nicotine delivery kinetics and abuse liability. J Consult Clin Psychol 1993; 61:1-8.
12. Thompson TI, Schuster CR. Behavioral Pharmacology. Englewood Cliffs, NJ: Prentice-Hall, 1968.
13. Goldberg SR, Spealman RD, Risner ME, Henningfield JE. Control of behavior by intravenous nicotine injections in laboratory animals. Pharmacol Biochem Behav, 1983; 19:1011-20.
14. United States Department of Health and Human Services (US DHHS), Public Health Services. Preventing Tobacco Use Among Young People: A Report of the Surgeon General. Atlanta, GA: US DHHS, Centers for Disease Control and Prevention, National Centers for Chronic Disease Prevention and Health Promotion, Office on Smoking and Health, 1994.
15. Benwell MM, Balfour DJ, Anderson JM. Evidence that tobacco smoking increases the density of (-)-[^{3}H] nicotine binding sites in human brain. J Neurochem 1988; 50:1243-47.
16. Srivastava ED, Russell MAH, Feyerabend C, Masterson JG, Rhodes J. Sensitivity and tolerance to nicotine in smokers and nonsmokers. Psychopharmacol, 1991;105:63-8.
17. Pickworth WB, Herning RI, Henningfield JE. Spontaneous EEG changes during tobacco abstinence and nicotine substitution. J Pharmacol Exp Ther 1989;251:976-82.
18. Hughes JR, Hatsukami DK. The nicotine withdrawal syndrome: a brief review and update. Int J Smoking Cessation 1992; 1:21-6.
19. West RJ, Russell MAH, Jarvis MJ, Feyerabend C. Does switching to an ultra-low nicotine cigarette induce nicotine withdrawal effects? Psychopharmacol 1984;84:120-3.
20. Robinson JH, Pritchard WS. The role of nicotine in tobacco use. Psychopharmacol 1992;108:397-407.
21. Warburton DM. Nicotine as a cognitive enhancer. Prog Neuro-Psychopharmacol Biol Psychiatry 1992;16:181-191.
22. Heishman SJ, Taylor RC, Henningfield JE. Nicotine and smoking: A review of effects on human performance. Exp Clin Psychopharm; in press.
23. George H. Gallup International Institute: Teen-age attitudes and behavior concerning tobacco. Report of the Findings. Princeton, New Jersey, September 1992.

24. Peto R, Lopez AD, Boreham J, Thun M, Heath C Jr. Mortality from tobacco in developed countries: Indirect estimation from national vital statistics. Lancet 1992;339:1268-78.
25. Anthony JC, Warner LA, Kessler RC. Comparative epidemiology of dependence on tobacco, alcohol, controlled substances, and inhalants: Basic findings from the National Comorbidity Survey. Exp Clin Psychopharm, In press.
26. Gilman AG, Rall TW, Taylor P. Goodman and Gilman's The Pharmacological Basis of Therapeutics. New York: Pergamon Press, 1990:180-181, 545-549.
27. Lasagna L, von Festinger JM, Beecher HK. Drug-induced mood changes in man. 1. Observations on healthy subjects, chronically ill patients, and "post-addicts" J Am Med Assoc, 1955;157:1006-20.
28. Fischman MW, Mello NK, editors. Testing for abuse liability of drugs in humans. NIDA Research Monograph No. 92. Washington, D.C.: Government Printing Office, 1989. (DHHS publication no. (ADM) 89-1613).

Effects of Nicotine on
Biological Systems II
Advances in Pharmacological Sciences

THE FUNCTIONAL CONCEPTION OF NICOTINE USE

David M. Warburton

Psychology, Reading University, Reading, RG6 2AL, UK

Summary: A large body of evidence shows that nicotine improves mood and enhances information processing capacity. These results can be interpreted in terms of a functional model of nicotine use. In this model, nicotine use can be seen as purposive, a behavior to obtain psychological resources.

Introduction

This paper will discuss the use of nicotine as a purposive activity, which represents an attempt of the person to cope with their environment. In this respect, the use of nicotine has features in common with other appetitive behaviors, such as eating, drinking and sex (1). This approach enables a functional analysis of the use of nicotine (2-4). It considers questions about why nicotine use occurs and examines the conditions that initiate and sustain nicotine use. The answers avoid explanations of nicotine use in terms of "need" and the "reward or reinforcement" derived from "satisfying" this need.

Instead, nicotine use is seen as an integration of sets of factors such as situation, psychological state and sensory factors as well as plasma nicotine level. None of these factors is considered more important than the others. The primary factor in nicotine use is the functional interrelationship between the factors. The causes of use can be both exogenous and endogenous. A smoker may have more than one motive for nicotine use and use may have different functions in different situations. In addition, nicotine use can be the outcome of not only the characteristics of the situation, but the personality of the individual.

The Addiction Conception of Nicotine Use

The functional view differs from the addiction conception of nicotine use (5). In this view, a "need for nicotine" has developed through exposure which manifests itself as a withdrawal syndrome during abstinence. Nicotine use is "reinforcing", because it "satisfies the nicotine need". In this model, the reasons for nicotine use are exclusively the result of an endogenous state.

The most salient feature of the supposed state is that it interferes with, or in the extreme case removes, the capacity for voluntary behaviour with respect to nicotine. The mechanism for the "nicotine addiction" state is said to be based on nicotine psychopharmacology (5), so that nicotine exposure compels nicotine use irrespective of, or against, the person's 'will'. Recently, there has even been an attempt to estimate the threshold dose for nicotine exposure, which will result in "nicotine addiction" (6).

However, the specification of nicotine use as being non-volitional, because it has a demonstrable effects on the nervous system is fallacious. Nicotine psychopharmacology is not the same as the psychopharmacology of deciding to use, or not to use nicotine voluntarily. Of course, personal experiences have their bases in the psychoactive effects of nicotine, but it cannot be concluded that because nicotine is psychoactive, that 'therefore' the person has no further decision-making capacity.

The view that drugs have inherent behavioural attributes has been widely criticised. For example, Hughes et al (7) pointed out that:- "If behavioral effects are inherent properties, then we should not expect to change with a new setting. As another example, use of the term property falsely encourages the belief that drug dependence arises from the pharmacological properties of drugs. If certain drugs have intrinsic reinforcing properties, then we should expect once used they will always produce dependence." (p 557). It is clear from the literature on adolescent experimentation with drugs that this is not the case.

Early Experiences with Nicotine

The claim that teenagers easily and almost inevitably get hooked if they smoke only a few cigarettes was made by Russell (8) in his paper "The nicotine addiction trap; a 40 year sentence for four cigarettes". His data were obtained from a representative sample of UK adults, who were asked to recall their past smoking experiences from the stage of first experimentation with smoking through to regular smoking (9).

Russell (8) calculated that of those who smoke one cigarette, 71 percent will smoke regularly for 5 years or more. Similarly, of those who smoke more than one cigarette 82 percent become regular smokers, and those who smoke four cigarettes have a 94 percent chance of becoming long-term regular smokers. Of course, it is possible that this figure is exaggerated by adult, non-smokers forgetting that they had tried a cigarette in adolescence or feeling that smoking the odd cigarette many years previously did not really count as having smoked (10). In addition, the retrospective information comes from a time prior to the Report of the Royal College of Physicians (11) which first warned British smokers about the dangers of smoking to health.

Modern information on the exposure issue can be derived from a survey which investigated children's experimentation with smoking (12). The population of the survey consisted of secondary school pupils in the age range 11-15 years who were interviewed three times, at the beginning of the second, third and fourth years of secondary education. While 61 percent tried smoking, only 36 percent had more than one and 14 percent became regular smokers.

A study in New Zealand (13) showed that at 9 years of age, 36 percent of the children reported that they had smoked on at least one occasion and this increased by seven percent a year until 80 percent had tried. However, daily smokers at age 15 years were equally likely to come from the 'infrequent' group and the 'never smoked' group at age 9 years.

Even for those who went on to smoke, it would be wrong to interpret such data in terms of the effect of smoking a single cigarette. It is certainly likely that trying a cigarette for the first time produces cognitive changes (e.g. changes in self-image, changes in beliefs about the effects of smoking) that will increase the probability of trying a second cigarette should the opportunity arise. It seems likely that those factors that increase the chances of trying a first cigarette by some given age also increase the likelihood of smoking again within a given subsequent time period (10). Thus, if you are prone to try a cigarette (because of personality, family, peers, stressors, etc.), then you will be similarly predisposed to smoke another cigarette. In our own study (14), we concluded that the predictors of smoking were not experimentation, but factors which result in problems for the individual, such as emotional or behavioural problems, trouble with the police, a broken home, etc. (13, 15).

Thus, the predictors of smoking were not exposure to nicotine, but factors which result in problems for the individual. Individuals choose to continue to smoke, because it provides nicotine which ameliorates the effects of stressors by its mood effects.

Smoking and Mood

The relaxing effects of smoking are the everyday experience of many smokers (16-21). However, anyone who has experienced a benzodiazepine or alcohol would agree that these effects of nicotine are mild in comparison with the other compounds. Our own studies (4, 22) compared the pleasurable relaxation and pleasurable stimulation from different substances. Tranquilizers and alcohol were significantly more relaxing than tobacco. Similarly, the mood elevating effects are much smaller for tobacco in comparison with alcohol. Using visual analogue scales (21), subjects rated themselves as calmer, more tranquil, more sociable, more friendly, more contented, more relaxed and happier with each puff on a cigarette. Non-nicotine cigarettes produced no mood changes, suggesting that the effects were due to nicotine. Mood changes were correlated with plasma nicotine; the mood changes increased as plasma nicotine increased throughout the cigarette (22). When a second cigarette was given 30 minutes later (i.e. no nicotine deprivation) the pre-smoking mood level was below that achieved at the end of the first cigarette, but it increased during smoking until it was above the level that was achieved at the end of the first cigarette, which is what one would expect if nicotine was responsible for the mood improvement.

When ad lib smoking was allowed prior to the test sessions, improvement was again found. Thus, the results could not have been due to a reversal of nicotine withdrawal. Other studies with minimal deprivation of 30 minutes have shown that nicotine abstinence was not required for the anxiolytic effects (23).

These findings strongly suggest that one of the functions of nicotine is to reduce feelings of anxiety and anger and so ameliorate the effects of stressors. Certainly, people increase their nicotine intake when undergoing stressful testing (24-26). The smoking cessation literature consistently records that smokers report that they miss these positive effects of nicotine when they quit smoking (27).

Nicotine Exposure and Mood State

It could be argued that the mood changes that occur after quitting are due to the results of exposure to nicotine and that the positive mood effects with smoking or nicotine replacement are the result of a reversal of "withdrawal effect", the consequence of changes in the brain as a result of chronic exposure to nicotine.

We have looked at the question of whether exposure to nicotine induces negative mood states by analyzing the mental health of smokers, ex-smokers and current smokers, including cigar and pipe smokers. We used information which was derived from The Health and Lifestyle Survey (28), which contains data from a sample of 9,003 individuals aged 18 and over and living in Great Britain. For our analysis, the respondents were classified into one of four groups, depending on their smoking habits. Current smokers consisted of those people who were smoking at least one cigarette per day at the time of the survey. People who had smoked at least one cigarette per day for a minimum of six months but no longer did so, were classified as ex-smokers. Never smokers were people who had not smoked at least one cigarette per day for a minimum of six months. Percentages were calculated for the three smoking groups and age standardised.

When we examined the incidence of depression or nervous illness, current smokers were more likely ($p < 0.001$) to have suffered than never smokers, but ex-smokers were intermediate between never smokers and current smokers and were not significantly different from either group ($p>0.1$). However, there were clear sex differences; for women respondents, current and ex-smokers were more likely to have experienced depression or nervous illness than never smokers ($p<0.001$ for both groups), while for men ex-smokers and never smokers differed significantly from current smokers ($p<0.001$).

When the ex-smokers were subdivided into number of years since stopping, there was no clear pattern; those who had quit 20 or more years or less than 10 years had a significantly greater incidence of depression or nervous illness ($p<0.05$ and $p<0.01$ respectively), while those who had stopped between 10 and 19 years did not differ significantly in incidence from never smokers, $p>0.1$. A comparison of ex-smokers, by years of quitting with current smokers showed those who had quit 20 or more years or less than 10 years had no smaller incidence than a current smokers ($p>0.1$), while those who had stopped between 10 and 19 years had significantly less depression or nervous illness ($p<0.01$). From these age-adjusted data, there is no simple relationship between duration of abstinence and mental health.

A separate analysis was carried out on the smokers. They were classified on the basis of consumption into less than 10 cigarettes per day, between 10 and 20 cigarettes and more than 20 cigarettes per day. There were no significant differences between the three groups in terms of the incidence of depression or nervous illness, i.e no relation with the level of daily nicotine exposure.

Of particular interest was the comparison between cigar, pipe smokers and never smokers. There was no significant difference between the groups in the incidence of depression or nervous illness, 7.4 percent, 7.5 percent and 8.8 percent respectively. It is known that pipe and cigar smoking can produce pharmacological doses of nicotine even if the smoke is not inhaled, because of the larger dose entering the mouth and the alkalinity of the smoke (29). In addition, many pipe and cigar smokers have switched from cigarettes and do inhale so that they obtain cigarette-like plasma nicotine levels (29). Once again, there seems to be no relation between the incidence of illness and the level of daily nicotine exposure.

A second method of examining the question was to examine sleep habits. Sleeping less than seven hours was a characteristic of those who were "under strain", but not associated with "worrying" or "depression" in the population (30). A similar set of analyses were performed on the population. Current smokers were more likely ($p < 0.001$) to sleep less than never smokers, but ex-smokers were not significantly different from current smokers ($p>0.1$). When the ex-smokers were subdivided into number of years since stopping, there were no differences between those who had quit 20 or more years, those who had stopped between 10 and 19 years or those who had quit less than 10 years before ($p>0.1$), which did not differ significantly in incidence from never smokers, $p>0.1$. A comparison of ex-smokers, by years of quitting with current smokers showed those who had quit 20 or more years were slightly more likely to sleep less than seven hours ($p<0.1$), in spite of age-adjustment.

A separate analysis was carried out on the smokers. They were classified on the basis of consumption into less than 10 cigarettes per day, between 10 and 20 cigarettes and more than 20 cigarettes per day. There were no significant differences between the first two groups, but those who smoked more than 20 per day had a higher incidence of sleeping less than seven hours. Once again, the comparison between cigar, pipe smokers, cigarette smokers and never smokers was interesting. There was no significant difference between the groups in the

incidence of sleeping less than seven hours, 32.8 percent, 24.4 percent, 35.6 percent and 33.3 percent respectively. Once again, there seems to be no relation between the level of daily nicotine exposure and sleeping less than seven hours.

In summary, there is little support for the idea that exposure to nicotine has adverse effects on mental health.

Conclusions

This paper has given evidence in support of a functional conception of nicotine use. It considers questions about how smoking occurs and examines the conditions that initiate and sustain nicotine use. It shows that use is not determined by exposure which manifests itself as a withdrawal syndrome during abstinence, so that nicotine is only used because it "satisfies the nicotine need".

References

1. Warburton DM. The appetite for nicotine. In: Appetite: Neural and Behavioural Bases. Legg CR, Booth DA, editors. Oxford: Oxford University Press, in press, 1994.
2. Warburton DM. The functions of smoking. In: Tobacco Smoking and Nicotine: A Neurobiological Approach. Martin WR, Van Loon GR, Iwamoto ET, Davis DL, editors. New York: Plenum Press, 1987: 51-61.
3. Warburton DM. The functional use of nicotine. In: Nicotine, Smoking and the Low Tar Programme. Wald N, Froggatt P, editors. Oxford University Press, 1988: 182-99.
4. Warburton DM, Revell A, Walters AC. Nicotine as a resource. In: The Pharmacology of Nicotine. Rand MJ, Thurau K, editors. Oxford: IRL Press, 1988: 359-73.
5. U.S. Department of Health and Human Services; The health consequences of smoking: nicotine addiction. A report of the Surgeon-General. DHHS Publication Number (CDC) 88-8406, U.S. Department of Health and Human Services, Office of the Assistant Secretary for Health, Office on Smoking and Health, Rockville, MD., 1988.
6. Benowitz NL, Henningfield JE. Establishing a nicotine threshold for addiction. N Engl J Med 1994; 331:123-125.
7. Hughes JR, Higgins ST, Bickel WK. Behavioral "properties" of drugs. Psychopharmacology 1988; 96:557.
8. Russell MAH. The nicotine addiction trap: a 40-year sentence for four cigarettes. Brit J Addict 1990; 85:293-300.
9. McKennell AC, Thomas RK. Adult's and adolescent' smoking habits and attitudes. London: Her Majesty's Stationary Office, 1967.
10. Sutton SR. Is taking up smoking a reasoned action? Brit J Addict 1992; 87:21-24.
11. Royal College of Physicians. Smoking and Health. London: Royal College of Physicians, 1962.
12. Goddard E. Smoking among Secondary School Children in England in 1988. London: Her Majesty's Stationary Office, 1990.
13. Stanton WR, Silva PA, Oei TPS. The origins and development of an addictive behaviour: a longitudinal study of smoking. The Dunedin Multidisciplinary Health and Development Research Unit, Dunedin, New Zealand, 1989.
14. Warburton DM, Revell AD, Thompson DH. Smokers of the future. Brit J Addict 1991; 86:621-25.

15. Sieber MF, Angst J. Alcohol, tobacco and cannabis: 12-year longitudinal associations with antecedent social context and personality. Drug Alcohol Depend 1990; 25:281-92.
16. McKennell AC. Smoking motivation factors. Brit J Soc Clin Psychol 1970; 9:8-22.
17. McKennell AC. A comparison of two smoking typologies. Research Paper No. 12. Tobacco Research Council, London, 1973.
18. Tomkins SS. Psychological model of smoking behaviour. American J of Public Health 1966; 56:17-20.
19. Tomkins SS. A modified model of smoking behaviour. In: Smoking, Health and Behaviour. Borgatta E, Evans R, editors. Chicago: Aldine, 1968: 165-6.
20. Russell MAH, Peto J, Patel UA. The classification of smoking by factorial structure of motives. J Roy Stat Soc A, 1974; 137:313-46.
21. Warburton DM, Wesnes K. Individual differences in smoking and attentional performance. In: Smoking Behaviour. Thornton RE, editor. Edinburgh, Churchill-Livingstone, 1978: 19-43.
22. Warburton DM. The puzzle of nicotine use. In: The Psychopharmacology of the Addiction. Lader M, editor. Oxford University Press, 1988b: 27-49.
23. Pomerleau OF, Turk DC, Fertig JB. The effects of cigarette smoking on pain and anxiety. Addict Behav 1984; 9:265-71.
24. Ashton H, Stepney R. Smoking: Psychology and Pharmacology. Cambridge: Cambridge University Press, 1982.
25. Cherek DR. Effects of smoking different doses of nicotine on human aggressive behavior. Psychopharmacology 1981; 75:339-45.
26. Wesnes K, Warburton DM, Revell A. Work and stress as motives for smoking. In: Smoking and the Lung. Cumming G, Bonsignore G, editors. New York: Plenum, 1984: 233-48.
27. Shiffman SM, Jarvik ME. Smoking withdrawal symptoms in two weeks of abstinence. Psychopharmacology 1976; 50:35-9.
28. Cox BD, and eleven others. The health and lifestyle survey. The Health Promotion Research Trust, 1987.
29. Russell MAH. In: Tobacco Smoking and Nicotine: a Neurobiological Approach. Martin WR, Van Loon GR, Iwamoto ET, Davis DL, editors. New York: Plenum Press, 1987: 25-50.
30. Blaxter M. Health and Lifestyle. London: Routledge, 1990.

Effects of Nicotine on
Biological Systems II
Advances in Pharmacological Sciences

NICOTINE IS ADDICTIVE: THE ISSUE OF FREE CHOICE

R West

St George's Hospital Medical School, Cranmer Terrace,
London SW17 0RE, UK

Summary: The issue of whether nicotine should be regarded as addictive is important because it relates to the way that social, legal and medical institutions treat smokers and those with smoking-related diseases. The core feature of addiction is compulsion or inability to exercise free choice. Lack of free choice is evidenced by inconsistency between sincerely expressed desire and actual behaviour and by behaviour that violates the principle of 'bounded rationality'. On both of these criteria, smoking behaviour appears to operate under compulsion rather than free choice. For example, a majority of smokers shows evidence of sincerely wishing they could stop but they fail to do so when they try. Also, factors associated with success of attempts to stop appear to be unrelated to the perceived costs and benefits of smoking but are related to dependence mechanisms.

Introduction

The debate over whether nicotine is addictive has largely been resolved in the academic and clinical community and the issues have been considered at length in many places (1,2,3). Nevertheless, there are still dissenting voices (4). In an issue as complex as this, new findings can still be brought to bear and the significance of existing findings can always be re-evaluated. This paper attempts to do this by focusing on competing views of smoking as an addiction on the one hand and a free choice on the other. From the perspective of this paper, the addictive quality of smoking is defined in terms of compulsive use or failure to exercise free choice.

Evidence for the role of free choice in behaviour

Exercise of free choice may be evidenced by two main kinds of information. First of all, there should be a concordance between the sincerely expressed wish of an individual to behave in a particular way, and his or her actual behaviour given the opportunity. Secondly, the behaviour should conform to the principles of 'bounded rationality' that operate in normal human decision making (5).

Considering first of all the issue of conflict between desire and behaviour, there are some important points to note. The first concerns timing. Clearly if an individual changes his or her mind this need not imply failure to exercise free choice. However, if one observes an individual engaging in an action while *at the same time* sincerely expressing a desire not to, the imputation should be that there is an element of compulsion. Then there is the question of what constitutes a *sincerely* expressed desire. The views people express often vary with the context, and pressure towards conformity may lead individuals to express views to which they do not subscribe. In deciding on sincerity it is prudent to consider how it manifests itself in behaviour and how it varies with context. Finally, a distinction should be drawn between *desire* and a feeling of moral imperative. There are many situations where people feel they 'ought' to behave in a particular way but do not do so because they 'want' to do something else. There is no failure to exercise free choice here; there is simply a conflict between duty and pleasure.

Turning to the second feature of free choice, that it would be expected to follow broadly rational principles, while people all deviate from pure rationality to an extent, there are boundaries beyond which sane individuals rarely if ever stray (5). Thus, in a simple choice between a long and happy life and short painful one few would opt for the latter unless there were some other factor compelling them to do so. Similarly, if offered the simple choice between handing over £20 per week or nothing, few would opt for the former unless compelled to do so. In more complex situations, individuals may make errors of judgement, exhibit biases, or vary in the values that they attach to particular outcomes. However, when behaviours show clear evidence of being self-destructive or counterproductive and there is no suggestion that the individual concerned wishes to harm himself or herself, it is sensible to look for pressures compelling the individual to behave in a manner that he or she would not wish to given a free choice.

The role of free choice in smoking behaviour

Applying these ideas to the smoking situation, population surveys in the UK consistently find that the majority of smokers say they would like to give up. Yet clearly they have not put this desire into effect. For example, the 1992 General Household Survey in the UK found that 67 per cent of smokers reported that they would like to give up altogether (6). Of even greater significance is the fact that heavier smokers were more likely to say they wanted to give up (70 per cent in those smoking 10 or more cigarettes per day compared with 58 per cent among those smoking up to 9 cigarettes per day). Therefore among smokers there is an inverse relationship between expressed desire and actual behaviour. In a large health and lifestyle survey carried out in the UK at the same time, 48 per cent of smokers of 16 or more per day expressed a 'strong' desire to stop compared with 41 per cent of smokers of 15 or fewer per day (7).

As to the sincerity of the expression of desire not to smoke, it is noteworthy that a recent survey in the UK found that 80 per cent of regular smokers had tried to stop in the past and 58% reported having tried more than once (8). If one focuses just on those who do make an attempt to give up, studies consistently show that only a very small minority last more than a week (9). One can go further and examine how far reversion to smoking represented a conscious desire to return to a cherished habit. In fact, very few smokers show any evidence of having simply decided not to continue with abstinence and most still express a desire to give up at some point in the future (8). In fact when relapse occurs, smokers generally do not decide to return to smoking, they decide to have at most several cigarettes and then resume abstinence yet in almost all cases, this leads to full blown relapse.

Taken as a whole the above facts are not consistent with a view of smoking as the result of a free choice. There is clear evidence of a conflict between sincerely expressed desire and actual behaviour.

We now consider how far smoking deviates from the bounded rationality that characterises free choice behaviour. Smokers in the UK now almost all accept that smoking is likely to lead to an early death and serious illness. For example, a recent health and lifestyle survey reported that 90 per cent of smokers believed that smoking increased risk of lung cancer, which was similar to the figure for non-smokers (94%) (8); a slightly lower figure of 84 per cent believed that smoking caused chronic bronchitis (compared with 91 per cent of former smokers); and 78 per cent of smokers believed that smoking caused heart disease (compared with 84 per cent of ex-smokers). A recent survey of student nurses also found little difference between the beliefs of current smokers and ex-smokers in terms of the health risks from smoking, with 80 per cent of smokers believing that smoking caused premature death compared with 87 per cent of ex-smokers (10).

Apart from possible health costs, the material cost of cigarettes is high in the UK, and places a burden particularly on poor families (11). Thus, families of smokers with no adult in work spend on average 12 per cent of their disposable income on cigarettes. This has a substantial effect in exacerbating their hardship (11).

If smoking resulted from a free choice, the benefits of smoking should be perceived as outweighing these acknowledged costs. Smokers certainly do perceive benefits. In varying proportions they report smoking to be pleasurable, relaxing and stimulating (12). Many also see advantages to the reduction in body weight conferred by smoking.

This raises the question as to whether smokers may simply weigh the short-term benefits of smoking against the costs as one might with, for example, eating high-fat foods or rock climbing. In fact there are two issues here. The first concerns whether smokers are correct in believing that smoking confers benefits; the second concerns whether, whatever the accuracy of their beliefs, their cost-benefit analysis is reflected in their behaviour.

The issue of whether smoking confers genuine benefits has been discussed at length elsewhere (13). In summary: the putative stress relieving properties of smoking are probably (though not definitely) little more than relief of withdrawal symptoms; smoking may aid concentration although the evidence is inconsistent on this matter; smoking does help keep weight down in most people. As to whether smoking provides pleasure, the issue is not so much whether this is true (because it seems reasonable on this matter to take smokers' self-reports at face value), but whether the pleasure represents relief from withdrawal discomfort or a simple positive hedonic value.

Even if smokers were wholly incorrect in believing that smoking conferred benefits, they would still be exercising free choice if their perception was that the benefits outweighed the costs. To stop people smoking would simply involve convincing them that the benefits were an illusion. Violation of free choice could only be considered to occur if the perception of the costs and benefits bore no clear relationship with actual smoking behaviour. The primary test would be an examination of whether likelihood of persisting in an attempt to give up smoking was related to prior beliefs about the benefits and hazards of smoking. Many studies have been carried out looking at factors associated with making an attempt to quit and factors associated with the success of a given attempt. In general it is clear that making an attempt to quit is quite well described by the theory of planned behaviour which is a modified version of the cost-benefit model (14). Under this theory, intentional behaviour is an analysis of the likely costs and benefits of the behaviour in question (labelled 'attitude to the behaviour') plus concerns about the views of other people (labelled 'perceived norm') plus a belief that one will be able to put the behaviour into effect (labelled 'perceived control'). Studies have shown that these factors are associated with the desire to give up smoking (15).

However, the key issue concerns the likelihood of success of any attempt. Here the results are different. Applying the theory of planned behaviour, one study found that perceived behavioural control emerged as the main variable discriminating smokers from non-smokers both in a general population sample and a sample of pregnant women (16). Focusing specifically on smoking cessation, there is no evidence that success in giving up is linked with perceived costs and benefits.

A recent trial of a 5-HT antagonist as an aid to smoking cessation provided an opportunity to examine relationships between smoker characteristics and ability to maintain abstinence. In the event the drug under test proved no more effective than a placebo so that data from 97 subjects who attempted abstinence were used irrespective of the drug condition to which they had been assigned. The sample consisted of smokers attending a smokers clinic using a group-treatment format similar to that developed at the Maudsley (17). The sample, methods and findings are described in detail elsewhere (18) and are only summarised here. There were 39 men and 58 women. Out of 97 smokers attempting abstinence, 67 managed complete abstinence for the four weeks of treatment, validated by saliva cotinine, while 30 failed to do so. The results showed that smokers' reports of how much they would miss the pleasure of smoking using a simple 10-point rating were unrelated to likelihood of remaining abstinent for the four weeks of treatment. Nor was abstinence related to how much the smokers said they wanted to give up at this attempt using a 10-point rating, nor their rating of their current health status on a 4-point scale (see Figure 1). This conflicts with the view that smokers who failed to manage abstinence were simply deciding that they were prepared to accept whatever costs may be incurred in smoking because of the pleasures or gains to be had.

Success in maintaining abstinence was related, however, to how confident they felt that would be able to give up (rated on a 10-point scale), how difficult they thought it would be to give up (rated on a 10-point scale), their saliva cotinine concentration measured two weeks rior to attempting to give up smoking (ng/ml), their prior cigarette consumption, how long they went before lighting up each day when smoking, to what extent they felt they were going through a stressful time (10-point scale), and how dizzy or light-headed they felt when smoking the first cigarette of the day (rated on a 10-point scale) (see Figure 2). Note that all the measures were taken before they had attempted to give up smoking. Differences between the abstainers and non-abstainers were significant at $p<0.05$ by t-test.

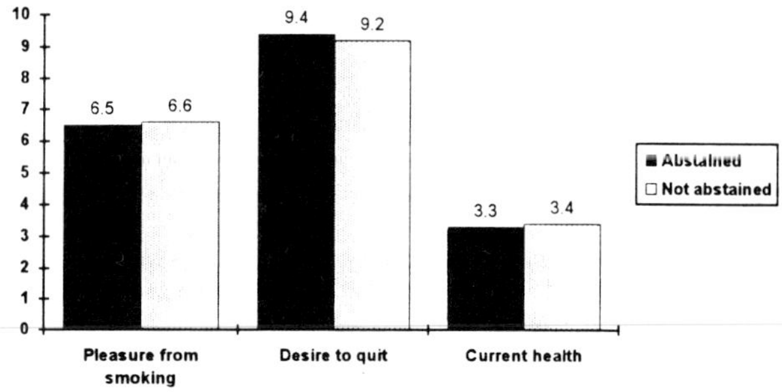

Figure 1. Similarities between those who abstained with help of group treatment for four weeks and those who did not in St George's smoking trial

The relationship between dizziness after the first cigarette of the day and subsequent ability to sustain abstinence is interesting. It suggests that smokers with greater chronic tolerance to this particular nicotine effect found it more difficult to abstain.

Examining the litany of factors that have been found at some time to be associated with success of an attempt to give up smoking, the pattern is generally one of lower chances of success being linked with higher prior levels of smoking, greater dysthymia, more severe withdrawal discomfort, lower psychological endurance, and less social support (19-24). Therefore, while decision making processes are apparently brought to bear in making attempts to stop smoking, factors that would be expected to be related to dependence are important in determining the success of these attempts.

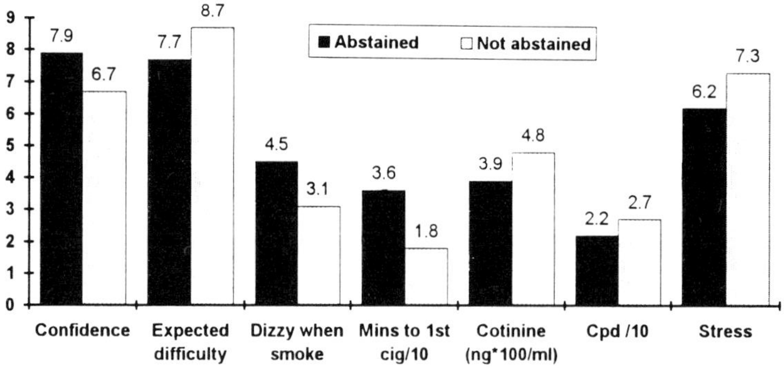

Figure 2. Significant differences between those who abstained with help of group treatment for four weeks and those who did not in St George's smoking trial

Conclusions

The current view of addiction is that its core feature is compulsive use. Compulsive use involves a failure to exercise free choice. Thus to say smoking is addictive is to say that a substantial proportion of smokers are smoking against their will; they are no longer able to exercise free choice in this aspect of their lifestyle. If this is true, it has important legal and social implications. The great majority of researchers working in the field believe it to be true but there are still a few who remain unconvinced and so the debate continues.

This paper has proposed that two features of a given behaviour would constitute evidence against the view that it resulted from the exercise of free choice. The first would be continuation of the behaviour despite sincere expressions of a desire to stop. The second would be a significant deviation from the bounded rationality that underpins human decision making. Evidence reviewed in this paper indicates that both of these features apply to smoking and therefore conflicts with the hypothesis that smoking behaviour in the main results from a free choice.

References

1. Hughes JR. Smoking is a drug dependence: a reply to Robinson and Pritchard. Psychopharm 1993;113:282.
2. USDHHS. The Health Consequences of Smoking: Nicotine Addiction.Rockville, MD: 1988
3. West R. Nicotine addiction: a re-analysis of the arguments. Psychopharm 108;408-410.
4. Robinson JH, Pritchard W. The role of nicotine in tobacco use. Psychopharm 1992;108:397-407.
5. Eiser JR, Van der Plight J. Attitudes and Decisions.London: Routledge, 1988
6. Thomas M, Goddard E, Hickman M, Hunter P. 1991 General Household Survey.London: HMSO, 1993
7. Cox BD, Huppert FA, Whichelow MJ. The Health and Lifestyle Survey: Seven Years on. Aldershot: Dartmouth, 1993
8. Health Education Authority. Health and Lifestyle Survey.London: HMSO, In Press.
9. Hughes JR, Gulliver SB, Fenwick JW, et al. Smoking cessation among self-quitters. Health Psychology 1992;11:331-334.
10. West R, Hargreaves M. Factors associated with smoking in student nurses. Psychology and Health In Press;.
11. Marsh A, McKay S. Poor Smokers.Oxford: Policy Studies Institute, 1994
12. Shiffman S. Assessing smoking patterns and motives. J Cons Clin Psychol 1993;61:732-742.
13. West R. Beneficial effects of nicotine: fact or fiction? Addiction 1993;88:589.
14. Ajzen I. From intention to actions: a theory of planned behaviour. In: Kuhl J, Beckmann J, ed. Action-control: From Cognition to Behavior. Heidelberg: Springer, 1985: 11-39.
15. Marsh A. Smoking and illness: what smokers really believe. Health Trends 1985;17:7-12.
16. Godin G, Valois P, Lepage L, Desharnais R. Predictors of smoking behaviour: an application of Ajzen's theory of planned behaviour. Br J Addiction 1992;87:1335-1343.
17. West R, Hajek P. Factors associated with outcome of an attempt to give up smoking. Manuscript in preparation.
18. Horwitz MB, Hindi-Alexander M, Wagner T. Psychosocial mediators of abstinence, relapse and continued smoking: a one-year follow-up of a minimal intervention. Add Behav 1985;10:29-39.
19. Burns BH. Chronic Chest disease, personality and success in stopping cigarette smoking. Br J Prev Soc Med 1969;23:23-27.

20. Glass A. Blue mood, blackened lungs: depression and smoking. J Am Med Assoc 1990;264:1583-1584.
21. Gritz ER, Carr CR, Marcus AC. Unaided smoking cessation: great American Smokeout and New Year's Day quitters. J Psychosocial Oncology 1989;6:217-234.
22. West R, Hajek P, Belcher M. Severity of withdrawal symptoms as a predictor of outcome of an attempt to quit smoking. Psychol Med 1989;19:981-985.
23. Hajek P. Individual differences in difficulty quitting smoking. Addiction 1991;86:555-558.
24. Hanson BS, Isacsson SO, Janzon L. Social support and quitting smoking for good. Is there an association? Add Behav 1990;15:221-230.

Effects of Nicotine on
Biological Systems II
Advances in Pharmacological Sciences

DIFFERENTIATING HABITS AND ADDICTIONS: THE EVIDENCE THAT NICOTINE IS NOT 'ADDICTIVE'

John H. Robinson and Walter S. Pritchard

Psychophysiology Laboratory, R.J. Reynolds Tobacco Company,
P.O. Box 2959, Winston-Salem, NC 27102

Summary: We review the data supporting our position that nicotine and smoking are more accurately labeled as habit than addiction. The case is presented that a meaningful definition of addiction should allow researchers to distinguish the pharmacologic and behavioral effects of substances like nicotine and caffeine from those of addicting drugs such as heroin and cocaine. A key attribute that has differentiated habits from addictions through the decades, and is still valid today, is that addicting drugs result in behavioral intoxication of the user. The pleasurable sensory experience, as well as the positive emotional and cognitive effects that smokers report, are discussed as an alternative explanation as to why people smoke.

The opportunity to contribute to this volume on the topic of nicotine 'addiction' is very timely. Recent events in the United States have again focused national attention on the debate concerning whether tobacco products and the nicotine contained in tobacco products are 'addictive'. Media coverage of U.S. Congressional and Food and Drug Administration hearings has served to promote emotional issues surrounding the debate, while doing little to add to the scientific discussion. This manuscript will demonstrate that the same data used by the nicotine-addiction proponents to support their view that nicotine is addictive, also readily support another view that clearly distinguishes the physiologic, pharmacologic and behavioral effects of substances like nicotine and caffeine from addicting drugs such as alcohol, heroin and cocaine.

In order to achieve an open dialogue in this debate, we ask that the reader put aside the emotional issues involved with smoking and nicotine, including the health risks that have principally been ascribed to the 'tar' fraction of smoke, and focus on a critical and objective evaluation of the data that will be presented.

This manuscript will briefly summarize our previously published challenge (1) to the conclusions in the 1988 Surgeon General's Report (2) that tobacco products, and the nicotine contained within tobacco products, are as addictive as heroin and cocaine. In our previous manuscript, we argued from the viewpoint presented in the 1964 Surgeon General's Report (3) that smoking and nicotine are more accurately labeled as *habit* than addiction, since the effects of nicotine, like caffeine, are fundamentally different from those of addicting drugs.

As expected, our position has generated a significant amount of discussion (4-6) and we welcome this opportunity to continue to discuss the meaning of addiction, why intoxication is critically important to defining an addicting drug, and an alternative explanation as to why people smoke.

The Meaning of Addiction

The determination of whether any substance should be labeled as 'addictive' obviously depends on one's definition. Unfortunately, no universally accepted definition of addiction currently exists. Instead, researchers and policy makers choose the important elements of their particular definition from a wide range of components (3-7).

One common 'definition' offered by the nicotine-addiction proponents seems to be a very simplistic one - any behavior which some people may find difficult to stop. If one accepts this simplistic definition, then smoking would indeed be labeled an 'addiction'. Unfortunately, adopting this definition leads to a wide variety of substances and activities being labeled as addictions, including estrogen-replacement therapy (8), caffeine in coffee (9), and even carrots (10).

Part of our responsibility as scientists is to explore, differentiate, and dissect the underlying 'reasons why'. As behavioral scientists, we find it unfortunate that a definition that should serve as a foundation for conducting meaningful scientific studies apparently cannot distinguish the pharmacological and behavioral effects of crack smoking from coffee drinking, and cocaine from colas.

Instead of adopting a language that recognizes obvious differences in this wide array of behaviors, the addiction proponents choose to anchor all pleasurable, repetitive behaviors at one end of the scale by labeling everything an addiction. This has not always been true, however. When a scientifically meaningful definition of addiction did exist, important distinctions were made between the effects of habituating substances like nicotine in cigarette smoke or caffeine in coffee and addicting drugs such as heroin, cocaine and alcohol.

Smoking and coffee drinking were classified as habits, based on the critical attribute that differentiated all addicting drugs: *Behavioral intoxication of the user*. In response, the nicotine-addiction proponents argue that the concept of 'habituation vs. addiction' is no longer 'recognized' or that these definitions are 'outdated'. Stolerman and Jarvis (6) have suggested that it is time to confine the term habituation "to the archives of history".

Equating vastly differing behaviors under one nebulous term should be unacceptable to all of us as scientists. In contrast, by labeling nicotine and caffeine as habituating we quickly highlight important features that separate these compounds from addicting drugs such as alcohol, heroin and cocaine.

This suggests that instead of being confined to the archives, perhaps it is time to re-establish the concept of habituation, or perhaps develop a new term, as a construct that adds significant meaning to the discussion of addicting drugs.

Intoxication

A critical issue in the debate centers on our argument that, through the decades, behavioral intoxication has been a defining feature of addicting drugs. We use the term 'behavioral intoxication' as described in DSM-IV (7) to mean disturbances of perception, attention, wakefulness, thinking, judgment and interpersonal relations that drastically alter the person's behavior, lifestyle and value system. These disruptions limit the addict's ability to think or reason clearly, work or drive safely or even to perform the simplest of manual tasks normally. Anyone who has experience with a friend or relative with a drug or alcohol problem knows exactly what we are describing, and consequently why intoxication is important to defining an addicting drug.

The addiction proponents respond to this by simply dismissing or ignoring our view, stating that intoxication is not a defining attribute of addicting drugs. However, the published literature seems to be clearly on our side with regard to this issue. As stated earlier, the 1964 Surgeon General's Report (3), quoting the WHO definitions of addiction and habituation, used the clearly defined feature of intoxication to distinguish nicotine and caffeine from addicting drugs. In 1978, WHO defined "psychotropic" (dependence producing or addicting) drugs as those that produced "hallucinations or disturbances in motor function or thinking or behavior or perception or mood" (WHO, Technical Report Series, No. 618, 1978, p 8). In 1984, NIDA defined "psychoactive" (addicting) substances as those that produced "a distortion of the perception of time, space and the location of objects within space" along with disruptions in "physical coordination or psychomotor functioning" (DHHS Publication No. (ADM)85-1374, 1984, pp. 19-20).

It was not until the publication of the 1988 Surgeon General's Report (2) that the criterion of intoxication, previously a critical component of the definition of a 'psychoactive drug', was dropped. The reason for this is obvious. *Smokers simply do not smoke to become intoxicated.*

We would also point out that the concept of intoxication as a principal feature of addicting drugs is valid even today, but is ignored when it comes to smoking and nicotine. The American Psychiatric Association has recently published the fourth edition of its Diagnostic and Statistical Manual of Mental Disorders (DSM-IV, 7). In DSM-IV, 11 classes of substances that produce disorders are discussed, and behavioral intoxication plays a critical role in these substance disorders. In fact, pick any of the substance disorders listed in DSM-IV and you will find it produces some form of behavioral intoxication -- every substance except nicotine of course. Nicotine intoxication is dismissed with the statement, "This category does not apply to nicotine" (p. 183). We therefore find it somewhat surprising that our critics deny that intoxication has been and still is a principal feature of addicting drugs (4-6).

In contrast to the distinct and scientifically testable definitions we review, the nicotine-addiction proponents offer vague definitions and analogies in response. The most persistent argument they offer centers on the 'highly controlled or compulsive use' of tobacco products. We are told that smokers have lost control of their ability to decide to stop smoking and that they experience powerful urges, uncontrollable cravings and irresistible desires when they attempt to stop smoking that they are apparently powerless to fight.

Much of the evidence to support these claims relies on smokers' assertions that they would like to quit, but cannot (4). Yet, how do we measure someone's desire to quit? What constitutes a loss of control or an irresistible urge? In the United States, over 40 million smokers have quit smoking without any help. The smoker's ability to think or reason clearly is *not* diminished when making the decision to quit or continue smoking. In short, this is not a behavior that the smoker has lost control over.

In response to this, the addiction proponents say that even those smokers who quit may require several serious attempts before they are successful. This is not surprising. Most smokers who quit report that they were successful when they finally became truly motivated to quit, when the pluses for them no longer outweighed the minuses.

We would like to be very clear on one point. Our argument that behaviors like smoking and coffee drinking are more accurately labeled as habits than addictions does not suggest that everyone would easily be able to quit. Successfully changing a well-ingrained, pleasurable behavior can be extremely difficult. One need only look at those who continually fight to lose weight, only to regain it and start the process over again, to verify that pleasurable behaviors, behaviors we say we want to change, may be very difficult to modify.

Why People Smoke

If smokers are not addicted, why then do they continue to engage in a behavior that has such well-publicized risks associated with it? The simple answer is smokers obviously enjoy smoking. The work of Dr. Rose, Professors Bättig, Adlkofer (this volume) and that at other labs, including our own, indicates that while a significant portion of this pleasure is associated with the sensory aspects of smoking, the 'taste', the smell, the 'feel' of the smoke, we also recognize that nicotine can produce mild pharmacological effects that are important to some smokers.

Professors Hindmarch and Warburton (this volume) indicate that smoking and nicotine can result in positive emotional, cognitive and performance benefits to the smoker. Drs. Levin, Joseph and Newhouse (this volume) discuss the positive effects nicotine can have on learning and memory.

In short, smoking is a pleasurable activity that seems to result in specific personal benefits to the smoker that probably serve to reinforce the habit. It should not be surprising, therefore, that some people report difficulty in quitting; but difficulty quitting does not serve as evidence that nicotine is an addicting drug on a par with alcohol, heroin and cocaine.

Drug addiction is the result of a complex interaction of behavioral and pharmacological variables. It cannot be defined or characterized by any single activity or effect. Simply stated, the determination of whether or not a substance is addictive should be related to the total behavioral effects of that substance. With this in mind, we argue in direct opposition to the 1988 Surgeon General's Report.

The physiologic, pharmacologic and behavioral effects of smoking and nicotine are fundamentally different from those of addicting drugs such as heroin and cocaine. To conclude otherwise may actually do a disservice to those attempting to quit by providing a crutch suggesting it may be too difficult to even try. It could actually encourage experimenting with illicit drugs by suggesting they are no different than cigarettes.

Finally, equating smoking and nicotine with heroin and cocaine endangers our credibility as scientists with the average layman who can readily distinguish the effects of smoking or coffee drinking from the tragedies of cocaine and heroin addiction.

References

1. Robinson JH, Pritchard WS. The role of nicotine in tobacco use. Psychopharmacology. 1992; 108: 397-407.
2. USDHHS. The health consequences of smoking: Nicotine addiction. A report of the Surgeon General. Washington, DC. US Government Printing Office, 1988.
3. USDHEW. Smoking and Health. Report of the advisory committee to the Surgeon General of the Public Health Service. Washington, DC. US Government Printing Office, 1964.
4. West R. Nicotine addiction: a re-analysis of the arguments. Psychopharmacology. 1992; 108: 408-410.
5. Hughes J. Smoking is a drug dependence: a reply to Robinson and Pritchard. Psychopharmacology. 1993: 113: 282-283.
6. Stolerman IP, Jarvis MJ. The scientific case that nicotine is addictive. Psychopharmacology, In press.
7. American Psychiatric Association: Diagnostic and statistical manual of mental disorders, Fourth Edition. Washington, DC. American Psychiatric Association, 1994.
8. Bewley S., Bewley TH. Drug dependence with oestrogen replacement therapy. Lancet. 1992, 339: 290-291.
9. Griffiths RR, Woodson PP. Reinforcing effects of caffeine in humans. J Pharmacol Exp Ther. 1988. 246:21-29.
10. Cerney L, Cerney K. Can carrots be addictive? An extraordinary form of drug dependence Br J Addict. 1992, 87:1195-1197.

Effects of Nicotine on
Biological Systems II
Advances in Pharmacological Sciences

DIFFERENCES IN RESPONSE TO NICOTINE ARE DETERMINED BY GENETIC FACTORS

J.A. Stitzel, S.F. Robinson, M.J. Marks, and A.C. Collins

Institute for Behavioral Genetics, University of Colorado, Boulder, CO, USA

Summary: Genetic influences on several behavioral and physiological responses elicited by acute injections of nicotine have been detected in a screen of inbred mouse strains. Some (35-40%) of the between strain variability may be due to differences in the number of brain nicotinic receptors (nAChR), and, given that restriction fragment length polymorphism (RFLP) analysis has detected polymorphisms for several of the known nicotinic receptor genes, variability may also arise as a result of strain specific nAChR alleles. Acute sensitivity to nicotine also seems to predict the propensity for developing tolerance to nicotine and oral self administration. These genetically-determined differences seen in the mouse might help to explain individual differences in smoking-related behaviors in humans.

Introduction

Potential explanations for the chronic use of tobacco have focused on the addiction issue, especially since the editors of the 1988 United States Surgeon General's Report (1) argued that nicotine is just as addicting as cocaine and heroin. However, Robinson and Pritchard (2) have argued that smokers are not addicted to nicotine and that chronic nicotine use is a habit. These highly disparate views may reflect differences in definitions applied to addiction. Alternatively, it may be that some, but not all, tobacco smokers are addicted to nicotine. Differences between individuals are routinely detected in studies of smoking and might arise for many reasons. Individual differences may be due to genetic factors since many studies (see 3 for a review) have demonstrated that identical twins are highly concordant for smoking. Eysenck (4) has argued that genetic factors influence amount and duration of tobacco use and might therefore influence whether an individual becomes addicted to nicotine after the smoking habit is initiated.

Our research efforts have been directed towards ascertaining whether genetic factors influence the responses of mice to nicotine. The responses that are measured, we believe, may be components of the dependence or addiction process. Specifically, we have studied potential genetic influences on sensitivity to a first dose of nicotine, on the development of tolerance to nicotine, and on a variant of nicotine self administration (oral preference). Potential relationships between genetically-determined differences in these nicotine-related behaviors and brain nicotinic receptor binding have been investigated.

Methods

A major effort compared the sensitivities of 19 inbred strains to nicotine by constructing dose-response curves for nicotine effects on respiratory rate, acoustic startle, Y-maze crossing and rearing activities, heart rate and body temperature (5), and seizures (6). All of these responses to nicotine, except seizures, were measured in each animal as part of a test battery. The doses required to evoke standard effects (ED_{50}-like values) were calculated for each strain.

Preference for nicotine-containing solutions was also determined in six of the inbred mouse strains. Male mice were placed singly in a cage and provided with two bottles that contained water or a nicotine solution. Baseline fluid consumption was established prior to the addition of nicotine to one of the bottles. The positions of the two bottles were switched daily and every fifth day the nicotine concentration was increased to a maximum of 0.2 mg/ml. Daily consumption was measured.

Nicotinic receptor binding was measured in regionally dissected tissue obtained from the 19 inbred mouse strains using L-[^{3}H]nicotine and [^{125}I]-α-bungarotoxin (α-BTX) as the ligands as described previously (7).

Studies have been initiated to determine whether strains that differ in sensitivity to nicotine are polymorphic at one or more of the nicotinic receptor loci. Southern analysis was performed on genomic DNA from DBA and C3H mice that had been digested with a panel of 21 restriction enzymes. [^{32}P]labelled cDNAs for the rat α3, α4, α5, α7, ß2 or ß4 nicotinic receptor subunits were used as probes. Resultant autoradiograms of the probed membranes were analyzed for restriction fragment length polymorphisms (RFLPs).

Results

Nicotine produced dose-dependent changes for all measures tested. The left panel of Figure 1 depicts the effects of nicotine on Y-maze crossing activity in six of the inbred mouse strains, the middle panel presents the effects on body temperature, and the right panel presents the latency to nicotine-induced seizures following i.v. infusion of a 2 mg/kg/min dose. The rank order of strain sensitivities was virtually identical for the locomotor and temperature depressant effects of nicotine but differed from the seizure tests, suggesting that different genes regulate seizures and the two other responses to nicotine.

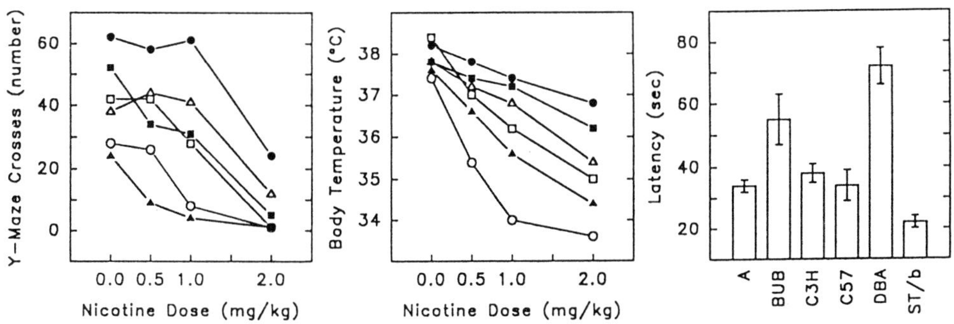

Figure 1. A comparison of responses to nicotine of six inbred mouse strains. Y-maze activity was measured for a 3-min period starting 5 min after injection (left panel) and body temperature was measured in the same animals 15 min after injection (middle panel). Latency to seizure (right panel) was measured during infusion with nicotine 2 mg/kg/min. Strain designations are A (○), BuB (●), C3H (▵), C57 (▴), DBA (□), ST/b (■).

Figure 2 illustrates the results of the experiment where male mice of six inbred mouse strains were tested for their oral ingestion of nicotine-containing solutions. Note that the animals were tested for their

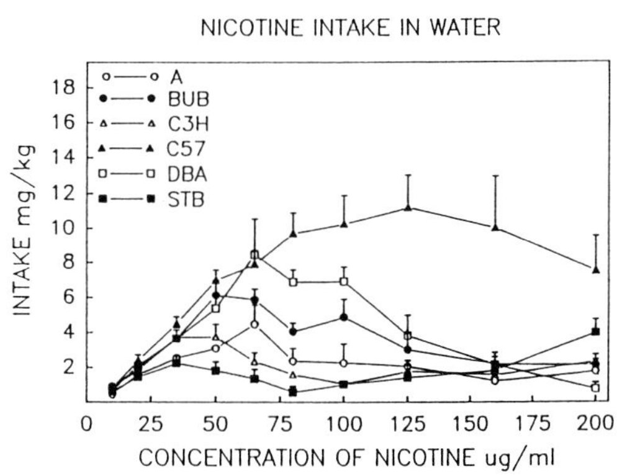

Figure 2. A strain comparison of oral intake of nicotine. Mice were presented with a choice of tap water or nicotine and daily preference measured as described in the methods.

intake of nicotine when a choice (tap water) was available. Some of the strains totally avoided nicotine (ST/b) whereas others (C57, DBA) ingested relatively large doses.

Correlations between sensitivity to acute doses of nicotine, as measured by effective doses (ED values), and maximal nicotine intake were calculated. Only one significant correlation was obtained: latency to seizures and maximum intake are highly correlated ($r = 0.89$, $p<0.01$).

Nicotinic receptor binding was measured in eight brain regions in each of the strains. Figure 3 depicts a scattergram that compares α-BTX binding in hippocampus and sensitivity to nicotine-induced seizures. This analysis suggests that increased sensitivity to nicotine-induced seizures (decreased latency to seizures) is negatively correlated with α-BTX binding. This was confirmed in the complete analysis carried out in 19 inbred strains (8). Significant correlations between nicotine binding and other responses to nicotine were not found using six inbred mouse strains, but, when more strains were tested, significant correlations were obtained (8).

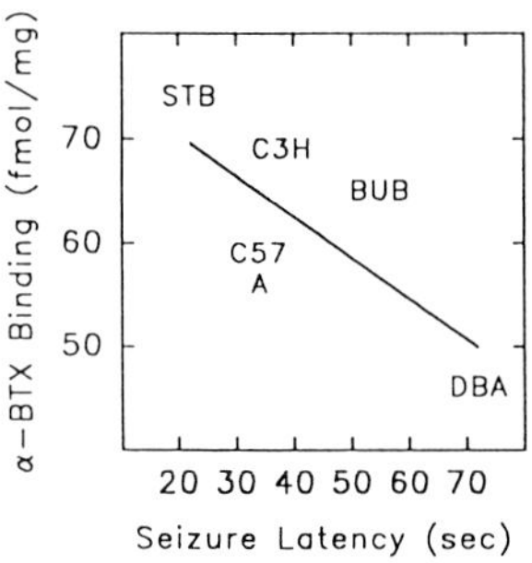

Figure 3. Correlation between seizure sensitivity and hippocampal α-BTX binding. The binding of α-BTX to hippocampal membranes was measured and compared with latency to seizures following i.v. nicotine infusion. The strain names indicate mean values.

The number of nicotinic receptors correlates with the responses to nicotine that were tested but may explain only about 35-40% of the variance in sensitivity to nicotine. Consequently, we have begun to explore whether mouse strains that differ in their responses to nicotine are polymorphic for nicotinic receptor genes. Identification of polymorphisms at a given nAChR locus will allow us to explore potential relationships between nAChR genotype and nicotine response phenotype. To date, we have identified polymorphisms for the α3, α5, α7, and ß4 subunits and are investigating whether polymorphisms for the α7 nicotinic receptor gene (the gene that codes for the α-BTX binding site) co-segregate with differences in seizure sensitivity.

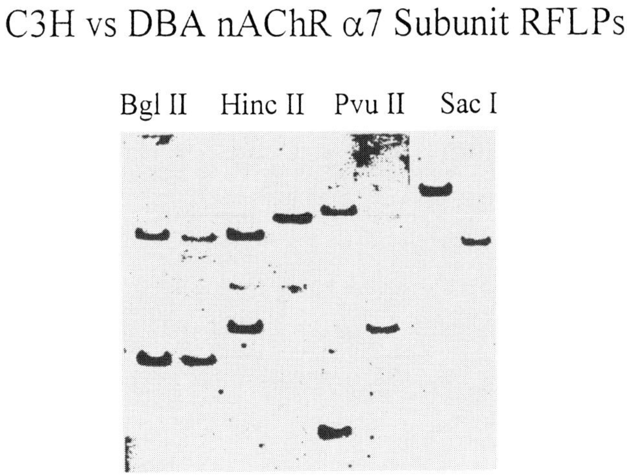

Figure 4. shows the results of an analysis for potential RFLPs of the α7 gene: three of the four restriction enzymes shown detected polymorphisms between the C3H and DBA strains.

Discussion

The data presented here demonstrate that inbred mouse strains differ in sensitivity to acute doses of nicotine. Some of this variability in sensitivity seems to be explained by variance in the number of brain nicotinic receptor binding sites: Those mouse strains that are most sensitive to nicotine (lowest ED values) seem to have the highest number of brain nicotinic receptors. Some specificity was seen in that sensitivity to nicotine-induced seizures was correlated with the number of α-BTX binding sites whereas other effects (activity, temperature) seem to be associated with the [^{3}H]-nicotine binding site. Studies in progress are attempting to assess whether polymorphisms associated with nicotinic receptor subunit genes covary with behavioral and physiological responses to nicotine.

Initial sensitivity to nicotine's seizure-inducing effects seems to predict whether an animal will self-administer nicotine, as measured by oral ingestion. An apparent relationship between seizure sensitivity and intake ($r = 0.89$, $p<0.001$) emerged with seizure resistant strains tending to drink more nicotine. If this finding predicts human tobacco use, it may be that those individuals who are less affected by the stimulant effects of nicotine are more likely to become heavy smokers.

First dose sensitivity to nicotine also seems to be a predictor for the development of tolerance to nicotine. A study that used the same six strains used in the studies described here revealed that mouse strains vary in the minimum infusion dose required to evoke measurable changes in sensitivity to nicotine (7). The correlation coefficient between the minimum infusion dose required to produce detectable tolerance to a given nicotine effect and first dose sensitivity for that effect ranged between 0.75 and 0.94.

Clearly, caution should be used when extrapolating results obtained with animals to the human condition. Nonetheless, results obtained with the mouse unambiguously demonstrate that genetically determined variability exists for several measures that may be related to the addiction process. If this same variability exists in humans, it may be that understanding the mechanisms that underlie this variability will help to explain the apparent genetic influences on tobacco use by humans.

Acknowledgements: *This work was supported by DA-03194 and DA-05131. JAS is supported by a training grant from NIAAA (AA-07464) and ACC is the recipient of a Research Scientist Award (DA-00197).*

References

1. The Health Consequences of Smoking: Nicotine Addiction. A report of the Surgeon General. Rockville, MD, 1988.
2. Robinson JH, Pritchard WS. The role of nicotine in tobacco use. Psychopharmacol 1992; 108:397-407.
3. Collins AC. Genetic influences on tobacco use: a review of human and animal studies. Int J Addictions 1990-91; 25:35-55.
4. Eysenck HJ. The Causes and Effects of Smoking. London: Temple Smith, 1980.
5. Marks MJ, Stitzel JA, Collins AC. Genetic influences on nicotine responses. Pharmacol Biochem Behav 1989; 33:667-678.
6. Miner LL, Collins AC. Strain comparison of nicotine-induced seizure sensitivity and nicotinic receptors. Pharmacol Biochem Behav 1989; 33:469-475.
7. Marks MJ, Campbell SM, Romm E Collins AC. Genotype influences the development of tolerance to nicotine in the mouse. J Pharmacol Exp Therap 1991; 259:392-402.
8. Marks MJ, Romm E, Campbell SM, Collins AC. Variation of nicotinic binding sites among inbred strains. Pharmacol Biochem Behav 1989; 33:679-689.

Nicotine and human diseases

Effects of Nicotine on
Biological Systems II
Advances in Pharmacological Sciences

RELATIONSHIP BETWEEN SMOKING, NICOTINE AND ULCERATIVE COLITIS

GAO Thomas BSc MRCP and J Rhodes MD FRCP

Department of Gastroenterology, University Hospital of Wales
Heath Park, Cardiff, Wales, UK. CF4 4XW

Summary: There is strong epidemiological evidence that ulcerative colitis is a disease of non-smokers. Recent studies suggest nicotine is the active ingredient responsible for this effect and may be of therapeutic value in this disease. Possible mechanisms responsible for this remain purely speculative but may involve mucosal eicosanoids and intestinal mucus.

Introduction

Ulcerative colitis is a chronic inflammatory disease affecting the large bowel and commonly presents with diarrhoea, abdominal cramps, urgency of defaecation with the passage of blood, pus and mucus. Inflammation is limited to the mucosa and usually affects the rectum (proctitis) but then often extends proximally to involve part or all of the colon. This contrasts with Crohn's disease which may affect any part of the small and large bowel and extends beyond the mucosa through all layers of the bowel wall. Both conditions are encompassed by the term idiopathic inflammatory bowel disease. The cause of ulcerative colitis is unknown. Incidence figures vary between 5 - $10/10^5$ each year with a prevalence in the region of $100/10^5$. Males and females are equally affected and although it may occur at any age, the peak incidence occurs in the third and fourth decade. Enviromental factors which have been implicated in the cause include infectious agents, diet and stress, but the most consistent epidemiological association has been with smoking status.

Smoking and ulcerative colitis

The initial observation in 1981 that ulcerative colitis was largely a disease of non-smokers was made almost by chance when a population of patients with colitis were used as controls in a nutritional study in which smoking habits were considered (1). Only 8 percent of 230 patients with colitis were current smokers compared with 44 percent of controls. There are now over 15 case-control studies of hospital and community based populations in different countries which have consistently identified the association. In a meta-analysis of data from 9 suitable studies of ulcerative colitis available at the time, Calkins showed a remarkably consistent association of smoking with ulcerative colitis in terms of direction. The risk for the disease was reduced in current-smokers and increased in non-smokers with the greatest risk in ex-smokers (2). In one study the relative risks in smokers, never smokers and ex-smokers were 0.6, 1 and 2.5 respectively with a high figure of 4.4 for ex-smokers who smoked > 11 cigarettes daily (3). The Surgeon General's (4) criteria for causality were applied and all suggested the findings were consistent with a causal relationship. Additional support has come from the Boston Drug Surveillance Programme which showed a relative deficiency of smokers in patients with ulcerative colitis (5). Mortality figures from a large group of patients with ulcerative colitis showed a relative absence of smoking related diseases, particularly cardiovascular and respiratory illness in men (6). A further study of a large cohort of 17,000 women registered on a Family Planning Programme and subsequently followed- identified 31 who developed ulcerative colitis; smoking habit on entering the programme had been documented and showed a relative absence of smokers amongst those with colitis (7). The time at which patients stop smoking in relation to the onset of colitis has been of particular interest since over two-thirds of patients who are ex-smokers develop their disease after giving up smoking, with a particularly high incidence in the first few years (8,9,10). Furthermore the graphical plots for cumulative incidence of colitis in male life-long non-smokers compared with previous smokers show a shift to the right for ex-smokers. Fifty percent of those who had never smoked developed colitis by the age of 25 compared with 42 years for ex-smokers (11). There is an apparent delay in the onset of colitis of 17 years in patients who have been smokers.

If smoking reduces the risk of ulcerative colitis and most previous smokers develop their disease soon after stopping smoking - one must question whether smoking may have a favourable effect on active disease and perhaps maintain clinical remission. It would be neither practical nor ethical to consider trials in which patients were asked to start and stop smoking in studies to resolve these questions. However, we questioned 30 intermittent smokers about their observations; half of them thought their colitis symptoms improved over a period of 6 weeks whilst smoking 20 cigarettes daily (12).

Nicotine and ulcerative colitis

The strong negative association between smoking and colitis has stimulated trials with nicotine. Initial studies with nicotine gum were inconclusive and uncontrolled (13). With the introduction of transdermal nicotine it has been possible to test this compound more formally in controlled trials. In a pilot study, 12 of 16 patients with active left sided disease improved over 4 weeks with nicotine, but 6 of these deteriorated in the subsequent 4 weeks with placebo patches (14). In a recent randomised controlled double-blind trial of 72 patients with active left sided disease patients were treated with either transdermal nicotine patches or placebo for six weeks. Incremental doses of nicotine were given; most patients tolerated doses of 15 to 25 mg per 24 hours. All patients had been taking mesalamine, and 12 were receiving low dose glucocorticoids; these medications were continued without change during the study. Clinical, sigmoidoscopic, and histologic assessments were made at base line and at the end of the study. Improvement occurred in both groups but was significantly greater in those treated with nicotine for both clinical and histological measures. Seventeen of the 35 patients in the nicotine group had complete remissions, as compared with 9 of the patients in the placebo group. More patients in the nicotine group had side effects, the most common of which were nausea, light-headedness, headache and sleep disturbance. We conclude that the addition of transdermal nicotine to conventional therapy improves symptoms in patients with ulcerative colitis (15).

Mechanisms

Although it would seem very plausible that nicotine is the active ingredient in smoking responsible for the clinical effect in ulcerative colitis, the mechanism of action remains elusive. Strong epidemiological associations with a disease should point toward causative factors and should help to elucidate mechanisms responsible for the pathological process - it would be a great encouragement if these simple principles applied to ulcerative colitis! Somehow or other, despite the most intense scrutiny, the cause remains elusive, which leaves much opportunity for speculation.

Smoking produces a range of compounds which are pharmacologically active, of which nicotine is thought to be the most important. Some of the actions of nicotine could be relevant to the inflammation which occurs in ulcerative colitis. Smoking influences cellular immunity (16) and there is a suggestion that heavy smokers may have reduced levels of IgA in both saliva (17) and intestinal secretion (18). It also appears to reduce the skin responses to challenges from lauryl sulphate and ultraviolet radiation (19). It has been shown to reduce the levels of some mucosal eicosanoids including prostaglandins in man (20) whilst infusions of nicotine in animals may change eicosanoid tissue levels and the thickness of adherent surface mucus in the rectum (21). Smoking reduces blood flow in the rectum (22) and decreases gut permeability (23). Which, if any, of these mechanisms are relevant to its effect in ulcerative colitis remains purely speculative.

Conclusions

Ulcerative colitis is a disease of non-smokers and ex-smokers, and smoking appears to have a favourable effect on the disease. Patients with ulcerative colitis may benefit from transdermal nicotine as therapy but further clinical trials would be required to substantiate this suggestion. Elucidation of the mechanisms involved will probably give us a clearer insight into the pathophysiology of the disease.

In contrast the relationships between Crohn's disease and smoking are opposite to those found in colitis; smoking increases the risk of Crohn's disease and appears to have an unfavourable effect on the course of the disease .The conundrum of 'striking opposites' in the epidemiological link with smoking for two very similar inflammatory diseases of bowel is most intriguing and may be the key to a fuller understanding of the cause and pathogenesis for these diseases.

Since smoking is bad for Crohn's disease but may help ulcerative colitis, those with Crohn's should be strongly dissuaded from smoking. Those with colitis who improve while smoking but deteriorate when they quit will make their own decisions about smoking, but will need to know the facts. **No one however, should be encouraged to smoke!**

References

1. Harries AD, Baird A, Rhodes J . Non-smoking: a feature of ulcerative colitis. Br Med J 1982; 284 : 706.
2. Calkins BM. A meta - analysis of the role of smoking in Inflammatory bowel disease. Dig Dis Sci 1989; 34: 1841 - 54.

3. Lindberg E, Tysk C, Andersson K, Jarnerot G. Smoking and Inflammatory bowel disease. A case - control study. Gut 1988; 29: 352-7.
4. The health consequences of smoking: cancer. A report of the Surgeon General. US Department of Health and Human Services, Public Health Service, Office on Smoking and Health, Rockville, Maryland, 1982.
5. Jick H, Walker AM. Cigarette smoking and ulcerative colitis. N Eng J Med 1983; 308 : 261-3.
6. Gyde SN, Prior P, Dew MJ, Saunders V, Waterhouse JAH, Allen RN. Mortality in ulcerative colitis. Gastroenterology 1982; 83: 36-43.
7. Vessey M, Jewell D, Smith A, Yeates D, McPherson K. Chronic inflammatory bowel disease, Cigarette smoking, and use of oral contraceptives: findings in a large cohort study of women of childbearing age. Br Med J 1986; 292: 1101-3.
8. Motley RJ, Rhodes J, Ford GA et al. Time relationship between cessation of smoking and onset of ulcerative colitis. Digestion 1987; 37: 125-7.
9. Boyko EJ, Koepsell TD, Perera DR, Inui TS. Risk of ulcerative colitis among former and current cigarette smokers. N Eng J Med 1987; 316: 707-10.
10. Motley RJ, Rhodes J, Kay S, Morris T. Late presentation of ulcerative colitis in ex-smokers. In J Colorect Dis 1988; 3: 171-5.
11. Motley RJ, Rhodes J, Kay S, Morris T. Late presentation of ulcerative colitis in ex-smokers. In J Colorect Dis 1988; 3: 171-5.
12. Rudra T, Motley RJ, Rhodes J. Does smoking improve colitis? Scand J Gastroenterol 1989; 170 (supp): 61-3.
13. Perera DR, Janeway CM, Field A, Ylvisaker JT, Belic L, Jick H. Smoking and ulcerative colitis. Brit Med J 1984; 288:1533.
14. Srivastava ED, Russell MAH, Feyerabend C, Williams GT, Masterson JG, Rhodes J. Transdermal nicotine in active colitis. Eur J Gastroenterol 1991; 3: 815-18.
15. Pullan RD, Rhodes J, Ganesh S et al. Transdermal nicotine for active ulcerative colitis. N Engl J Med 1994; 330: 811-15.
16. Miller LG, Goldstein G, Murphy M, Ginns LC. Reversible alterations in immunoregulatory T cells in smoking. Chest 1982; 5: 527-9.
17. Barton JR, Raid MA, Gaze MN, Maran AGD, Ferguson A. Mucosal immunodeficiency in smokers and in patients with epithelial head and neck tumours. Gut 1990; 31: 378-82.
18. Srivastava ED, Barton JR, O'Mahoney S et al. Smoking, humoral immunity, and ulcerative colitis. Gut 1991; 32: 1016-19.
19. Mills CM, Hill SA, Marks R. Altered inflammatory responses in smokers. Br Med J 1993; 307: 911.
20. Motley RJ, Rhodes J, Williams G, Tavares IA, Bennett A. Smoking, eicosanoids and ulcerative colitis. J Pharm. Pharmacol 1990; 42: 288-9.
21. Zijlstra FJ, Srivastava ED, Rhodes M et al. Effect of nicotine on rectal mucus and ucosal eicosanoids. Gut 1994; 35: 247-251.
22. Srivastava ED, Russell MA, Feyerabend C, Rhodes J. Effect of ulcerative colitis and smoking on rectal blood flow. Gut 1990; 31: 1021-4.
23. Prytz H, Benoni C, Tagesson C. Does smoking tighten the gut? Scand J Gastroenterol 1989; 24:1084-8.

Effects of Nicotine on
Biological Systems II
Advances in Pharmacological Sciences

TRANSDERMAL NICOTINE IN TOURETTE'S SYNDROME

A.A. Silver, R.D. Shytle, M.K. Philipp and P.R. Sanberg

Departments of Psychiatry, Surgery, Pharmacology and Neurology
Division of Neurological Surgery, University of South Florida College of Medicine
Tampa, Florida 33612, USA

Summary: In an open clinical trial, transdermal nicotine patch significantly ameliorated the frequency and severity of tics in patients with Tourette's syndrome who were receiving dopamine receptor blockers. This effect, noted within 3 hours after application of the patch, persisted for a variable period of time after removal of the patch. In 2 patients, initially with incapacitating Tourette's symptoms, the effect persisted from 3 weeks to 4 1/2 months without further administration of nicotine.

Introduction

The significant potentiation of neuroleptic-induced catalepsy in rats by systemic and intracaudate injection of nicotine suggested that nicotine might increase the efficacy of neuroleptics in hyperkinetic motor disorders (1,4,7). In an open, non-blind clinical study, McConville and his colleagues (2,3) reported that tic frequency and severity diminished during a 30-minute chewing of Nicorette gum (each containing 2 mg of nicotine) in 10 patients with Tourette's syndrome who were treated with haloperidol. The improvement in symptoms along with concomitant decrease in attention and concentration problems, persisted during video-taped observations for 1 hour after nicotine gum in 9 of the 10 patients. Placebo gum had no effect on tics, while nicotine gum alone may have had a minor effect in diminishing tics. The effects of nicotine plus haloperidol, however, were transitory. Moreover, the bitter taste of Nicorette gum and symptoms of gastro-intestinal irritation made compliance questionable. This study reports on the use of transdermal nicotine patches on 11 patients with Tourette's syndrome, all treated with a dopamine receptor blocker and further extends our earlier report (8).

Method

This was an open-field clinical treatment, not blind, not placebo-controlled to explore the effects of transdermal nicotine patches in 11 patients under treatment in the Tourette's Syndrome Clinic at the University of South Florida. The subjects were selected only because they had appointments at the clinic at the time the nicotine open-trial was to be conducted and they and their parents had signed consent to participate in the clinical treatment. The 3 young adults in the group (R.P., M.H., and G.Z.) signed consent on their own authority. The age of the sample ranged from 11 years, 5 months to 32 years (mean of 16 years). There were 10 males and 1 female (L.A.). Because of the random selection of subjects, they differed in medication they were receiving at the time of the trials, in dose of medication, in the effectiveness of the medication, and in the duration of time they had been receiving their medication. Six of the subjects were receiving haloperidol. All 6, however, were also receiving additional medication; 3 of the 6 (M.M., M.H., D.B.) clonidine in addition to haloperidol, 2 of the 6 pimozide along with haloperidol (J.B., L.A.), and of the 6 (R.P.) clomipramine with haloperidol. Pimozide alone was used in 2 subjects (G.Z., A.A.), pimozide plus clonidine in 2 (J.E.B. and R.M.), and perfenazine and clomipramine in one (J.F.). The location, frequency and severity of the Tourette's symptoms also varied, spanning a wide range from the most severe (J.F.) who had been hospitalized with almost continuous head thrusting and barking vocalization with 1250 tics measured over a ten minute segment of video tape to 3 subjects who clinically were virtually symptom free (L.A., M.H., D.B.). Three subjects were heavy smokers (R.P., G.Z., R.M.), each smoking at least 2 packs of cigarettes daily. The remaining 8 subjects were non-smokers and their parents also non-smokers. All patients received a transdermal nicotine patch (Marion Merrell Dow); for 8 patients the patch was titrated to deliver 7 mg of nicotine in 24 hours, for 3 patients (M.M., R.P. and J.B.) a 14 mg patch was used. Each patch was removed in 24 hours. With the exception to be mentioned below, each patient received only one patch.

The localization, frequency and severity of the tics were determined by video tapes each approximately 30 minutes in duration, with taping done at baseline (before application of transdermal nicotine patch), 3 hours after application of the patch, 24 hours after application of the patch and a fourth video-taping at variable weeks after baseline. Location and frequency of the tics were determined by two independent observers counting the tics in each location recorded during the middle 10 minutes of each video. The number of tics in all anatomic locations was summed for the purpose of analysis. Data obtained to date will be reported here.

Severity of tics was scored by the clinician using a 5-point scale (5-most severe, cannot function in school, work, or socially; 4-daily living performed with difficulty, sensitivity to reaction of peers, strangers, anxiety on leaving own home, attends school in special classes, cannot keep jobs; 3-can function in daily living but not to potential, aware of but can overcome reaction of peers, anxiety present but under control; 2-can function in daily living, can realize potential, patient and family aware of symptoms but not disruptive of normal activity; 1-functions to potential, tics mild so that they may not be considered as deviant by peers, may have mild obsessions and compulsions which can be disguised in action). Because of the variability in location, frequency and severity of tics among subjects, change in symptoms after nicotine was evaluated as percentage change over baseline for each subject.

Table 1. Effect of Transdermal Nicotine on Frequency and Severity of Tics

PATIENT	AGE (Yrs-Mos)	CURRENT MEDICATION (Daily Dose)	BASELINE		0 - 3 HOURS		% OF CHANGE	
			# of Tics	Severity	# of Tics	Severity	# of Tics	Severity
M.M.	12.66	Haldol, 6 mg Clonidine, 0.3 mg	70	5	15	2	78	60
J.B.	14.66	Pimozide, 4 mg Haldol, 0.5 mg	158	5	7	1	96	80
L.A.	12.42	Haldol, 2 mg Pimozide, 2 mg	1	1	1	1	0	0
M.H.	19.25	Haldol, 2 mg Clonidine, 0.1 mg	2	2	0	1	100	50
D.B.	15.17	Haldol, 5 mg Clonidine, 0.25 mg	2	1	2	1	0	0
J.E.B.	11.58	Pimozide, 2 mg Clonidine, 0.25 mg	121	3	48	1	60	67
J.F.	15	Trilifon, 3 mg Anafranil, 25 mg	1250	5	585•	3	53	40
R.M.+	17	Pimozide, 4 mg Clonidine, 0.3 mg	27	2	42	2	-55	0
R.P.+	22	Haldol, 4 mg Anafranil, 200 mg	108	4	45	2	58	50
G.Z.+	32	Pimozide, 5 mg	247	4	108	3	56	25
A.A.	13	Pimozide, 4 mg	234	2	59	2	75	0
Medians	15.1		114.5	2.5	43.5	1.5	57	32.5
Means	16.79		201.82	3.09	82.91	1.73	47.36	33.82
SEMs	1.79		108.17	0.48	51.18	0.24	14.20	9.11

• plus 4 days
+ smokers

Results

As shown in Table 1, it appears that frequency and severity were significantly reduced 3 hours after application of the transdermal nicotine patch when compared to baseline. This was confirmed by the nonparametric matched-pairs Wilcoxon signed-ranks test (Number of tics, $Z=-2.43$, $p<0.05$; Severity, $Z=-2.39$, $p<0.05$). Review of the data suggests that improvement with nicotine was less obvious in patients whose Tourette's symptoms were adequately controlled with medication while those with disturbing symptoms often exhibited dramatic improvements. Further post-hoc analyses revealed a significant correlation between baseline tic severity and percent change at 3 hours post-patch application (Spearman rho=0.732, $p<0.05$).

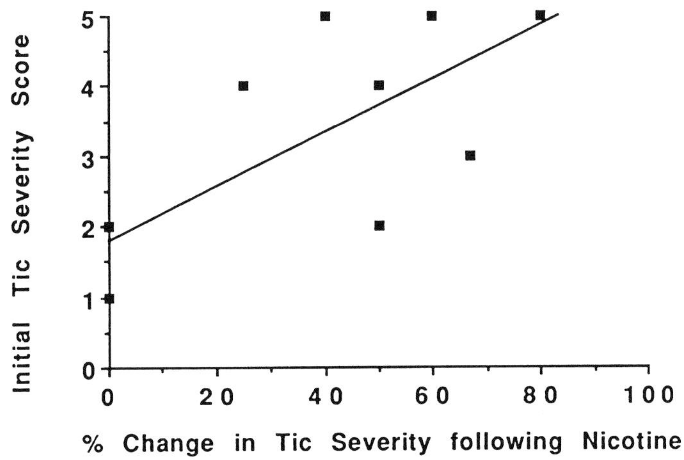

Figure 1. Relationship between Initial Tic Severity and Percent Change following application of the nicotine patch.

For example, two patients (L.A. and D.B.) with minimal Tourette's symptoms before application of the patch had minimal symptoms after the patch, with no significant change in severity. On the other hand, severely incapacitated patients had significant reduction in tic frequency and severity and two cases, J.B. and J.F., along with significant tic reduction and tic severity also had improvement in visual-motor function. Two of the three smokers also had significant improvement (R.P. with a 58% reduction in tics and a severity rating from 4 to 2; G.Z. approximately 56% reduction in tics and severity rating from 4 to 3). However, one of the 3 smokers (R.M.) experienced an increase in tics with no change in severity 3 hours after transdermal nicotine.

Two subjects, the most incapacitated in the group, were followed beyond three hours after initial application of the patch. J.F., a 15 year old boy diagnosed with Tourette's syndrome at age 10 years, developed an acute exacerbation of his symptoms approximately 2 weeks before referral to us. During those two weeks he had been hospitalized receiving perfenazine, 3 mg daily, and clomipramine, 25 mg daily, without appreciable relief. When we first saw him he had severe motor and vocal tics, with head thrusts and jaw extension in synchrony with loud barking noises. There was coprolalia and a complex hand clapping. He was anxious and frightened, although able to talk appropriately. The frequency of tics over a ten minute segment of video tape was 1250 with severity of 5. Three hours after application of a 7 mg transdermal nicotine patch, symptoms had virtually disappeared and the patient was eager to attend a friend's birthday party. The patch was removed in 24 hours. He was not seen until 4 days after application of the patch, when the only symptom was slapping of his thigh. This was done in a rhythmic fashion with a frequency of 585 times over a ten minute video tape. A second 7 mg transdermal nicotine tape was then applied. Again within 3 hours all symptoms had disappeared. No further nicotine was used. Three weeks later, no symptoms were evident on video-taping. Follow-up three months after initial transdermal nicotine found J.F. had returned to school with minimal slapping of his thigh when stressed. Medication remained at 3 mg perfenazine and 25 mg clomipramine daily.

J.B., age 14 years, 8 months, had been diagnosed with Tourette's syndrome at age 5 years. Followed by his neurologist, he had been under intermittent control with a variety of medication and was receiving pimozide, 4 mg, daily and haloperidol, 0.5 mg daily, when he was referred to us because for the previous 2 weeks, he had been unable to attend school. He had been thrusting his head back, violently, so that his neck was twisted and painful. This was accompanied by coprolalia, punching his knees, biting his clothes, and hitting his head against the wall. His tic frequency count at baseline was 158 with a severity of 5. Within 3 hours after application of a 14 mg transdermal nicotine patch, tics were reduced to a frequency of 7 with a severity of 1. His mother reported that symptom relief was noted in one hour and that, in spite of nausea, he was able to do his written school work with greater legibility and ease than he could even before the exacerbation of symptoms. Twenty four hours after application of the patch, there was only an occasional head tic (frequency 2 in 10 minutes, with a severity of 1). Four weeks later only an occasional tic was noted and the patient felt well. Follow-up 4 1/2 months after initial application of transdermal nicotine revealed an infrequent tic with a severity of only 1 to 2. He was still receiving pimozide, 4 mg, and haloperidol, 0.5 mg, daily.

Side effects: Ten of the twelve subjects complained of itching at the site of the patch. This, however, was transient and did not require removal of the patch. In two adolescents, nausea was significant, in one of these accompanied by headache. In each of the two, the patch was removed in approximately 2 hours. The one female subject felt that the nausea and headache was so disturbing that she did not wish any further nicotine. On the positive side, most subjects felt more relief with the nicotine, and objectively were less restless within 3 hours after application of the patch. The transdermal nicotine patch appears to be well tolerated.

Discussion

In conclusion, transdermal nicotine patch appeared to be effective in obtunding symptoms of Tourette's syndrome in adolescents and young adults whose Tourette's symptoms had not been satisfactorily controlled with dopamine receptor blockers. In contrast to previous studies where only haloperidol was used as the dopamine blocker, (2,3,6-8) significant improvement was seen here when pimozide or perfenazine was the dopamine blocker. In smokers, improvement in Tourette's symptoms was less obvious than in non-smokers, although even in two of the three smokers in this sample, significant improvement in tic frequency was noted and severity was decreased.

The onset of symptom relief is rapid, within three hours, suggesting that the absorption rate of transdermal nicotine is rapid. Pharmacokinetic data supplied by Marion Merrell Dow indicate that after Nicoderm application, plasma concentration rises rapidly reaching a plateau within 2 to 4 hours and then slowly declines during 4 to 24 hours with a maximum concentration of nicotine in the 5-12 microgram/milliliter range (average 4-10 micrograms/milliliter). The half-life of nicotine is 1-2 hours, but the primary metabolite of nicotine, cotinine, whose action is less active than nicotine, has a half-life of 15-20 hours. Nevertheless, with only 1 application of transdermal patch, the symptom relief may extend beyond the 24 hour period and, in 2 patients followed, the effect persisted beyond 3 weeks without further nicotine. Shortly after completion of this open-trial, we learned that similar long-term therapeutic benefits of the nicotine patch have now been independently confirmed in three patients with Tourette's syndrome from another clinical laboratory (5). However, how frequently the patch must be applied to control symptoms is yet to be determined. Overall, these results suggest the need for double-blind placebo controlled studies to test the validity of these findings and to determine the effect of nicotine alone in the symptoms of Tourette's syndrome.

Acknowledgments: *We appreciate the generous support of the Smokeless Tobacco Research Council, Inc. and Marion Merrell Dow Inc.*

References

1. Emerich DF, Zanol MD, Norman AB, McConville BJ, Sanberg PR. Nicotine potentiates haloperidol-induced catalepsy and hypoactivity. Pharmacology, Biochemistry and Behavior 1991; 38:875-880.
2. McConville BJ, Fogelson MH, Norman AB, Klykylo WM, Manderschlid PZ, Parker KW, Sanberg PR (1991). Nicotine potentiation of haloperidol in reducing tic frequency in Tourette's disorder. Am J Psychiatry 1991; 148:793-794.
3. McConville BJ, Sanberg PR, Fogelson MH, King J, Cirna R, Parker KW, Norman AB. The effects of nicotine plus haloperidol compared to nicotine only and placebo nicotine only in reducing tic severity and frequency in Tourette's disorder. Biol Psychiatry 1992; 31:832-840.
4. Moss DE, Mandersherd PZ, Mongomery SP, Norman AB, Sanberg PR. Nicotine and cannabinoids as adjuncts to neuroleptics in the treament of Tourette Syndrome and other motor disorders. Life Sciences 1989; 44:1521-1525.
5. Reveley MA, Bird R, Stirton RF, Dursun SM. Microstructural analysis of the symptoms of Tourette's Syndrome and the effects of a trial use of transdermal nicotine patch. J Psychopharmacol Supplement. 1994; A30:117.
6. Sanberg PR, Fogelson HM, Manderscherd PZ, Parker KW, Norman AB, McConville BJ. Nicotine gum and haloperidol in Tourette's Syndrome. The Lancet 1988; 1:592.
7. Sanberg PR, McConville BJ, Fogelson HM, Manderscheid PZ, Parker KW, Blythe MM, Klykle WM, Norman AB. Nicotine potentiates the effects of haloperidol in animals and patients with Tourette syndrome. Biomed Pharmacother 1989; 43:19-23.
8. Silver AA, Sanberg PR. Transdermal nicotine patch and potentiation of haloperidol in Tourette's Syndrome. The Lancet 1993; 342:182.

Effects of Nicotine on
Biological Systems II
Advances in Pharmacological Sciences

NICOTINE AND NEUROPSYCHIATRIC DISORDERS: SCHIZOPHRENIA

John R. Hughes and Pauline McHugh

University of Vermont, 38 Fletcher Place,
Burlington, Vermont 05401-1419, USA

Summary: Smoking is 2-3 times more prevalent and quitting is 2-3 times less prevalent in psychiatric patients than in the general population. Whether this is due to neurotransmitter or behavioral deficits, attempts to combat medication side-effects, modeling or adjunctive behavior is unknown. Studies of the association of smoking and schizophrenia illustrate how smoking may modulate symptoms of a disorder and interfere with the efficacy and acceptability of treatment of the disorder. Thus, psychiatric clinicians and researchers need to record and adjust for the smoking behavior of their patients/subjects.

Introduction

This review documents the high prevalence of smoking in psychiatric patients, illustrates several possible reasons for this high prevalence and then uses schizophrenia to illustrate several ways in which smoking can interact with the diagnosis and treatment of psychiatric disorders. Given recent regulations for smoke-free psychiatric units and data that acute or chronic smoking or nicotine withdrawal can interfere with scientific experiments and clinical diagnosis (1,2), clinicians and scientists need to be more aware of the smoking status of their psychiatric patients and the psychiatric status of their smoking patients.

In the general population in the US, approximately 25% are current smokers, 25% are exsmokers and 50% are neversmokers (3). In contrast, among patients with alcohol/drug dependence, mania or schizophrenia, approximately 80% are current smokers, 10% are exsmokers and 10% are neversmokers (1). Among patients with other psychiatric disorders, 45-60% are current smokers, 10% are exsmokers and 30-45% are neversmokers (1). These high rates of smoking and lower rates of cessation among psychiatric patients are not due to differences in education, employment or use of alcohol or caffeine between psychiatric patients and controls (4).

Whether the association of smoking and psychiatric disorders is causal or not is unclear (2). In terms of smoking causing psychiatric disorders, the onset of smoking precedes the onset of the large majority of psychiatric disorders; however, a biological mechanism for how smoking might induce a disorder is not clear. In terms of depression causing smoking, adolescents who have depressed moods are more likely to take up smoking (5,6); however, whether such moods are precursors of a major depression or other psychiatric disorders is unclear. Another possibility is that smoking does not cause psychiatric disorders or vice versa but rather that a third variable causes both smoking and the psychiatric disorder. For example, one twin study suggests the association of smoking and depression is due to shared genes (7). In addition, it is plausible that a lack of reinforcers, a multiplicity of aversive events, low self-esteem, poor assertiveness, etc. could lead to both smoking and depression (8). Finally, psychiatric patients have poor cognitive, psychological and social resources; thus, other events may initiate smoking and then a psychiatric disorder intervene to undermine cessation efforts.

Several reasons why psychiatric patients are more likely to smoke can be hypothesized, but none of these hypotheses have been well tested (4). For example, patients with several disorders are thought to have deficits in dopamine, noradrenaline, etc. Since smoking increases these neurotransmitters (9), psychiatric patients may smoke to correct these deficits. Psychiatric patients also often have behavioral deficits; e.g., poor ability to concentrate, relaxation skills and anger control. Since nicotine improves these (10), perhaps psychiatric patients smoke to obtain help with these problems. Many psychiatric medications are sedating and this can be aversive. Since smoking increases the clearance of many of these drugs (1) and, as a stimulant, offsets the sedating effects of such drugs, perhaps psychiatric patients use smoking to combat side-effects of psychiatric drugs. Modeling is a common initiator of behaviors. Since so many psychiatric patients smoke, perhaps just being around other psychiatric patients initiates smoking. Finally, since psychiatric patients often have large chunks of time with few activities, perhaps smoking serves as a behavioral filler; i.e., its simply something to do, or what has been termed adjunctive or schedule-induced behavior (11).

Smoking and Schizophrenia

Due to space limitations, this article will review only the association of smoking and schizophrenia (12) to illustrate several of the ways smoking can interact with a psychiatric illness.

Smoking and the Etiology of Schizophrenia

Newer theories of the pathophysiology of schizophrenia have posited a functional dopamine excess in the mesolimbic system causes the positive symptoms (e.g., hallucinations, delusions, thought disorder) and a dopamine deficiency in the frontal areas causes the negative symptoms (e.g., anhedonia, apathy, ambivalence) of schizophrenia (13). Since acute injections of nicotine increase dopamine (14), one possibility is that some schizophrenics smoke to combat the negative symptoms of schizophrenia. Another possibility is that early negative symptoms place the schizophrenic at special risk for taking up smoking. Prior to the first psychotic break, schizophrenics typically have poor social skills that diminish their ability to access social reinforcers plus they are hypohedonic such that perhaps they are responsive only to predictable, immediate reinforcers. Thus, schizophrenics may be especially likely to smoke because nicotine via smoking is an easily accessible, reliable and immediate reinforcer. Direct tests of the above hypotheses of why so many schizophrenics smoke have not been reported; however, two studies have looked for an association of smoking and negative symptoms. In one study, negative symptoms did not differ between smoking and nonsmoking schizophrenics (15). In the other study, smokers had more negative symptoms than nonsmokers; however, the same was also true for positive symptoms (16).

Smoking and the Treatment of Schizophrenia

Smoking increases the metabolism of several neuroleptics; e.g., clozapine, fluphenazine and haloperidol (1). In fact, in some patients, smoking and smoking cessation have dramatic effects on haloperidol levels (17). Interestingly, seven of nine studies have found that schizophrenics who smoke receive twice as much neuroleptics as nonsmoking schizophrenics (15,16,18-24). Although this may be because smoking schizophrenics are more symptomatic (15,16), it also is likely due to increased metabolism of neuroleptics by smoking (1).

Smoking may also influence side-effects from neuroleptics. Two studies have compared akathesia in smoking and nonsmoking schizophrenics (16,24). Both found smokers had more akathesia; however, neither of these studies corrected for the increased neuroleptic dosing of smoking schizophrenics. Six studies have examined parkinsonian side-effects from neuroleptics (16,21,25-28). Five found that smokers had less parkinsonian symptoms even though they received higher doses of neuroleptics. Finally, six studies have examined tardive dyskinesia (16,23-25,27-29). Four found more tardive dyskinesia in smoking schizophrenics and three failed to find this association. Theories to explain the above findings have focused on 1) the fact that chronic dosing of nicotine may decrease rather than increase dopamine, and 2) by increasing noradrenaline and adrenaline, smoking may protect dopamine neurons from free radicals or from MPTP damage to the mesolimbic system (30). None of these hypotheses has a body of evidence to support it.

Conclusion

At present, most psychiatric and alcohol/drug abuse clinicians and researchers ignore the smoking status of their patients/subjects (31). This brief review of smoking and schizophrenia illustrates that smoking can influence the symptoms of a psychiatric disorder, the efficacy of pharmacological treatments of the disorder and the severity of side-effects from the pharmacotherapy. For these reasons, both clinicians and researchers need to note and adjust for the smoking status of their patients.

Acknowledgements: *Writing of this article was funded by Research Scientist Development Award DA-00109 (Dr. Hughes) and Institutional Training Grant (T32 DA07242) (Ms. McHugh) from the National Institute on Drug Abuse.*

References

1. Hughes JR. Possible effects of smoke-free inpatient units on psychiatric diagnosis and treatment. J Clin Psychiatr 1993; 54:109-114.
2. Glassman AH. Cigarette smoking: Implications for psychiatric illness. Am J Psychiatr 1993; 150:546-553.
3. Cigarette smoking among adults - United States, 1991. Morbidity and Mortality Weekly Report 1993.
4. Hughes JR, Hatsukami DK, Mitchell JE, Dahlgren BA. Prevalence of smoking in psychiatric outpatients. Am J Psychiatr 1986; 143:993-997.
5. Kandel DB, Davies M. Adult sequelae of adolescent depressive symptoms. Arch Gen Psychiatr 1986; 43:255-262.
6. Lerner JV, Vicary JR. Difficult temperament and drug use: Analyses from the New York longitudinal study. J Drug Education 1984; 14:1-9.
7. Kendler KS, Neale MC, MacLean CJ, Heath AC, Eaves LJ, Kessler RC. Smoking and major depression. Arch Gen Psychiatr 1993; 50:36-43.
8. Hughes JR. Clonidine, depression and smoking cessation. JAMA 1988; 254:2901-2902.
9. Pomerleau OF, Pomerleau CF. Neuroregulators and the reinforcement of smoking: Towards a biobehavioral explanation. Neuroscience and Biobehavioral Reviews 1984; 8:503-513.

10. US Department of Health and Human Services. Effects of nicotine that may promote tobacco use. In: Benowitz NL, Grunberg NE, Henningfield JE, Lando HA, eds. The Health Consequences of Smoking. Nicotine Addiction, Washington, USDHHS, Pub No (CDC) 88-8406:A Report of the Surgeon General, 1988:377-458.
11. Falk JL. Schedule-induced drug self-administration. In: Van Haaren F, ed. Methods in Behavioral Pharmacology. Amsterdam:Elsevier Science Publishers, 1993:301-328.
12. Lohr JB, Flynn K. Smoking and schizophrenia. Schizophrenia Research 1992; 8:93-102.
13. Davis KL, Kahn RS, Ko G, Davidson M. Dopamine in schizophrenia: A review and reconceptualization. Am J Psychiatr 1991; 148:1474-1486.
14. Clarke PBS. Recent advances in understanding the actions of nicotine in the central nervous system. J Natl Cancer Inst Monograph 1992; 2:229-238.
15. Ziedonis DM, Kosten TR, Glazer WM, Frances RJ. Nicotine dependence and schizophrenia. Hosp Comm Psychiatr 1994; 45:204-206.
16. Goff DC, Henderson DC, Amico E. Cigarette smoking in schizophrenia: Relationship to psychopathology and medication side effects. Am J Psychiatr 1992; 149:1189-1194.
17. Stimmel G, Falloon I. Chlorpromazine plasma levels, adverse effects, and tobacco smoking: Case report. J Clin Psychiatr 1983; 44:420-421.
18. Vinarova E, Vinar O, Kalvach Z. Smokers need higher doses of neuroleptic drugs. Biol Psychiatr 1984; 19:1265-1268.
19. Jann MW, Saklad SR, Ereshefsky L, Richards AL, Harrington CA, Davis CM. Effects of smoking on haloperidol and reduced haloperidol plasma concentrations and haloperidol clearance. Psychopharm 1986; 90:468-470.
20. Miller DD, Kelly MW, Perry PJ, Coryell WH. The influence of cigarette smoking on haloperidol pharmacokinetics. Biol Psychiatr 1990; 28:529-531.
21. Decina P, Caracci G, Sandik R, Berman W, Mukherjee S, Scapicchio P. Cigarette smoking and neuroleptic-induced parkinsonism. Biol Psychiatr 1990; 28:502-508.
22. Sandyk R, Kay SR. Tobacco addiction as a marker of age at onset of schizophrenia. Intern J Neuroscience 1991; 57:259-262.
23. Yassa R, Samarthji L, Korpassy A, Ally J. Nicotine exposure and tardive dyskinesia. Biol Psychiatr 1987; 22:67-72.
24. Menza MA, Grossman N, Van Horn M, Cody R, Forman N. Smoking and movement disorders in psychiatric patients. Biol Psychiatr 1991; 30:109-115.
25. Wagner B, Wolf GK, Ulmar G. Does smoking reduce the risk of neuroleptic parkinsonoids? Pharmacopsychiatry 1988; 21:302-303.
26. Yassa R, Lal S, Korpassy A, Ally J. Nicotine exposure and tardive dyskinesia. Biol Psychiatr 1987; 22:67-72.
27. Binder RL, Kazamatsure H, Nishimura T, McNiel DE. Smoking and tardive dyskinesia. Biol Psychiatr 1987; 22:1281-1282.
28. Youssef HA, Waddinton JL. Morbidity and mortality in tardive dyskinesia: Associations in chronic schizophrenia. Acta Psychiatr Scand 1987; 75:74-77.
29. Chiles JA, Cohen S, Maiuro R, Wright R. Smoking and schizophrenic psychopathology. The American Journal on Addictions 1993; 2:315-319.
30. Kirch DG, Alho AM, Wyatt RJ. Hypothesis: A nicotine-dopamine interaction linking smoking with parkinson's disease and tardive dyskinesia. Cellular and Molecular Neurobiology 1988; 8:285-291.
31. Hughes JR, Howard TS. Nicotine don't get no respect. JAMA 1994; 271:585.

Effects of Nicotine on
Biological Systems II
Advances in Pharmacological Sciences

NICOTINIC RECEPTORS AND THE PATHOPHYSIOLOGY OF SCHIZOPHRENIA

R. Freedman, S. Leonard, L. Adler, P. Bickford, W. Byerley, H. Coon, C. Miller, V. Luntz-Leybman, M. Myles-Worsley, H. Nagamoto, G. Rose, K. Stevens, M. Waldo

Departments of Psychiatry and Pharmacology, Denver VAMC and University of Colorado Health Sciences Center, Denver, CO 80262 and Department of Psychiatry, University of Utah, Salt Lake City, UT 84132

Summary: Nicotine normalizes several physiological abnormalities found in schizophrenia, including a deficit in the inhibitory gating of the P50 evoked response to repeated auditory stimuli. Animal models suggest the involvement of $\alpha7$ nicotinic receptor subunits on hippocampal interneurons. The deficit is distributed in the families of schizophrenics in a pattern consistent with autosomal dominant transmission, and there is some evidence suggestive of linkage at the chromosomal locus of the $\alpha7$ gene.

Introduction

Although cigarette smoking is a common behavior, people who have schizophrenia have long been observed to have a higher prevalence of smoking and to smoke more heavily than the general population. One reason suggested for this behavior is the possibility that nicotine normalizes some aspect of the pathophysiology of the illness (1). This chapter reviews clinical evidence that nicotine normalizes a specific electrophysiological deficit in schizophrenics, pharmacological studies in animal models pointing to the involvement of the $\alpha7$ nicotinic receptor subunit, and genetic studies suggestive of linkage of the physiological defect to the chromosomal location of the gene for the $\alpha7$ subunit.

Inhibitory Gating of Sensory Response in Schizophrenia

Auditory stimulation evokes a series of waves that can be recorded from the scalp surface of humans. As the stimuli are repeated, many of these potentials rapidly diminish in amplitude. This habituation of response is a manifestation of the activity of inhibitory neuronal gating mechanisms. Such habituation permits filtering of repetitive sounds in the environment, so that the brain can focus its attention on more important, changing elements. Schizophrenics have a diminished capacity to filter out unimportant features of their environment, so that their attention is drawn capriciously to many details that normal persons would ignore (2). Their preoccupation with these details can lead to misperception of their environment and perhaps to the hallucinations and delusions characteristic of their illness. For example, schizophrenic patients are often unable to filter out the sound of a refrigerator motor, a sound that normals can hear, but generally ignore; the cycling of the motor can sometimes become part of the patient's delusional beliefs, taking on significance as a sign from an external force. We have used the P50 wave, a midlatency auditory-evoked potential, to characterize differences in filtering of auditory information between schizophrenics and normals (3). The auditory stimuli are delivered in pairs, so that the first stimulus activates inhibitory gating mechanisms that are responsible for diminishing the response to the second stimulus. The ratio of amplitudes of the second P50 wave to the first is a measure of the activity of the inhibitory gating mechanism. Normal subjects have significantly lower ratios (17.8 ± 17.0%; mean ± sem, $n = 43$) than schizophrenics (100.1 ± 31.8%; $n = 37$, $t = 14.6$, df 78, $p < 0.001$). These data suggest that schizophrenic patients have a defect in an inhibitory gating mechanism. This defect is not normalized by treatment with the typical neuroleptic drugs used to treat schizophrenia, such as haloperidol. About half the first-degree relatives of schizophrenics share this defect in the inhibitory gating of the P50 wave. The defect may thus represent part of the genetic risk for schizophrenia, which produces clinical illness in only some family members, in combination with additional genetic and environmental risk factors, such as defects in the structure and volume of various brain regions and alterations in the activity of other neurotransmitters, such as dopamine.

Effects of Nicotine on Sensory Dysfunction in Schizophrenia

Nicotine administration transiently normalizes the defect in the inhibitory gating of the P50 wave in schizophrenics. Because many schizophrenic patients have a history of smoking and take anticholinergic medications as part of the treatment of their illness, the initial tests of nicotine were performed in relatives of schizophrenics (4). Six relatives who had the defect in inhibitory gating of the P50 wave, but no history of smoking or exposure to psychotropic drugs, received nicotine-containing gum in a double-blind, placebo-controlled experiment (Figure 1). There was no significant change in the P50 ratio after placebo, but after nicotine administration the ratio fell from 97.7 ± 38.4% to 44.6 ± 9.6% ($t = 2.42$, df 10, $p < 0.05$).

A subsequent experiment was conducted with 10 schizophrenic patients who smoked, who were studied after overnight abstinence from cigarettes (5). Their initial recordings showed the defect in inhibitory gating of P50 previously described. They were then allowed to smoke cigarettes for 15 minutes, smoking as many as they wished until they felt satisfied. Generally, they smoked three cigarettes in 15 minutes. After smoking, their mean P50 ratio fell to below 25%, a value within the normal range. By contrast, a group of 10 smokers with no psychiatric illness had normal P50 ratios before smoking and a small increase in the ratio after smoking. The effect for the schizophrenics was lost within 30 minutes of smoking. Cigarette smoking has also been reported to normalize another psychophysiological deficit in schizophrenics, their inability to perform smooth pursuit eye movements (6).

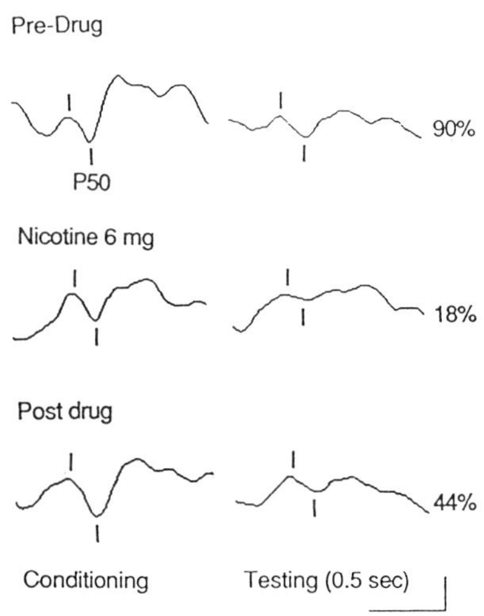

Figure 1. Auditory evoked potentials of the relative of a schizophrenic, showing abnormal gating of the P50 wave in the baseline period, normalization of the deficit after the administration of nicotine-containing gum, and return towards abnormal 15 minutes later. Each potential is the average of 16 responses to paired stimuli, with 0.5 seconds between the first or conditioning and second or testing stimuli. A computer algorithm measured the P50 wave and computed the percentage ratio of the test to conditioning amplitudes. Vertical calibration is -5μV; horizontal is 50 ms. From ref. 4.

Animal Models of Sensory Gating Deficits

The rat P20-N40 wave, recorded from the CA3-CA4 region of the hippocampus shows suppression of response to repeated stimuli similar to the human P50 wave. This animal model was used to identify possible mechanisms of the deficit observed in schizophrenia. In animals, suppression of response is lost after lesion of the fimbria-fornix, the pathway between the septal nuclei to the hippocampus, which contains, among other tracts, the cholinergic afferents to the hippocampus. To determine if a cholinergic mechanism is involved in the inhibitory gating of the P20-N40 response to repeated stimuli, we administered various cholinergic antagonists intraventricularly while recording the hippocampal evoked response. Scopolamine and mecamylamine did not alter the inhibitory gating of response; however, d-tubocurarine blocked this effect. α-bungarotoxin, in doses as low as 50 nM, also blocked the inhibitory gating of P20-N40, producing a physiological defect similar to that observed with the P50 wave in schizophrenics (7). κ-bungarotoxin had no effect. These pharmacological data suggest that receptors containing the α7 subunit may be critical to the inhibitory gating of hippocampal auditory response. [^{125}I]-α-bungarotoxin binds diffusely throughout the hippocampus, but there is particularly intense labeling of a small population of neurons in CA3 and CA4, which are a subpopulation of the GABA-containing interneurons of this region (8). Cholinergic afferents, through α-bungarotoxin-sensitive receptors containing α7 subunits, may excite interneurons, which then discharge in prolonged fashion to produce a long-lasting inhibition of the response of pyramidal neurons to afferents from the perforant path.

Whether or not a deficit in nicotinic cholinergic neurotransmission accounts for the abnormality in the inhibitory gating of sensory response in schizophrenia remains unknown. In post mortem brain tissue from schizophrenics, the cholinergic innervation of the hippocampus appears to be intact (9), so that the defect may be post synaptic. It is possible that there are abnormalities in the GABAergic target neurons or in the cholinergic receptor itself. We have found diminished α-bungarotoxin labeling of apparent interneurons, as well as decreased binding of [^{3}H]-cytisine in post mortem samples of hippocampus from schizophrenics, compared to tissue samples from non-schizophrenics matched for age, sex, and smoking history (10).

Genetic Studies of Sensory Gating Deficits in Schizophrenia

Because the deficit in inhibitory gating is distributed in families of schizophrenic probands in a pattern consistent with autosomal dominant transmission, it is possible to determine if there is linkage to a specific chromosomal location. We have recorded P50 evoked potentials in 9 moderate-sized, multi-generational families, containing 32 schizophrenics and 88 unaffected relatives. Thirty of the schizophrenics had P50 ratios over 50%, as did about half of all family members. The overall linkage results were not significant, with no lod scores over 3.0 for either P50 ratio or schizophrenia as a phenotype (11). Because schizophrenia may be a genetically heterogenous illness, each family was analyzed individually to determine if any showed evidence for linkage, and, because of the small size of the pedigrees, simulations were made to determine the 95% confidence limits of lod scores for unlinked markers. Scores over this limit for any family would indicate an area of possible interest, with only a 5% probability of a false positive. For the 9 families, there were four locations at which there was a lod score for the P50 ratio phenotype over the 95% confidence limit, which could be verified by a flanking marker and which also had a positive lod score for the schizophrenia phenotype. For one family, the location showing this evidence for linkage was on chromosome 4, for another it was on chromosome 7, for a third it was on chromosome 11, and for two families, the location was on chromosome 15 at band q14. Although the lod scores at these locations exceed the 95% confidence limits for unlinked markers, they are still well below 3.0, so that they do not unambiguously indicate linkage.

Nonetheless, the 15q14 location is of interest because it is the site of the gene for the α7 nicotinic receptor subunit (12). Thus, two independent lines of evidence point to the possibility that the α7 receptor contributes to the pathophysiology of schizophrenia: (a) the animal modeling of a physiological deficit found in schizophrenics, and (b) linkage analysis of the deficit in the families of schizophrenics. Further investigation is required, however, to determine whether there are mutations in the α7 gene in schizophrenics that are responsible for the linkage signal and that produce a functional abnormality in the receptor, resulting in the pathological sensory processing observed in the illness.

References

1. Goff DS, Henderson DC, Amico E. Cigarette smoking in schizophrenia: relationship to psychopathology and medication side effects. Am J Psychiatry 1992;149:1189-1194.
2. McGhie A, Chapman JS. Disorders of attention and perception in early schizophrenia. Br J Med Psychol 1961;34:103-117.
3. Adler LE, Pachtman E, Franks R, Pecevich M, Waldo MC, Freedman R. Neurophysiological evidence for a defect in neuronal mechanisms involved in sensory gating in schizophrenia. Biol Psychiatry 1985;17:1284-1296.
4. Adler LE, Hoffer LD, Griffith J, Waldo MC, Freedman R. Normalization by nicotine of deficiient auditory sensory gating in the relatives of schizophrenics. Biol Psychiatry 992;32:607-616.
5. Adler LE, Hoffer LD, Wiser A, Freedman R. Cigarette smoking normalizes auditory physiology in schizophrenics. Am J Psychiatry 1993;150:1856-1861.

6. Klein C, Andersen B. On the influence of smoking upon smooth pursuit eye movements of schizophrenics and normal controls. J Psychophysiol 1991;5:361-369.
7. Luntz-Leybman V, Bickford P, Freedman R. Cholinergic gating of response to auditory stimuli in rat hippocampus. Brain Res 1992;587:130-136.
8. Freedman R, Wetmore C, Stromberg I, Leonard S, Olson L. Alpha bungarotoxin binding to hippocampal interneurons: immunocytochemical characterization and effects on growth factor expression. J Neurosci 1993;13:1965-1975.
9. Karson CN, Casanova MF, Kleinman JE, Griffin WST. Choline acetyltransferase in schizophrenia. Am J Psychiatry 1993;150: 454-459.
10. Leonard S, Rollins Y, Ogel J, Drebing C, Dill D, Hall M, Freedman R. Expression of the alpha7 nicotinic acetylcholine receptor subunit in normal and schizophrenic postmortem brain. Schizophr Res 1993;9:223.
11. Coon H, Plaetke R, Holik J, Hoff M, Myles-Worsley M, Freedman R, Byerley W. Use of a neurophysiological trait in linkage analysis of schizophrenia. Biol Psychiatry 1993;34:277-289.
12. Chini B, Raimond E, Elgoyhen A, Moralli D, Balzaretti M, Heineman S. Molecular cloning and chromosomal localization of the human alpha7-nicotinic receptor subunit gene (CHRNA7). Genomics 1994;19:379-381.

Effects of Nicotine on
Biological Systems II
Advances in Pharmacological Sciences

THE EPIDEMIOLOGY OF CIGARETTE SMOKING AND PARKINSON'S DISEASE

John A. Baron
Dartmouth Medical School, Hanover, NH 03755 USA

Summary: Numerous studies have reported an inverse association between cigarette smoking and Parkinson's disease (PD); these consistently indicate that smokers have about half the risk of PD of non-smokers. The association appears to be causal: it is consistent, graded, and found in most population subgroups. It is possible that a "pre-Parkinson's personality" could explain the association, but this seems unlikely. The relationship between smoking and other dopamine-related processes provides further support for a causal relationship.

Introduction

Parkinson's disease (PD) is a relatively common neurological disorder characterized by degeneration of the pars compacta of the substantia nigra. The resulting loss of dopaminergic neurons in the nigrostriatal tracts result in the bradykinesia, tremor, and rigidity which characterize the disease. The pathology and clinical expression of PD are thus closely associated with an alteration in local neurotransmitter function; approximately 80% of striatal dopamine depletion typically must have occurred before symptoms become clinically apparent (1).

Parkinsonian symptoms may be caused by drugs, or occur as a consequence of encephalitis, but these secondary causes are relatively uncommon; the etiology of idiopathic parkinsonism, i.e. PD, remains obscure. Genetic factors appear to exert only a minor influence, but there are relatively few clues regarding possible environmental determinants of PD. It has been proposed that some exposure associated with rural living (particularly pesticides) increases risk, but this remains controversial (2). Cigarette smoking, however, has repeatedly been found to be inversely related to the risk of PD. Indeed, never smoking is by far the strongest environmental risk factor yet identified for the disease (3,4).

The Epidemiology of Cigarette Smoking and Parkinson's disease

Over 30 studies from North America, Europe and Asia have published data regarding the association between cigarette smoking and PD (3 22). Virtually all of them are consistent with an inverse association. The relative risk for ever smoking has generally been found to be about 0.5; that for current smoking about is probably somewhat lower (3,4). In several studies the PD risk decreased with increasing amount or duration of smoking (5-11). (In some of these studies the trends were not statistically significant, however.)

A variety of research designs have been employed in this research, all leading to the same conclusions. Various case-control studies have utilized prevalent or incident cases drawn from hospitals, clinics, or the general population; controls have been patients with other diseases, or individuals selected from populations at risk. Follow-up studies have focussed on both incident and fatal cases. Research among twins has indicated that in discordant pairs, the smoking twin tends to be protected (20-22). Some studies have been relatively informal, but even the more formal and careful ones have found the apparent protective effect.

The inverse association also seems to be present in all population subgroups in which it has been investigated; there is little indication that the impact of smoking varies by age, gender, or race. A few investigations have suggested that the smoking effect is absent or attenuated among early-onset cases (15,17). However, in these studies the differences in relative risks between age groups were not statistically significant, and other investigations have shown no differences with age, or a stronger effect among younger cases (8,16). Several studies have reported an association that is stronger in men than in women (8,9,12), although one large study reported the opposite (14). The smoking-PD association has been reported in several ethnic and racial groups: North American whites, Northern Europeans, Southern Europeans, and Japanese.

Interpretation of the Epidemiological Data

There is no doubt that there is an association between cigarette smoking and Parkinson's disease; the relevant issue is whether the relationship is causal. Several considerations indicate that it is. Foremost is the consistency and volume of the data, generated over 3 decades in several countries by studies having very different research designs. This pattern of results indicates that the inverse association cannot be explained by response or selection biases in case-control investigations, changes in smoking status after a base-line evaluation in cohort studies, a tendency for the motor deficits of the disease to impair the physical ability to smoke, or the social context of smoking in a particular culture. The decreasing risk over the amount or duration of smoking provides further support for a causal association.

Some studies have reported that intake of coffee or alcohol is inversely related to PD (8,9,11,14,19,22), but typically the association has not been as strong as for cigarette smoking, and other investigations have not found these inverse associations at all (9,10,14,21). Little is known about the independent effects of coffee or alcohol (after taking into account smoking); what data there are (11) indicates that at least some of the association of these beverages with PD is due to the cigarette smoking that often accompanies them. Thus the smoking effect is not due to coffee or alcohol intake, and these other habits seem not to have similarly strong protective relationships with PD risk as does smoking.

Several studies have found a smoking-PD relationship that was not statistically significant, and these "negative" results have at times been described as contradicting a smoking-PD association. Virtually all of these reports were consistent with an inverse association, however, and had estimated relative risks less than 1.0. This actually suggests a protective effect, and does not provide evidence against an association.

Another source of confusion derives from the finding that the age at onset of smoking Parkinson's disease cases seems not to be older than that of non-smokers. Some investigators have taken this to indicate that smoking cannot reduce the risk of PD. However, differences in the age of onset of PD in smokers and non-smokers cannot be used to provide information about the rate at which these two groups get PD, unless the age distribution of the groups is known and accounted for (23). For example, it certainly would be accepted that working as a parachute soldier is more hazardous than being in nursery school. Yet the average age at death of nursery school pupils is lower than that of parachute soldiers; the differences in the age distributions of the two populations make a comparison of the ages at death an invalid comparison of death rates (24). Similar problems will distort inferences from differences in age at onset of PD in smokers vs. non-smokers.

Attempting to reach conclusions about the effect of smoking by comparing PD risk in groups with different smoking prevalences (e.g. men and women) is also not effective. Smoking is not the only factor influencing PD risk; unless the other influences are taken into account, the prevalence of PD in various groups may well not correlate with smoking. This is an example of the difficulties associated with ecological studies (25).

Another factor that has led to confusion is the high mortality caused by smoking. This has prompted some investigators to postulate that the epidemiological findings are an artifact of survival effects. However, even early-onset PD seems to be inversely related to smoking (8,10,16), a finding that cannot be explained by mortality differentials. Moreover, the increased mortality could distort the epidemiological findings for the older cases only if those smokers who die from other causes (e.g. cancer, cardiovascular disease, or lung disease) are largely those who might develop Parkinsonism had they lived. There is certainly no rationale or evidence for this, and such competing risks do not prevent the separate detection of smoking-related risks from coronary heart disease and chronic lung disease among smokers.

If smoking PD cases died at a higher rate than other smokers, the case-control studies of prevalent cases could conceivably be explained as an artifact. However, this would not affect case-control studies of incident cases, or follow-up studies, which have come to the same conclusion as the case-control studies.

Smoking and Parkinson's disease -- a non-causal association?

The problem of indirect associations -- confounding -- is a central concern of epidemiology. In the case of PD, the worry is that the smoking - PD relationship might be due to some other factor -- closely tied to smoking -- which actually causes the reduced PD risk. It is true that most of the PD studies have not adjusted for many covariables. However, since there are so few risk factors for PD other than age, it is not clear what other factors should be adjusted for.

Some commentary has focussed on the notion of a "pre-Parkinson personality," a neurochemical trait that predisposes both to PD and to avoidance of cigarette smoking. This is a possible scenario, since it is plausible that genetic factors or early environmental exposures might determine both neurochemical characteristics related to smoking uptake (typically in the teen-age years) and also to PD (typically much later in life).

In support of this hypothesis, there are indications that Parkinson's disease patients have a tendency to be conservative, rigid, and depressed -- even years before the diagnosis is made (26). However, this is itself controversial, and to explain the smoking-PD relationship, such traits must be very closely associated both with smoking and with PD. If ever smoking confers a relative risk of 0.5, then the Pre-Parkinsonian personality must itself confer a relative risk of at least 2, and have an ever-smoking prevalence no more than half that of individuals with other personalities, a strong association indeed. If there were such a factor, it could not be based on a genetic predisposition, or to early life exposures: findings from twin studies indicate that the smoking-PD association persists even after taking such factors into account (20-22). In addition, despite the close relationship to smoking, the pre-Parkinson personality must not lead to as strong an aversion to coffee or alcohol as to cigarettes. Otherwise, these habits would be as strongly (inversely) related to PD as smoking is.

Smoking, Other Movement Disorders, and Psychiatric Disease

The relationship of smoking with other neuro-psychiatric processes may have implications for the interpretation of its relationship to PD. Issues such as the pre-Parkinson personality will generally not be a concern for these disorders. Several studies have suggested that individuals who take neuroleptic drugs and who also smoke have a lower risk of (drug-induced) parkinsonism than non-smokers (27-30). On the other hand, patients on these drugs who smoke may have higher risk of tardive dyskinesia (30-32), which is believed to be a consequence of excess striatal dopaminergic stimulation. These two relationships support the smoking-PD association in the sense that they indicate that smoking enhances or preserves striatal dopaminergic functioning. Effects on Tourette's syndrome (33) do not fit into this pattern, but do support an impact on dopamine-related processes in the extrapyramidal motor system.

The association between smoking and schizophrenia and major depression also have implications for the hypothesis that smoking facilitates dopaminergic events (34,35). Both these psychiatric disorders have been linked to disturbances in dopaminergic pathways, and it is possible that the high rate of smoking is related to this, or is even self-medicating.

Conclusions

Several aspects of the epidemiological data regarding cigarette smoking and PD suggest that the association is a causal one. The lowered relative risk reflects a substantial inverse association that is graded, consistent over studies, and maintained in various subgroups. It is conceivable that a "pre-Parkinson personality" could explain the relationship, but the available data make this seem unlikely. Preliminary information regarding the association of smoking with other dopamine-related processes in the central nervous system supports the concept of a genuine effect of smoking on PD. Given these data, it is likely that something in cigarette smoke - presumably nicotine - either prevents degeneration of nigrostriatal neurons, or enhances their functioning. A trial of nicotine administration in early PD is certainly justified.

References

1. Hornykiewicz O, Kish SJ. Biochemical pathophysiology of Parkinson's disease. in Yahr MD, and Bergmann KJ (eds). Adv Neurology 1986; 45:19-34.
2. Tanner CM, Langston JW. Do environmental toxins cause Parkinson's disease? A critical review. Neurology 1990; 40 (Suppl 3):17-30.
3. Baron J. Cigarette smoking and Parkinson's disease. Neurology 1986; 36:1490-1496.
4. Graves A. Does smoking reduce the risks of Parkinson's and Alzheimer's diseases? J Smoking Related Diseases, 1994 (in press).
5. Kahn HA. The Dorn study of smoking and mortality among U.S. veterans: report on eight and one-half years of observation. In: Haenszel W, ed. Epidemiological approaches to the study of cancer and other chronic diseases, U.S. Department of Health, Education and Welfare, National Cancer Institute Monograph 19, 1966.
6. Nefzger MD, Quadfasel FA, Karl VC. A retrospective study of smoking in Parkinson's disease. Am J Epidemiol 1968; 88:149-58.
7. Kessler II, Diamond EL. Epidemiologic studies of Parkinson's disease. I. Smoking and Parkinson's disease: a survey and explanatory hypothesis. Am J Epidemiol 1971; 94:16-25.
8. Godwin-Austen RB, Lee PN, Marmot MG, Stern GM. Smoking and Parkinson's disease. J Neurol Neurosurg Psychiatry 1982; 45:577-81.
9. Jimenez-Jimenez FJ, Mateo D, Gimenez-Roldan S. Premorbid smoking, alcohol consumption, and coffee drinking habits in Parkinson's disease: a case-control study. Movement Disorders 1992; 7:339-344.
10. Butterfield PG, Valanis BG, Spencer PS, Lindeman CA, Nutt JG. Environmental antecedents of young-onset Parkinson's disease. Neurology 1993; 43:1150-1158.
11. Grandinetti A, Morens DM, Reed D, MacEachern D. Prospective study of cigarette smoking and the risk of developing idiopathic Parkinson's disease. Am J Epidemiol 1994; 139:1129-1138.
12. Kessler II. Epidemiologic studies of Parkinson's disease. III. A community-based survey. Am J Epidemiol 1972;96:242-54.
13. Marttila RJ, Rinne UK. Smoking and Parkinson's disease. Acta Neurol Scand 1980; 62:322-325.
14. Haack DG, Baumann RJ, McKean HE, et al. Nicotine exposure and Parkinson's disease. Am J Epidemiol 1981; 114:191-200.
15. Stern M, Dulaney E, Gruber SB, Golbe L, Bergen M, Hurtig H, Gollmp S, Stolley P. The epidemiology of Parkinson's disease. A case-control study of young-onset and old-onset patients. Arch Neurol 1991; 48:903-907.
16. Zuber M, Verdier-Taillefer M-H, Alperovitch M, de Recondo J. Smoking and Parkinson's disease: differences according to age at disease onset. Neuroepidemiolgy 1991; 10:103-4.
17. Rajput AH, Offord KP, Beard M, Kurland LT. A case-control study of smoking habits, dementia, and other illnesses in idiopathic Parkinson's disease. Neurology 1987; 37:226-232.
18. Semchuk KM, Love EJ, Lee RG. Parkinson's disease: a test of the multifactorial etiologic hypothesis. Neurology 1993; 43:1173-1180.
19. Hubble JP, Ramachandran V, Hassenein RES, Gray C, Koller WC. Personality and depression in Parkinson's disease. J Nervous and Mental Dis 1993; 181:657-662.
20. Duvoisin RC, Eldridge R, Williams A, Nutt J, Calne D. Twin study of Parkinson disease. Neurology (NY) 1981;31:77-80.

21. Ward CD, Duvoisin RC, Ince SE, Nutt JD, Eldridge R, Calne DB. Parkinson's disease in 65 pairs of twins and in a set of quadruplets. Neurology (Cleveland) 1983;33:815-24.
22. Bharucha NE, Stokes L, Schoenberg BS, Ward C, Ince S, Nutt JG, Eldridge R, Calne DB, Mantel N, Duvoisin R. A case-control study of twin pairs discordant for Parkinsons' disease: a search for environmental risk factors. Neurology 1986; 36:284-288.
23. Baron JA, Greenberg ER. letter to the Editor: Parkinson's disease and smoking. Arch. Neurol. 44:1110-1, 1987.
24. Rothman KJ. Left-handedness and life expectancy. N Engl J Med 1991; 325:1041 (letter).
25. Piantadosi S, Byar DP, Green SB. The ecological fallacy. Am J Epidemiol 1988; 127:893-904.
26. Paulson GW, Dadmehr N. Is there a premorbid personality typical for Parkinson's disease? Neurology 1991; 41 (Suppl 2):73-76.
27. Reshef A, Rabey JM, Bar P, Schlosberg A, Korczyn AD. Smoking and drug-induced parkinsonism. Neurology 1987; 37:121.
28. Wagner B, Wolf GK, Ulmar G. Does smoking reduce the risk of neuroleptic parkinsonoids. Parmacopsychiat 1988; 21:302-303.
29. Decina P, Caracci G, Sandik R, Berman W, Mukherjee S, Scapicchia P. Cigarette smoking and neuroleptic-induced parkinsonism. Biol Psychiatry 1990; 28:502-508.
30. Yassa R, Lal S, Korpassy A, Ally J. Nicotine exposure and tardive dyskinesia. Biol Psychiatry 1987; 22:67-72.
31. Menza MA, Grossman N, Van Horn M, Cody R, Forman N. Smoking and movement disorders in psychiatric patients. Biol Psychiatry 1990; 30:109-115.
32. Binder R, Kazamatsuri H, Nishimura T, McNiel DE. Smoking and tardive Dyskinesia. Biol. Psychiatry 1987; 22:1280-1282.
33. McConville BJ, Fogelson MH, Norman AB, Klykylo WM, Manderscheid PZ, Parker KW, Sanberg PR. Nicotine potentiation of haloperidol in reducing tic frequency in Tourette's disorder. Am. J. Psychiatry 1991; 148:793-794.
34. Newhouse PA, Hughes JR. The role of nicotine and nicotinic mechanisms in neuropsychiatric disease. Br J Addiction 1991; 86:521-526.
35. Glassman AH. Cigarette smoking: implications for psychiatric illness. Am J Psych 1993; 150:546-553.

Effects of Nicotine on
Biological Systems II
Advances in Pharmacological Sciences

NICOTINE AND ANIMAL MODELS OF PARKINSON'S DISEASE

A. M. Janson[1], A. Møller[2], P. B. Hedlund[1], G. von Euler[1] and K. Fuxe[1]

[1]Department of Neuroscience, Karolinska Institute, S-171 77 Stockholm, Sweden.
[2]NeuroSearch, 26 Smedeland, DK-2600 Glostrup and
Stereological Research Laboratory, DK-8000 Aarhus, Denmark

Summary: The effect of chronic continuous (-)nicotine treatment in lesioned nigrostriatal dopamine (DA) systems was studied in rats with a partial unilateral mesodiencephalic lesion. (-)Nicotine significantly counteracted both the lesion-induced reduction in total number of nigral tyrosine-hydroxylase-like (TH-li) immunoreactive neurons and the lesion-induced upregulation of the high-affinity binding sites of the striatal DA D_2 receptors analyzed with [^{3}H]N-propylnorapomorphine. The suggested mechanism for the neuroprotection is a functional desensitization of the nicotinic cholinoceptors located on DA nerve cells. In a different lesion model, (-)nicotine either attenuated or enhanced 1-methyl-4-phenyl-1,2,3,6-tetrahydropyridine (MPTP)-induced neurotoxicity in mice depending on the dose-schedule. Here, we suggest an action of MPTP on the nicotinic cholinoceptors preventing functional desensitization from taking place.

Introduction

Parkinsonism is linked to a degeneration of DAergic neurons, particularly targeting those in the substantia nigra pars compacta (1). The mechanisms underlying these neurodegenerative changes still remain to be elucidated. Epidemiological evidence indicates that the incidence of Parkinson's disease is lower in cigarette smokers compared to non-smokers (2). The hypothesis that nicotine exerts neuroprotective effects in lesioned DA systems has been addressed in several studies (3-9). Here, we summarize some of our recent studies on animal models of parkinsonism, which attempt to elucidate the differential effects of various (-)nicotine treatment protocols on mechanical and neurotoxic lesions of the nigrostriatal DA system.

Materials and Methods

Animal and tissue treatments. Male Sprague-Dawley rats (B&K Universal, Stockholm, Sweden) were partially hemitransected at the right mesodiencephalic junction (3,9,10). (–)Nicotine ((-)nicotine-hydrogen (+)tartrate, BDH Chemicals, Poole, UK, 0.125 mg/kg/h dissolved in 0.9 % saline) was delivered for two weeks via subcutaneously implanted osmotic minipumps. To allow prompt high concentrations in the brain, four (-)nicotine injections (0.5 mg/kg, ip, 30 min time intervals) were given starting immediately following the lesion. Control animals received saline instead of (-)nicotine. Two weeks after the partial unilateral transection, the animals used for quantitative immunohistochemistry were perfused transcardially with saline followed by 4 % paraformaldehyde and 0.2 % picric acid in 0.1M PBS as described previously (8). The rats used for the binding experiment were decapitated using a guillotine two weeks after the unilateral lesion (9). In the mouse experiments male C57 Bl/6 mice (B&K Universal, Stockholm, Sweden) were killed by cervical dislocation. The neostriatal membranes were prepared as described earliet (11).

Quantitative immunocytochemistry. Systematically sampled sections from the midbrain were used for immunocytochemistry with antibodies against tyrosine hydoxylase (Dr Menek Goldstein, New York) and glial fibrillary acidic protein (Dakopatts, Glostrup, Denmark). The avidin-biotin-immunoperoxidase system (Vectastain, Burlingame, CA, USA) together with 3,3'-diaminobenzidine tetrahydrochloride (DAB, Sigma, St Louis, MO, USA) was used to visualize the cells, employing cresyl violet as a counterstain (8).

The stereological analysis (12) of the stained cells was performed using an Olympus CAST-system. The morphological measurements were carried out with this unbiased quantitative technique in a three-dimensional fraction of the entire volume of substantia nigra. The total number of four neuronal and non-neuronal cell populations were analyzed (see legend to Fig. 1) (8).

[^{3}H]N-propylnorapomorphine binding. Cryostat sections were incubated under equilibrium conditions (1 h, 23 °C) with 0.15-13 nM (-)N-propyl-[^{3}H]N-propylnorapomorphine ([^{3}H]NPA, 2.3 TBq/mmol, NEN, Boston, MA, USA) in a 50 mM Tris buffer (pH 7.4) containing 5 mM $MgCl_2$, 1 mM EDTA, 0.01 % L-ascorbic acid and 0.05 % bovine serum albumin. This agonist labels D_2 but not D_1 receptors *in vitro* (13). Saturation curves with ten concentrations of the ligand were performed, and the binding in the presence of the D_2 antagonist raclopride (10 μM;

Astra, Södertälje, Sweden) was defined as nonspecific. The intact and lesioned halves of the section were carefully wiped off with slightly moist GF/B filter papers (Whatman International, UK), and analyzed separately. The radioactivity of the filters was determined by liquid scintillation spectrometry (Beckman LS 1800, Irvine, CA, USA). The K_d and B_{max} values were calculated using a computer-assisted iterative non-linear regression analysis (9-14).

[^{3}H]methylcarbamylcholine binding. Neostriatal membranes were incubated under equilibrium conditions (1 h, 4°C) with 3.3 nM N-methyl-[^{3}H]methylcarbamylcholine iodide ([^{3}H]MCC, 3.2 TBq/mmol, NEN, Boston, MA, USA) in 50 mM Tris buffer (pH 7.4) containing 1 mM $MgCl_2$, 120 mM NaCl, 5.0 mM KCl and 2.0 mM $CaCl_2$ (11). This agonist selectively labels nicotinic cholinoceptors (15). Competitive-inhibition curves (each curve used tissue pooled from 10 mice) with 20 different concentrations (1 nM - 1 mM) of MPTP were analyzed. The binding in the presence of 10 μM (-)nicotine was defined as nonspecific. The radioactivity determination and curve analysis are described above.

Results

Quantitative immunocytochemistry. The partial mesodiencephalic hemitransection lead to a 65% decrease in the total number of TH-li + CV neurons, whereas neurons stained with CV alone showed a 20 % decrease (Fig. 1). In the non-neuronal cell populations a lesion-induced increase in the total number of cells was seen. (-)Nicotine significantly counteracted the lesion-induced decrease in total number of TH-li + CV neurons compared to saline-treated controls, but did not affect the other cell populations studied ipsilateral to the lesion. No significant effects of the (-)nicotine treatment were observed on the non-lesioned side in the four cell populations analyzed. See also Janson and Møller, 1993 (8).

[^{3}H]NPA binding. The partial mesodiencephalic hemitransection in saline-treated animals significantly increased the neostriatal B_{max} value of [^{3}H]NPA binding ipsilateral to the lesion, whereas the K_d value was not significantly affected. (-)Nicotine treatment significantly ($P<$ 0.05, two-tailed unpaired t-test) counteracted the lesion-induced increase in [^{3}H]NPA binding (Fig. 2).

[^{3}H]MCC binding. MPTP displaced the neostriatal [^{3}H]MCC binding (Fig. 3) with an IC_{50} value of 10.7 (6.1-18.9) μM (geometric mean, standard error limits), n = 3.

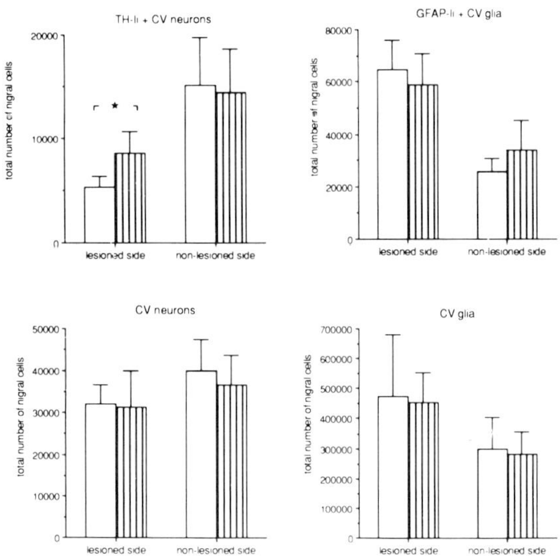

Figure 1. Effects of chronic continuous (-)nicotine treatment on the total number of nigral cells in rats with a partial mesodiencephalic hemitransection. Two populations of neuronal cells (upper and lower left panel, respectively) were counted ipsilaterally (lesioned side) and contralaterally (non-lesioned side) to the lesion in saline (open bars) and (-)nicotine-treated (striped bars) animals: those showing tyrosine-hydroxylase-like immunoreactivity with cresyl violet counterstaining (TH-li + CV) and those showing Nissl staining alone (CV neurons). Two non-neuronal cell populations (upper and lower right panel, respectively) were also quantified in the same animals: those showing glial fibrillary acidic protein-like immunoreactivity counterstained with cresyl violet (GFAP-li + CV glia) and those labelled with cresyl violet alone (CV glia). Means ± SD, n=5. Statistical analysis was performed comparing corresponding sides in saline- and (-)nicotine-treated rats with two-tailed unpaired t-test applying the Bonferroni procedure. * $P < 0.05$. The coefficient of error was 0.03 - 0.04. For details on treatment and procedures[17], see Materials and Methods. Modified from Janson and Møller, 1993 (8).

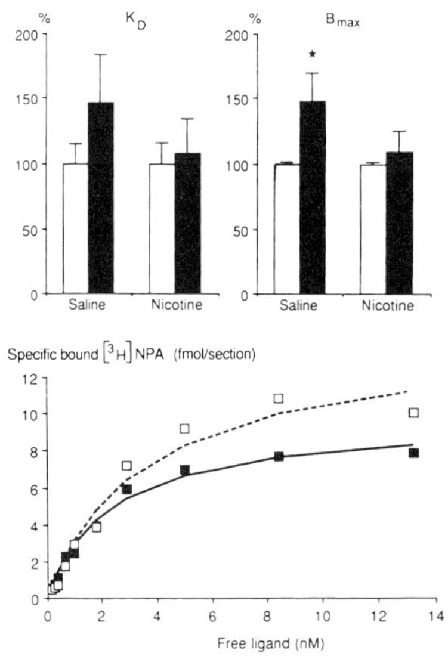

Figure 2. Effects of chronic continuous (-)nicotine treatment on changes in [^{3}H]NPA binding parameters induced by a partial mesodiencephalic hemitransection in rats. The K_d and B_{max} values on the lesioned side (filled bars) are expressed as a percentage (mean± S.E.M., n=6) of the intact side (open bars). The absolute K_d and B_{max} values on the intact side were 3.9 nM and 11.4 fmol/section in saline-treated rats and 3.4 nM and 9.9 fmol/section in (-)nicotine-treated rats respectively. Statistical analysis comparing the lesioned and non-lesioned side was performed with a two-tailed paired t-test applying the Bonferroni procedure. * $P<$ 0.05. For details on treatment and procedures[20], see Materials and Methods. Modified from Janson et al., 1994 (9).

Discussion

In this and earlier work we have studied the effects of (-)nicotine on nigrostriatal dopamine systems in rodents using two different animal models: a unilateral mechanical partial lesion and a neurotoxin-induced lesion using the selective neurotoxin MPTP in the mouse (7). The mechanical lesion spares the most medial portion of the nigrostriatal and striatonigral pathways (3,9-10) and has the advantage of leaving the contralateral nigrostriatal dopamine system virtually intact (3,5,8). In the MPTP-model we used a low dose (40 mg/kg, sc) of MPTP to allow compensatory mechanisms in surviving DA neurons (7).

The putatively neuroprotective actions by (-)nicotine in substantia nigra demonstrated in this and earlier morphological (3,8) as well as biochemical studies (4) have also been shown to involve neostriatal DA parameters evaluating TH-IR nerve terminals (3) as well as tissue (4) and *in vivo* extracellular levels of DA (5). The total number of neuronal and non-neuronal

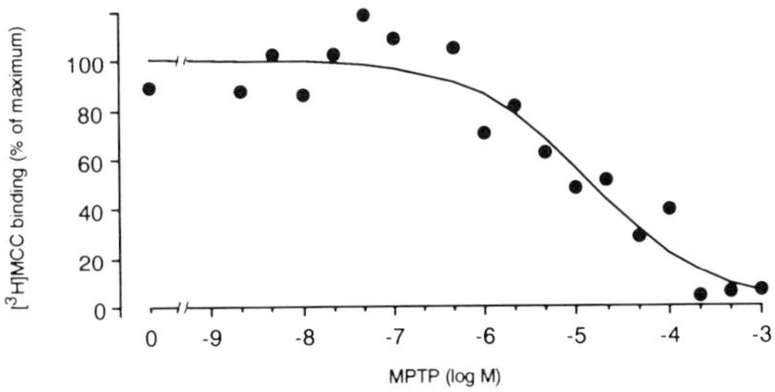

Figure 3. Representative competitive-inhibition curve showing the effect of MPTP on [^{3}H]MCC binding in neostriatal membrane preparations of the male C57 Bl/6 mouse. The IC_{50} value of this curve was 14.4 μM. For details on treatment and procedures, see Materials and Methods.

cell populations in substantia nigra have now been studied using a stereological approach which has the advantage of yielding unbiased estimates with a known coefficient of error. This quantitative analysis allows full resolution of individual cells in the densely populated substantia nigra and is independent of tissue shrinkage phenomena.

Compensatory phenomena on the lesioned side have been studied using a similar lesion protocol. Here, we demonstrate a lesion-induced upregulation of DA D_2 receptors in the neostriatum ipsilateral to the lesion. Previously, presynaptic compensatory mechanisms have been found including an increased burst firing in the DA neurons identified on the lesioned side *in vivo* (6). This firing pattern has been suggested to be the major determinant of their terminal release (16). Accordingly, ipsilateral to the lesion an increase in striatal DA utilization was observed (4). (-)Nicotine counteracts all these pre- and postsynaptical compensatory mechanisms, the latter probably reflecting an increased presence of DA on the lesioned side.

The suggested mechanism of action is probably linked to a functional desensitization of the nicotinic cholinoceptors (17) located on the DA nerve cell bodies, dendrites and terminals (3,9,18). Thus, chronic continuous (-)nicotine treatment could switch the low-affinity agonist-binding state of the nicotinic cholinoceptor with an open ion-channel towards a high affinity binding state with a high probability for a closed ion channel (19). As a corollary, the influx of cations including Na^{+} and Ca^{2+} in the neuron is reduced (20-21), which could lead to reduced energy demands to maintain ion homestasis. The reduction of Ca^{2+} influx could contribute to the putative increased survival of DA neurons (22).

In the MPTP lesioned mouse acute, but not chronic, (-)nicotine treatment has been shown to counteract the toxin-induced lesion if given 10 min before or at the same time as MPTP, but not 10 min later (7). These effects might be linked to an acute (-)nicotine-induced increased DA release, which interferes with the uptake of the neurotoxic metabolite MPP^{+} shown to utilize the DA transporter to enter the neuron (23). Chronically, however, (-)nicotine leads to a dose-dependent enhanced DA toxicity (7). These seemingly contradictory results might be linked to an action of MPTP or its metabolites on the nicotinic cholinoceptor interacting with the desensitization.

Here, we show that MPTP in itself may displace the agonist binding site with an IC_{50} of 10.7 μM. This is in line with the curare-like action of MPTP, demonstrated outside the central nervous system (24). Single channel recordings of nicotinic cholinoceptors in myocyte culture have shown that MPTP in the concentration range 10-500 μM appears to interact with both the ligand binding part of the receptor and its associated ionic channel (25). This effect of MPTP may interfere with the agonist binding to the nicotinic cholinoceptor preventing desensitization to take place.

In conclusion, chronic continuous (-)nicotine treatment has protective effects in lesioned dopamine systems, which may lead to the development of new neuroprotective therapies in Parkinson's disease. Further studies should be performed to elucidate the action of MPTP and its metabolites on the nicotinic cholinoceptor.

Acknowledgements: *The present study was supported by grants from the Swedish Tobacco Company and from the Groschinsky and Hierta foundations, the Swedish Medical Society Foundations and the Swedish Medical Research Council (12X-10816-02 and 14X-10377-03). We thank Beth Andbjer, Ulla-Britt Finnman, Ulla Hasselrot and Kioumars Delfanis for* *excellent technical assistance. The rabbit polyclonal tyrosine hydroxylase antiserum was kindly provided by Dr Menek Goldstein, Dept of Psychiatry, New York Univ. Med. Center, NY, USA.*

References

1. Marsden CD. Parkinson's disease. Lancet 1990; 335:948- 952.
2. Baron JA. Cigarette smoking and Parkinson's disease. Neurology 1986; 36: 1490-1496.
3. Janson AM, Fuxe K, Agnati LF, Kitayama I, Härfstrand A, Andersson K, Goldstein M. Chronic nicotine treatment counteracts the disappearance of tyrosine-hydroxylase-immunoreactive nerve cell bodies, dendrites and terminals in the mesostriatal dopamine system of the male rat after partial hemitransection. Brain Res. 1988; 455:332- 345.
4. Fuxe K, Janson AM, Jansson A, Andersson K, Eneroth P, Agnati LF. Chronic nicotine treatment increases dopamine levels and reduces dopamine utilization in substantia nigra and in surviving forebrain dopamine nerve terminal systems after a partial di-mesencephalic hemitransection, Naunyn-Schmiedeberg's Arch. Pharmacol. 1990; 341:171-181.
5. Janson AM, Meana JJ, Goiny M, Herrera-Marschitz M. Chronic nicotine treatment counteracts the decrease in extracellular neostriatal dopamine induced by a unilateral transection at the meso-diencephalic junction in rats: a microdialysis study. Neurosci. Lett. 1991; 134:88-92.
6. Grenhoff J, Janson AM, Svensson TH, Fuxe K. Chronic continuous nicotine treatment causes decreased burst firing of nigral dopamine neurons in rats partially hemitransected at the meso-diencephalic junction. Brain Res. 1991; 562:347-351.
7. Janson AM, Fuxe K, Goldstein M. Differential effects of acute and chronic nicotine treatment on MPTP-(1-methyl-4-phenyl-1,2,3,6-tetrahydropyridine) induced degeneration of nigrostriatal dopamine neurons in the black mouse, Clin. Investig. 1992; 70:232-238.
8. Janson AM, Møller A. Chronic nicotine treatment counteracts nigral cell loss induced by a partial mesodiencephalic hemitransection: an analysis of the total number and mean volume of neurons and glia in substantia nigra of the male rat. Neuroscience 1993; 57:931-941.
9. Janson AM, Hedlund PB, Fuxe K, von Euler G. Chronic nicotine treatment counteracts dopamine D_2 receptor upregulation induced by a partial meso-diencephalic hemitransection in the rat. Brain Res. 1994; 655:25-32.
10. Agnati LF, Fuxe K, Calza L, Benfenati F, Cavicchioli F, Toffano G, Goldstein M. Gangliosides increase the survival of lesioned nigral dopamine neurons and favour the recovery of dopaminergic synaptic function in striatum of rats by collateral sprouting, Acta Physiol. Scand. 1983; 119:347-363.
11. Von Euler G, Fuxe K, Finnman U-B, Agnati LF. Acute and subchronic corticosterone treatment differentially modulates subcortical limbic and neostriatal nicotinic cholinergic receptors. Brain Res. 1990; 526: 122-126.
12. Gundersen HJG, Bagger P, Bendtsen TF, Evans SM, Korbo L, Marcussen N, Møller A, Nielsen K, Nyengaard JR, Pakkenberg B, Sorensen FB, Vesterby A, West M. The new stereological tools: Disector, Fractionator, Nucleator and point sampled intercepts and their use in pathological research and diagnosis. Review Article. APMIS 1988; 96:857-881.
13. Von Euler G, Fuxe K, Benfenati F, Hansson T, Agnati LF, Gustafsson JÅ. Neurotensin modulates the binding characteristics of dopamine D_2 receptors in rat striatal membranes also following treatment with toluene, Acta Physiol Scand. 1989; 135:443- 448.
14. Hedlund P, von Euler G, Fuxe K. Activation of 5-hydroxytryptamine $_{1A}$ receptors increases the affinity of galanin receptors in di- and telencephalic areas of the rat. Brain Res. 1991; 560:251-259.
15. Abood LG, Grassi S. ^{3}H-methylcarbamylcholine: a new radioligand for studying brain nicotinic receptors. Biochem. Pharmacol. 1986; 35:4199-4202.
16. Gonon FG, Buda MJ. Regulation of dopamine release by impulse flow and by autoreceptors as studied by in vivo voltammetry in the rat striatum, Neuroscience 1985; 14:765-774.
17. Marks MJ, Burch JB, Collins AC. Effects of chronic nicotine infusion on tolerance development and nicotinic receptors. J. Pharmacol. Exp. Ther. 1983; 226:817-825.
18. Clarke PB, Pert A. Autoradiographic evidence for nicotine receptors on nigrostriatal and mesolimbic dopaminergic neurons. Brain Res. 1985; 348:353-358.

19. Heidmann T, Bernhart J, Neumann E, Changeux J-P. Rapid kinetics of agonist binding and permeability response analyzed in parallell on acetylcholine receptor rich membranes from Torpedo marmorata. Biochemistry 1983; 22:5452-5459.
20. Changeux J-P. The TiPS lecture. The nicotinic acetylcholine receptor: an allosteric protein prototype of ligand-gated ion channels. Trends Pharmacol. Sci. 1990; 11:485-492.
21. Noronha-Blob L, Gover R, Baumgold J. Calcium influx mediated by nicotinic receptors and voltage gated calcium channels in SK-N-SH human neuroblastoma cells. Biochem. Biophys. Res. Commun. 1989; 162:1230-1235.
22. Choi DW. Calcium-mediated cytotoxicity: relationship to specific channel types and role in ischemic damage. Trends Neurosci. 1988; 11:465-469.
23. Javitch JA, D'Amato RJ, Strittmatter SM, Snyder SH. Parkinsonism-inducing neurotoxin N-methyl-1,2,3,6-tetrahydropyridine: uptake of the metabolite N-methyl-4-phenylpyridine by dopamine neurons explains selective toxicity. Proc. Natl. Acad. Sci. U.S.A. 1985; 82:2173-2177.
24. Hsu KS, Fu WM, Lin-Shiau SY. Studies on the neuromuscular blocking actions of MPTP in the mouse phrenic nerve-diaphragm. Neuropharmacology 1993; 32:597-603.
25. Hsu KS, Fu WM, Lin-Shiau SY. Blockade by MPTP of the nicotine acetylcholine receptor channesl in embryonic Xenopus muscle cells. Neuropharmacology 1994; 33 35-41.

Effects of Nicotine on
Biological Systems II
Advances in Pharmacological Sciences

NICOTINE EFFECTS ON MEMORY PERFORMANCE

Edward D. Levin[1] and Diane Torry[2]

[1]Departments of Psychiatry and Pharmacology, Duke University Medical Center, Durham, NC 27710, USA, [2]Department of Biochemistry, University of Bath, UK.

Summary: Nicotine has been found by many but not all studies to improve and mecamylamine to impair performance on cognitive tasks in humans and experimental animals. In our laboratory, a very reproducible finding has been that chronic infusion of nicotine significantly improves memory performance in a win-shift version of the radial-arm maze. Nicotine attenuates the memory deficits caused by lesions of the fimbria-fornix or the medial basalocortical projection. Chronic co-administration of the nicotinic antagonist mecamylamine eliminates the nicotine effect. Some aspects of the cognitive deficit in Alzheimer's disease are attenuated by nicotine, suggesting promise for the therapeutic use of nicotine or other nicotinic ligands for cognitive dysfunction.

Nicotine Effects

Nicotine has been found to have a variety of cognitive enhancing properties in humans including improved attention and memory functioning (1). The most consistent finding is that nicotine and cigarette smoking improves vigilance and rapid information processing (1-3). Smoking improves vigilance performance by attenuating the decline in performance which normally occurs over time (3). However, in some cases it is difficult to determine how much the improved performance after smoking is due to relief of the deficits shown during nicotine withdrawal (4). Some studies have documented nicotine-induced improvements in memory function, but other investigators have found deficits, no facilitation or variable effects (for a review see (5)). Roth and co-workers (6) found that subjects who typically smoked their first cigarette within the first hour after getting up performed the memory tasks better after smoking than after deprivation. The reverse was true for subjects who smoked their first cigarette later.

A wide variety of animal studies have shown that nicotine administration improves learning and memory performance; however, some studies have found that nicotine administration impairs cognition or has no detectable effects (for a review see (5)). We have found with rats that acute nicotine injections (0.2 mg/kg) improve working memory in the win-shift radial-arm maze test (7, 8). This facilitation is blocked by the nicotinic antagonist mecamylamine and the muscarinic antagonist scopolamine (7). Acute intracerebroventricular (ICV) nicotine infusion improved radial-arm maze choice accuracy in rats with little training on the radial-arm maze but not in rats which were highly trained (9). However, even in the highly trained rats, ICV nicotine was effective in reversing the impairment caused by mecamylamine. Interestingly, infusion of nicotine into the ventral tegmental area (VTA) or substantia nigra (SN) impaired radial-arm maze memory performance (10). This was reversed by mecamylamine, which at a higher dose caused a deficit.

Chronic nicotine administration has also been found to improve memory performance. We have consistently found in eight experiments that chronic nicotine infusions of 12 mg/kg/day improves memory performance in the radial-arm maze over periods of 3-4 weeks (11-16). No tolerance to the memory enhancing effects was seen. In some studies we have found the improvement to persist well after nicotine withdrawal (11, 12, 17). The nicotine effect is reversed by concurrent administration of chronic mecamylamine (13). Acute challenge with mecamylamine caused an overall impairment in both control and nicotine groups, but the enhanced performance of the nicotine group relative to controls was preserved (11). In contrast, acute administration with the muscarinic antagonist scopolamine selectively eliminated the chronic nicotine-induced improvement in choice accuracy (11). Like acute nicotine effects on cognitive performance, chronic nicotine-induced working memory facilitation seems to be task dependent. The same dose of nicotine which consistently improved performance in the win-shift radial-arm maze working memory task had no significant effect on T-maze spatial alternation performance. This task has a greater component of proactive interference which may have detracted from the nicotine effect. Dunnett and colleagues (18) have found that nicotine increases the effects of proactive interference.

Mecamylamine Effects

Mecamylamine a nicotinic antagonist has been used as a probe of nicotinic involvement in memory function. It has been found to impair memory performance of rodents in several tasks including passive avoidance and the working memory version of the radial-arm maze (for a review see (5)). We have repeatedly found that acute mecamylamine impairs working memory in the win-shift radial-arm maze (for a review see (5)).

Some experiments have detected paradoxical improvements with mecamylamine administration. Moran (19) showed that acute mecamylamine (1-10 mg/kg) improved choice accuracy in a T-maze spatial alteration task when 30 second delays were used, whereas it impaired performance when no delays were used. Recently, we have found similar effects. Mecamylamine (2.5-10 mg/kg) caused a dose-related improvement in choice accuracy in a delayed matching to position radial-arm maze task which used delays of 15-60 seconds (Levin, unpublished data). Mecamylamine may have attenuated proactive interference during the delays. Paradoxical effects can also be seen with chronic administration. Chronic mecamylamine (3 mg/kg/day) produced a paradoxical improvement in win-shift radial-arm maze choice accuracy during the first week of administration (13). Similarly, the same mecamylamine dose caused a paradoxical improvement in choice accuracy during the first week at a short delay in a spatial alternation T-maze task (15).

Acute ICV mecamylamine impairs memory performance (9, 20). More specifically, memory deficits have been seen after mecamylamine infusions into the VTA, SN (10), hippocampus (21, 22) or amygdala (23). Nicotine reverses mecamylamine-induced deficits produced by ICV (9) or amygdala (23) infusion.

Mecamylamine effects on cognitive performance have not been widely investigated in humans. One study has shown that mecamylamine slows cognitive performance in humans (24). A recent study by Newhouse (25) showed that mecamylamine caused significant deficits in learning, increased response bias and slowed reaction time, with elderly subjects showing enhanced sensitivity.

Interactions with Dopaminergic Systems

We have found that nicotinic receptors have important interactions with dopamine (DA) systems with regard to working memory function. The choice accuracy impairment in the radial-arm maze caused by the nicotinic blocker mecamylamine is potentiated by a D_2 blocker (26) and is reversed by a D_2 agonist (27), whereas D_1 ligands had no interactive effects. In contrast, muscarinic receptors interact more closely with D_1 receptors. The radial-arm maze choice accuracy impairment caused by the muscarinic blocker scopolamine is reversed by a D_1 but not a D_2 antagonist (28). The D_1/D_2 agonist pergolide (29) and the D_2/D_3 agonist quinpirole (30) both have mutually potentiating effects when given together with acute doses of 0.2 mg/kg of nicotine. Interesting, we recently found that this dose of nicotine is effective in attenuating a choice accuracy deficit elicited by the D_1 agonist SKF 38393 (30).

The interactions of DA systems with chronic nicotine effects seems to be less apparent than with acute nicotine effects. In three recent studies we examined the effects of D_2 manipulations in rats chronically administered nicotine (16). Nicotine-induced improvements in working memory performance were seen in all three studies. Neither raclopride, a D_2 antagonist, or quinpirole, a D_2/D_3 agonist, altered the choice accuracy improvement caused by nicotine.

Nicotine-Induced Reversal of Cognitive Deficits

The evidence for nicotine-induced improvement in attentiveness and memory discussed above suggests that nicotine or other nicotinic ligands may be useful as therapies for cognitive dysfunction associated with syndromes such as Alzheimer's disease. A large number of studies have found substantial declines in cortical nicotinic receptor density in Alzheimer's disease (for a review see (5)). Acute nicotine improved choice accuracy performance of aged rats (31) and monkeys (32). Acute nicotine also counteracted the memory deficits seen after lesions of the nucleus basalis (33, 34), septum (35) and hippocampus (36). We recently studied the effects of chronic nicotine administration in reversing fimbria-fornix or basalocortical lesions (14). Chronic nicotine infusion (12 mg/kg/day) significantly improved radial-arm maze choice accuracy regardless of lesion status, such that the lesioned rats with nicotine administration were not different from untreated control rats.

Newhouse et al. (37) have found that acute intraveneous nicotine in a dose of 0.25 μg/kg/min. for one hour significantly improved performance on a verbal recall task in Alzheimer's patients. However, adverse effects of anxiety and depression were also seen. Sahakian, Jones and co-workers (38, 39) found that subcutaneous injections of nicotine (0.4, 0.6 and 0.8 mg/kg) in Alzheimer's patients caused a significant dose-related improvement in attention and rapid information processing but did not improve short-term memory as assessed by the digit span test.

Chronic administration is probably necessary for successful treatment of Alzheimer's disease, whether the effect of the drug is to alleviate the symptoms or to arrest the neurodegeneration. The necessity for chronic administration presents problems for most agonists which would cause receptor downregulation and diminished effect with chronic administration. The mixed agonist/antagonist effects of nicotine may impair its efficacy as a treatment for Alzheimer's disease (40), but this property may also be advantageous for continued efficacy with chronic administration. The deficit in nicotinic receptors in Alzheimer's patients may impair response to nicotine. We have found that blocking nicotinic receptors with chronic mecamylamine is effective in blocking the chronic nicotine-induced memory improvement (13). However, the preliminary clinical data show that even if there is diminished response to nicotine in Alzheimer's patients, nicotine treatment is still potent enough to have beneficial effects. The effectiveness of nicotine in attenuating cognitive deficits caused by aging and lesions is encouraging for the development of nicotinic-based treatments for cognitive impairment in humans.

Acknowledgments: *Funding from the Council for Tobacco Research-USA, RJ Reynolds, the Alzheimer's Association and the National Science Foundation helped make these studies possible.*

References

1. Warburton DM. Nicotine as a cognitive enhancer. Prog. Neuro-Psychopharmacol. Biol. Psychiat. 1992; 16: 181-919.
2. Warburton DM, Wesnes K. Mechanisms for habitual substance use: Food, alcohol and cigarettes. In: Gale A, Edwards JA eds., Physiological Correlates of Human Behaviour, Vol. 1: Basic Issues, London, Academic Press, 1984: 277-297.
3. Wesnes K, Warburton DM. Smoking, nicotine and human performance. Pharmacol. Ther. 1983; 21: 189-208.
4. Heishman SJ, Snyder FR, Henningfield JE. Effect of repeated nicotine administration in nonsmokers. NIDA research 1991; series 105: 314-315.
5. Levin E. Nicotinic systems and cognitive function. Psychopharmacology 1992; 108: 417-431.
6. Roth N, Lutiger B, Hasenfratz M, Bättig K, Knye M. Smoking deprivation in "early" and "late" smokers and memory function. Psychopharmacology 1992; 106: 253-260.
7. Levin ED, Rose JE. Nicotinic and muscarinic interactions and choice accuracy in the radial-arm maze. Brain Res Bull 1991; 27: 125-128.
8. Levin ED, Rose JE, Abood L. Effects of nicotinic dimethylaminoethyl esters on working memory performance of rats in the radial-arm maze. Pharmacol. Biochem. Behav. 1994; submitted.
9. Brucato FH, Levin ED, Rose JE, Swartzwelder HS. Interocerebroventricular nicotine and mecamylamine alters radial-arm maze performance in rats. Drug Dev. Res. 1994; 31: 18-23.
10. Levin ED, Briggs SJ, Christopher NC, Auman JT. Working memory performance and cholinergic effects in the ventral tegmental area and substantia nigra. Brain Res. 1994; in press.

11. Levin ED, Rose JE. Anticholinergic sensitivity following chronic nicotine administration as measured by radial-arm maze performance in rats. Behav. Pharmacol. 1990; 1: 511-520.
12. Levin ED, Lee C, Rose JE, Reyes A, Ellison G, Jaravik M, Gritz E. Chronic nicotine and withdrawal effects on radial-arm maze performance in rats. Behav. Neural. Biol. 1990; 53: 269-276.
13. Levin ED, Briggs SJ, Christopher NC, Rose JE. Chronic nicotinic stimulation and blockade effects on working memory. Behav. Pharmacol. 1993; 4: 179-182.
14. Levin ED, Christopher NC, Briggs SJ, Rose JE. Chronic nicotine reverses working memory deficits caused by lesions of the fimbria or medial basalocortical projection. Cog. Brain Res. 1993; 1: 137-143.
15. Levin ED, Christopher NC, Briggs SJ. Comparison of chronic nicotine and mecamylamine effects on working memory performance in the radial-arm maze and T-maze alternation. Soc Neurosci Abs 1994; in press.
16. Levin ED, Rose JE. Acute and chronic nicotinic interactions with dopamine systems and working memory performance. In: Lajtha A, Abood L eds., Functional Diversity of Interacting Receptors, New York:The New York Academy of Sciences, 1994, in press.
17. Levin ED, Briggs SJ, Christopher NC, Rose JE. Persistence of chronic nicotine-induced cognitive facilitation. Behav Neural Biol 1992; 58: 152-158.
18. Dunnett SB, Martel FL. Proactive interference effects on short-term memory in rats: 1. Basic parameters and drug effects. Behav. Neurosci. 1990; 104: 655-665.
19. Moran PM. Differential effects of scopolamine and mecamylamine on working and reference memory in the rat. Pharmacol Biochem Behav 1993; 45: 533-538.
20. Decker MW, Majchrzak MJ. Effects of systemic and intracerebroventricular administration of mecamylamine, a nicotine cholinergic antagonist, on spatial memory in rats. Psychopharmacology 1992; 107: 530-534.
21. Ohno M, Yamamoto T, Watanabe S. Blockade of hippocampal nicotinic receptors impairs working memory but not reference memory in rats. Pharmacol Biochem Behav 1993; 45: 89-93.
22. Blozovski D. Deficits in passive avoidance learning in young rats following mecamylamine injections in the hippocampo-entorhinal area. Exp. Brain Res. 1983; 50: 442-448.
23. Blozovski D, Dumery V. Development of amygdaloid choinergic mediation of passive avoidance learning in the rat. Exp. Brain. Res. 1987; 67: 70-76.
24. Stolerman IP, Goldfarb T, Fink R, Jarvik ME. Influencing cigarette smoking with nicotine antagonists. Psychopharmacology 1973; 28: 247-259.
25. Newhouse PA, Potter A, Corwin J, Lenox R. Modeling the nicotinic receptor loss in dementia using the nicotinic antagonist mecamylamine: Effects on human cognitive functioning. Drug Dev. Res. 1994; 31: 71-79.
26. McGurk SR, Levin ED, Butcher LL. Radial-arm maze performance in rats is impaired by a combination of nicotinic-cholinergic and D_2 dopaminergic drugs. Psychopharmacology 1989; 99: 371-373.
27. Levin ED, McGurk SR, Rose JE, Butcher LL. Reversal of a mecamylamine-induced cognitive deficit with the D_2 agonist, LY 171555. Pharmacol. Biochem. Behav. 1989; 33: 919-922.
28. Levin ED. Scopolamine interactions with D_1 and D_2 antagonists on radial-arm maze performance in rats. Behav Neural Biol 1988; 50: 240-245.
29. Levin ED. Interactive effects of nicotinic and muscarinic agonists with the dopaminergic agonist pergolide on rats in the radial-arm maze. In preparation 1994.
30. Levin ED, Eisner B. Nicotine interactions with D_1 and D_2 agonists: Effects on working memory function. Drug Dev. Res. 1994; 31: 32-37.
31. Widzowski DV, Cregan E, Bialobok P. Effects of nicotinic agonists and antagonists on spatial working memory in normal adult and aged rats. Drug Dev. Res. 1994; 31: 24-31.
32. Buccafusco JJ, Jackson WJ. Beneficial effects of nicotine administered prior to a delayed matching-to-sample task in young and aged monkeys. Neurobiol. Aging 1991; 12: 233-238.
33. Hodges H, Gray JA, Allen Y, Sinden J. The role of the forebrain cholinergic projection system in performance in the radial-arm maze in memory-impaired rats. In: Adlkofer F, Thurau K eds., Effects of Nicotine on Biological Systems, Basel: Birkhäuser Verlag, 1991: 389-399.
34. Tilson HA, McLamb RL, Shaw S, Rogers BC, Pediatikakis P, Cook L. Radial-arm maze deficits produced by colchicine administered into the area of the nucleus basalis are ameliorated by cholinergic agonists. Brain Res 1988; 438: 83-94.

35. Decker MW, Majchrzak MJ, Anderson DJ. Effects of nicotine on spatial memory deficits in rats with septal lesions. Brain Res. 1992; 572: 281-285.
36. Hodges H, Sinden J, Turner JJ, Netto CA, Sowinski P, Gray JA. Nicotine as a tool to characterise the role of the forebrain cholinergic projection system in cognition. In: Lippiello PM, Collins AC, Gray JA, Robinson JH eds., The Biology of Nicotine: Current Research Issues, New York:Raven Press, 1992: 157-182.
37. Newhouse PA, Sunderland T, Tariot PN, Blumhardt CL, Weingartner H, Mellow A, Murphy DL. Intravenous nicotine in Alzheimer's disease: A pilot study. Psychopharmacology 1988; 95: 171-175.
38. Jones GMM, Sahakian BJ, Levy R, Warburton DM, Gray JA. Effects of acute subcutaneous nicotine on attention, information processing and short-term memory in Alzheimer's disease. Psychopharmacology 1992; 108: 485-494.
39. Sahakian B, Jones G, Levy R, Gray J, Warburton D. The effects of nicotine on attention, information processing, and short-term memory in patients with dementia of Alzheimer type. Br. J. Psychiat. 1989; 154: 797-800.
40. Kellar KJ, Wonnacott S. Nicotinic cholinergic receptors in Alzheimer's disease. In: Wonnacott S, Russell MAH, Stolerman IP eds., Nicotine Psychopharmacology: Molecular, Cellular, and Behavioral Aspects,, Oxford: Oxford University Press, 1990: 341-373.

Effects of Nicotine on
Biological Systems II
Advances in Pharmacological Sciences

POSSIBLE MECHANISMS UNDERLYING BENEFICIAL EFFECTS OF NICOTINE ON COGNITIVE FUNCTION

M.H. Joseph, G. Grigoryan, H. Hodges and J.A. Gray

MRC Behavioral Neurochemistry Group and Department of Psychology
Institute of Psychiatry, De Crespigny Park, London SE5 8AF, UK

Summary: A number of studies, principally from our laboratories, are reviewed, showing that acute systemic nicotine can improve performance both in patients with Alzheimer's disease, and in experimental animals bearing lesions to the forebrain cholinergic projections. Nicotine acts to increase the release of dopamine (DA) and noradrenaline in the rat brain, transmitters associated with attention and vigilance. Lesions to, or pharmacological blockade of, the noradrenergic or the dopaminergic system indicate however that the ameliorative effects of nicotine in the rat model are independent of effects on noradrenaline, and at least partly independent of effects on DA. The evidence concerning a releasing effect of nicotine on dopamine in humans is reviewed.

Introduction

Alzheimer's disease is known to be accompanied by degeneration in the forebrain cholinergic projection systems (reviewed in 1), and by a substantial loss of forebrain nicotinic, but not muscarinic, cholinergic receptors (2, 3). This has led to a number of investigations into the acute cognitive effects of nicotine in Alzheimer patients. Studies from our group (4, 5) and others (6) have reported that subcutaneous nicotine can improve performance in vigilance, but not in memory tasks in patients with Alzheimer's disease. In related animal studies, our group has also been investigating the behavioural effects of selective lesions to the cells of origin of the forebrain cholinergic projection system (FCPS), and the effects of chronic alcohol administration. Both of these procedures cause marked depletions in forebrain acetylcholine, and may therefore provide partial animal models of Alzheimer's and of alcoholic dementia.

In both these models deficits occur in Jarrard's (7) version of the 8-arm radial maze (8, 9), and also in the water maze (see below). However these deficits are seen across all aspects of these tasks, rather than being confined to working or reference memory in the spatial or cue versions of the maze task. This suggests a deficit in a general cognitive function such as vigilance or attention, rather than a specific deficit in memory. In these animal models we have further found that subcutaneous nicotine is able to reverse the deficits observed in each of the four components, though most markedly in working memory (8, 10). This may correspond to the effects on vigilance described in our patient studies.

Nicotinic receptors are widely distributed in the brain, and occur both pre- and post-synaptically. Whether nicotine acts indirectly, by a presynaptic action increasing acetylcholine release, or directly at post-synaptic cholinergic receptors, these actions must impinge upon neurones using other transmitters. Nicotine has a wide variety of actions in the brain, and influences many transmitters indirectly; in particular it increases the release of the catecholamines (CAs) dopamine and noradrenaline. Since these transmitter systems are involved in arousal, attention and vigilance, it seemed plausible that at least part of the actions of nicotine in the lesioned animals might be mediated via the release of CAs. Accordingly we have studied the extent to which CA antagonists, or CA depleting lesions could reverse the ameliorative effects of nicotine in our animal models.

Actions of nicotine on noradrenaline release

Systemic nicotine increases the rate of hydroxylation of tyrosine in a variety of catecholaminergically innervated areas of rat brain (11), and releases CAs in vitro from slices or synaptosomal preparations from the brain (12). Further studies from our laboratory, using in vivo dialysis, have correspondingly shown that nicotine releases noradrenaline (NA) in the hippocampus whether applied systemically (13), locally via the dialysis probe, or when infused into the locus coerulus (14). However the NA releasing effect of systemic nicotine was blocked by the nicotinic antagonist mecamylamine only when it was infused into the locus coeruleus (cell body area), and not when it was infused into the hippocampus (terminal area) (14), indicating a site of action at the noradrenergic cell bodies. Repeated administration of nicotine results in a potentiation of the noradrenergic response. This too seems to depend upon an action at the noradrenergic cell bodies, increasing the synthesis of tyrosine hydroxylase, which is subsequently transported to the terminals (15-17). Accordingly, in the experiments described below we determined whether the ameliorative effects of nicotine in animals with forebrain cholinergic lesions were prevented either by of blockade of beta-adrenergic receptors with propranolol, or by lesions to the dorsal NA bundle, which substantially depletes the NA innervation of the hippocampus and cortex.

Are the effects of nicotine in the maze tasks mediated through noradrenaline?

As noted above, impairment in the radial maze is produced by an FCPS lesion resulting from s-AMPA injections into the medial septal area and the nucleus basalis of Meynert. A significant improvement in lesion-induced errors, especially in working memory is seen following nicotine (0.1 mg/kg). Grigoryan et al (18) have recently reported from our laboratory that while propranolol alone (0.5 and 5.0 mg/kg) dose-dependently increases working memory errors in the place task, combining nicotine with the low and high dose of propranolol returns the error rate to below baseline or to baseline respectively. Thus neither dose of propranolol is able to block the effect of nicotine. In a further experiment (19) control and FCPS lesioned animals were compared with animals receiving lesions both to the FCPS projection system and to the dorsal bundle (lesioning noradrenergic projections to the hippocampus and cortex). The results showed that the combined lesion group had no greater impairment than the FCPS lesion group in their rate of learning of the position of a concealed platform in a working memory version of the Morris water maze task, and that nicotine (0.1 mg/kg) was equally effective in improving the performance of both groups. The conclusion from these two experiments is that the effect of nicotine in the lesioned animals is not likely to be mediated via NA release.

Actions of nicotine on dopamine release

Among the principal dopaminergically innervated areas of rat brain, nicotine appears to have a stronger action in the mesolimbic than in the mesostriatal projections. Tyrosine hydroxylation is selectively increased here, as is dopamine release measured by in vivo dialysis. Again the effect is seen after either systemic or local administration (11, 20-22). Again, as for noradrenaline, a recent report (23) indicates that the DA releasing effect of systemic nicotine depends on an action at the dopaminergic cell bodies (in this case in the ventral tegmental area) and not at the terminals in the nucleus accumbens. We sought a functional correlate of the dopamine releasing actions of nicotine using an attentional paradigm, latent inhibition (LI), which is known to be disrupted by indirect DA agonists

(24,25) and believed to depend upon accumbal function. We found that systemic nicotine could indeed disrupt LI (26) at doses which we and others had previously shown to release DA in the accumbens in vivo. The disruptive effect was reversed by haloperidol, confirming that it was mediated by increased dopamine function. These results may be of particular relevance in the present context, since LI depends upon learning not to attend to familiar stimuli, a process which may be directly related to vigilance.

Are the effects of nicotine in the maze tasks mediated through dopamine?

The improvement caused by nicotine in the radial maze performance of rats bearing ibotenate or quisqualate-induced lesions of the FCPS is not mimicked by another dopamine releasing drug, amphetamine, at 0.5 or 2.0 mg/kg (10). However it should be noted that the effect of acute amphetamine on DA release is largely impulse independent (see, for example, 27) in contrast to that of nicotine. Conversely, we studied the effect of blockade of DA function with haloperidol. This drug at 0.25 mg/kg clearly reversed the ameliorative effect of nicotine in the water maze task (Grigoryan and Mitchell, unpublished). In a further experiment therefore (28), control and FCPS lesioned animals were compared with animals receiving 6-hydroxy-dopamine induced lesions of the dopamine terminals in the nucleus accumbens and animals receiving both lesions. While accumbens lesions did result in an additional impairment on some measures in the water maze task, nicotine (0.1 mg/kg) resulted in a significant improvement in the combined lesion group, albeit somewhat less than that in the FCPS-lesion only group. Thus, although the results with manipulation of the DA system are somewhat more equivocal, it is clearly the case that a significant part of the nicotine-induced improvement is independent of DA release, .at least in the nuclens accumbens.

Can DA release account for part of the cognitive effect of nicotine in humans?

It is relevent to ask whether any part of the cognitive effect of nicotine at the levels given to patients, or at the higher levels self-administered by smokers, could in principle be due to increased DA release. Many theorists believe that the rewarding effects of drugs of abuse are mediated through a common action in increasing DA release in the accumbens. Although it may be argued that animal studies suggest that higher doses than those tolerated by human non-smoker subjects may be required to do this, we should bear in mind that both patients and smokers may have alterations in receptor sensitivity, due in the first case to denervation, and in the second to chronic nicotine administration. Moreover, analogies with animals should not be taken too far; we need evidence from human studies, although such evidence will necessarily be indirect.

If the rewarding effect of nicotine derived from smoking is mediated by DA release, then the degree of reward should be reduced by concommitant administration of haloperidol. In the short term, this would be expected to increase the intake of nicotine. Dawe et al. (29) obtained a result consistent with this prediction; in overnight-abstinent smokers given placebo or haloperidol under blind conditions and allowed free access to cigarettes for one hour during a masking procedure, those pretreated with haloperidol had a small but significant increase in plasma nicotine over those given placebo. However they had smoked no more cigarettes, and a pharmacokinetic interpretation of this data has still to be ruled out.

In view of our robust finding (26, 30) that nicotine disrupts LI in rats, and the report of Gray et al.(31) that amphetamine disrupts LI in humans, we have sought to determine whether nicotine can disrupt LI in humans also. However, neither subcutaneous nicotine (0.3 or 0.6 mg), nor the smoking of one cigarette by overnight-abstinent smokers reduced LI in volunteers (32). In our laboratories, Lee, Russell and Gray (unpublished) also found that haloperidol was unable to reverse the ameliorative effect of 0.3 and 0.6 mg sc nicotine on a vigilance task. None of these observations would support the hypothesis of nicotine induced dopamine release in humans.

Conversely, Al-Adawi and Powell (unpublished) from our department studied the interaction between heightened motivation and smoking. They used a card sorting task under a basal, and under a finacially rewarded condition in control subjects, and in subjects abstaining from smoking during Ramadan. They found that the increase in speed under the rewarded condition was greater in smokers abstaining only during daylight hours for Ramadan after they had smoked a cigarette in the evening, while it was unchanged in control subjects, or in subjects abstaining for the whole of Ramadan, tested at the same times.

Clearly it is difficult to draw a single conclusion from these varied results. However we may suggest that nicotine, at the maximal doses tolerated acutely by non-smokers is probably insufficient to cause DA release in the human brain. The higher doses which could be given under conditions of peripheral nicotinic blockade might be sufficient. However in smokers, levels of nicotine may be achieved which can release DA in the brain, and this may have functional effects, including those on vigilance and reward. This difference could relate either to smokers' peripheral tolerance, permitting them to achieve higher plasma levels of nicotine, or to increases in the sensitivity of their brain nicotinic receptors.

Conclusions

Animal data clearly establish that nicotine can ameliorate the cognitive deficits produced by cholinergic lesions. They also clearly establish in intact animals that nicotine, at somewhat higher doses, can release NA in the hippocampus and DA in the accumbens by actions at the cell bodies in the locus coeruleus and ventral tegmental area respectively. The ameliorative effects of nicotine in lesioned animals do not appear to depend upon NA release, and are at most only partly dependent on DA release. This makes it probable that the ameliorative effects of nicotine in Alzheimer's patients are directly mediated in the denervated hippocampus or frontal cortex. However additional indirect effects via DA release could occur in human subjects taking nicotine, especially in smokers.

Acknowledgements: *MH Joseph is a Reader in Behavioural Neurochemistry, and a member of the UK Medical Research Council External Scientific Staff. Helen Hodges is a Wellcome Senior Research Fellow. Parts of our research programme have been supported by the MRC, the Wellcome Trust, R J Reynolds Tobacco Company, the Council for Tobacco Research, BAT and Schering AG, Berlin. We express our thanks to all of these bodies. We are grateful to our cited colleagues, and in particular to Samir Al-Adawi, Chuly Lee, Steve Mitchell, Scott Peters and Jane Powell for allowing us to discuss their unpublished data.*

References

1. Kopelman MD. The cholinergic neurotransmitter system in human memory and dementia: a review. Quart J Exp Psychol 1986; 38A: 535-573.
2. Kellar KJ, Whitehouse PJ, Martino-Barrows AM, Marcus K, Price DL. Muscarinic and nicotinic cholinergic binding sites in Alzheimer's disease cerebral cortex. Brain Res 1987; 436: 62-68.
3. Perry EK, Perry RH, Smith CJ, Dick DJ, Candy JM, Edwardson JA, Fairburn A, Blessed G Nicotinic receptor abnormalities in Alzheimer's and Parkinson's diseases. J Neurol Neurosurg Psychiat 1987; 50: 806-809.
4. Sahakian B, Jones G, Levy R, Gray J, Warburton D. Effects of nicotine on attention, information processing, and short-term memory in patients with dementia of the Alzheimer type. Br J Psychiat 1989; 154: 797-800.
5. Jones GMM, Sahakian BJ, Levy R, Warburton DM, Gray JA. Effects of acute subcutaneous nicotine on attention, information processsing and short-term memory in Alzheimer's disease. Psychopharmacology 1992; 198: 485-494.
6. Newhouse P, Sunderland T, Tariot P, Blumhardt C, Weingartner H, Mellow W. Intravenous nicotine in Alzheimer's disease: a pilot study. Psychopharmacology 1988; 95: 171-175.

7. Jarrard LE, Feldon J, Rawlins JNP, Sinden JD, Gray JA. The effects of intrahippocampal ibotenate on resistance to extinction after continuous or partial reinforcement. Exp Brain Res 1986; 61: 519-530.
8. Hodges H, Allen Y, Sinden J, Lantos PL, Gray JA. The effects of cholinergic-rich neural grafts on radial maze performance of rats after excitotoxic lesions of the forebrain cholinergic projection system: 2. cholinergic drugs as probes to investigate lesion-induced deficits and transplant-induced functional recovery. Neuroscience 1991; 45: 609-623.
9. Hodges H, Allen Y, Sinden J, Mitchell SN, Arendt T, Lantos PL, Gray JA. The effects of cholinergic drugs and cholinergic-rich foetal neural transplants on alcohol-induced deficits in radial maze performance in rats. Behav Brain Res 1991; 43: 7-28.
10. Turner JJ, Hodges H, Sinden JD, Gray JA. Comparison of radial maze performance of rats after ibotenate and quisqualate lesions of the forebrain cholinergic projection system: effects of pharmacological challenge and changes in training regime. Behavioural Pharmacology 1992; 3: 359-373.
11. Mitchell SN, Brazell MP, Joseph MH, Alavijeh MS, Gray JA. Regionally specific effects of acute and chronic nicotine on rates of catecholamine and indoleamine synthesis in rat brain. Eur J Pharmacol 1989; 167: 311-322.
12. Snell LD, Johnson KM. Effects of nicotinic agonists and antaonists on NMDA-induced ^{3}H-norepinephrine release and ^{3}H-TCHP binding in rat hippocampus. Synapse 1989; 3: 129-135
13. Brazell MP, Mitchell SN, Gray JA. Effect of acute administration of nicotine on in vivo release of noradrenaline in the hippocampus of freely moving rats: a dose-response and antagonist study. Neuropharmacology 1991; 30: 823-833.
14. Mitchell SN. Role of the locus coeruleus in noradrenergic response to a systemic administration of nicotine. Neuropharmacology 1993; 32: 937-949.
15. Smith KM, Mitchell SN, Joseph MH. Effects of chronic and subchronic nicotine on tyrosine hydroxylase activity in noradrenergic and dopaminergic neurons in rat brain. J. Neurochem 1991; 56: 989-997.
16. Smith KM, Joseph MH, Gray JA. A single dose of nicotine is sufficient to increase tyrosine hydroxylase in noradrenergic neurones. In: Adlkofer A, Thurau K editors. Effects of nicotine on biological systems. Advances in Pharmacological Sciences. Basel: Birkhauser Verlag, 1991: pp 345-350.
17. Mitchell SN, Smith KM, Joseph MH, Gray JA. Increases in tyrosine hydroxylase mRNA in the locus coeruleus after a single dose of nicotine are followed by time-dependent increases in enzyme activity and noradrenaline release. Neuroscience 1993; 56: 989-997.
18. Grigorian GA, Peters S, Gray JA, Hodges H. Interactions between the effects of propranolol and nicotine on radial maze performance of the rat with lesions of the forebrain cholinergic projection system Behav Pharmacol 1994; 5: 265-280
19. Gray JA, Mitchell SN, Joseph MH, Grigorian GA, Dawe S, Hodges H. Neurochemical mechanisms mediating the behavioural and cognitive effects of nicotine. Drug Development Res 1994; 31: 3-17. and Grigorian GA, Mitchell SN, Sinden JD and Gray JA. Are the cognitive enhancing effects of nicotine in the rat mediated by an interaction with the noradrenergic system? Pharmacol Biochem Behav 1994; (under revision for publication)
20. Brazell MP, Mitchell SN, Joseph MH, Gray JA. Acute administration of nicotine increases the in vivo extracellular levels of dopamine, 3,4-dihydroxyphenylacetic acid and ascorbic acid preferentially in the nucleus accumbens of the rat: comparison with caudate-putamen. Neuropharmacology 1990; 29: 1177-1185.
21. Imperato A, Mulas A, Di Chiara G. Nicotine preferentially stimulates dopamine release in the limbic system of freely moving rats. Eur J Pharmacol 1986; 132: 337-338.
22. Mifsud JC, Hernandez L, Hoebel BG. Nicotine infused into the nucleus accumbens increases synaptic dopamine as measured by in vivo microdialysis. Brain Res 1989; 478: 365-368
23. Nisell M, Nomikos GG, Svensson TH. Systemic nicotine-induced dopamine release in the rat nucleus accumbens is regulated by nicotinic receptors in the ventral tegmental area. Synapse 1994; 16: 36-44.
24. Weiner I. Neural substrates of latent inhibition: the switching model. Psychol Bull 1990; 108: 442-461.
25. Warburton EC, Joseph MH, Feldon J, Weiner I, Gray JA. Antagonism of amphetamine-induced disruption of LI in rats by haloperidol and ondansetron: implications for a possible antipsychotic action of ondansetron. Psychopharmacology 1994; 114: 657-664.
26. Joseph MH, Peters SL, Gray JA. Nicotine blocks latent inhibition in rats: evidence for a critical role of increased functional activity of dopamine in the mesolimbic system at conditioning rather than pre-exposure. Psychopharmacology 1993; 110: 187-192.

27. Hurd YL, Ungerstedt U. Ca++ dependence of the amphetamine, nomifensine and Lu 19-005 effect on in vivo dopamine transmission. Eur J Pharmacol 1989; 166: 261-269.
28. Grigorian GA, Hodges H, Mitchell SN, Sinden JD, Gray JA. Are the cognitive enhancing effects of nicotine in the rat mediated by an interaction with the noradrenergic system? 1994; (in preparation for publication)
29. Dawe S, Gerada C, Russell MAH, Gray JA. Nicotine intake in smokers increases following a single dose of haloperidol. Psychopharmacology 1994; in press
30. Joseph MH, Peters SL, Gray JA. Haloperidol, and selective D-1 and D-2 dopamine receptor antagonists, can block the disruption of latent inhibition by nicotine given only at conditioning in rats. 1994; (in preparation for publication).
31. Gray NS, Pickering AD, Hemsley DR, Dawling S, Gray JA. Abolition of latent inhibition by a single 5 mg dose of d-amphetamine in man. Psychopharmacology 1992; 107: 425-430.
32. Thornton JC, Dawe S, Lee C, Frangou S, Gray NS, Russell MAH, Gray JA. Effects of nicotine on latent inhibition in human subjects. Psychopharmacology 1994; in press

Effects of Nicotine on
Biological Systems II
Advances in Pharmacological Sciences

NICOTINIC MODULATION OF COGNITIVE FUNCTIONING IN HUMANS

Paul A. Newhouse[1], Alexandra Potter[1], Melissa Piasecki[1], Jennifer Geelmuyden[1], June Corwin[2] and Robert Lenox[1]

[1]Clinical Neuroscience Research Unit, Department of Psychiatry, University of Vermont College of Medicine, 1 South Prospect St, Burlington, VT 05401

[2]Department of Psychiatry, New York University School of Medicine and Manhattan VA Medical Center, NY, NY

Summary: The loss of central nicotinic receptors is a neurochemical hallmark of several degenerative brain disorders, notably Alzheimer's (AD) and Parkinson's disease (PD). However, uncertainty has remained about the significance of this loss for the cognitive symptomatology of these disorders. Investigations of the effects of nicotinic agents in both normal and diseased individuals have led to important information about the role of nicotinic systems in cognitive functioning. These studies suggest that amelioration of some of the cognitive disturbances in AD and PD may be possible utilizing nicotinic compounds.

Introduction

Prior to the development of the ability to image and map central nervous system (CNS) nicotinic receptors (1), cholinergic deficits in Alzheimer's disease were thought to be exclusively related to muscarinic receptor changes. However interest in the role of nicotinic systems in the cognitive disorder of Alzheimer's disease increased after Whitehouse and colleagues (2) and others (e.g. 3) showed that there is a marked reduction in nicotinic receptor density in the brains of Alzheimer's disease patients compared to age-matched controls.

However, the demonstration of a decline in receptor number does not in and of itself indicate that these changes are responsible for the cognitive symptomatology of the disorder. It is necessary to show functional or cognitive consequences from the loss of these receptors or their associated cell processes in vivo. We review here a series of studies on the effects of nicotinic antagonists on cognitive functioning in young and older normal humans and studies of the acute effects of nicotinic stimulation on cognitive functioning in Alzheimer's disease patients.

Nicotinic Antagonist Studies

The use of antagonists allows the identification of cognitive operations affected by the interruption of agonist neurotransmission. The effects may indicate which cognitive domains are affected by lesions to the system of interest (4). These studies examined the effects of a temporary blockade of central nicotinic receptors by the drug mecamylamine on aspects of cognition. Mecamylamine is a centrally active non-competitive antagonist of nicotine (and presumably acetylcholine) at C6 (ganglionic) type central and peripheral nicotinic receptors (5).

Methods. Detailed description of experimental methodology is provided in Newhouse et al (6). In general, all normal subjects were nonsmokers, healthy, and cognitively intact as verified by physical exam, laboratory, and cognitive tests. Alzheimer's disease subjects fulfilled diagnostic criteria for probable Alzheimer's disease and were nonsmokers.

Cognitive testing consisted of a computer battery, one oral memory test, and behavioral ratings. Tests included: the Repeated Acquisition Test (RAT), which tests a subject's ability to retrieve previously acquired information as well as the ability to learn new information (7); the High-Low Imagery Test (a recognition memory test with high and low imagery words) (8); a choice reaction time test and a manikin spatial rotation test, both from the Walter Reed PAB (9); and the Selective Reminding Test.

All drugs were administered double-blind. The doses administered were 5, 10, and 20 mg of mecamylamine and placebo. Drug was administered at 0900. Cognitive testing was conducted at 0, +60 and +120 minutes.

Results. Mecamylamine administration produced dose-related impairment of the acquisition of new information with group differences in sensitivity. This was most clearly demonstrated by the RAT task in which subjects learn a button-pushing sequence (Figure 1). The young normals showed a significant increase in errors after the 20 mg dose. By contrast, the elderly normals showed significant impairment after the 10 and 20 mg dose, and the Alzheimer's disease subjects showed impairment after all three active doses (5, 10, and 20 mg). In the retrieval condition (old learning), there were no significant dose-related impairments in any group.

The selective reminding task, which involves verbal learning, demonstrated a similar pattern. Here, the young normals showed a small dose-related decline in total recall, but no change in recall failure. The old normals showed a significant and substantial increase in recall failure however, after the 20 mg dose, and the Alzheimer's disease patients showed increased recall failure after the 10 and 20 mg doses.

For the High-Low Imagery task, a test of recognition memory, there was a dose-related decline in discrimination for both normal groups with the elderly normals showing a greater effect than the young. Perhaps more interestingly, in the old normals, mecamylamine produced a dose-related change in response bias with a significant liberal shift after the 20 mg dose.

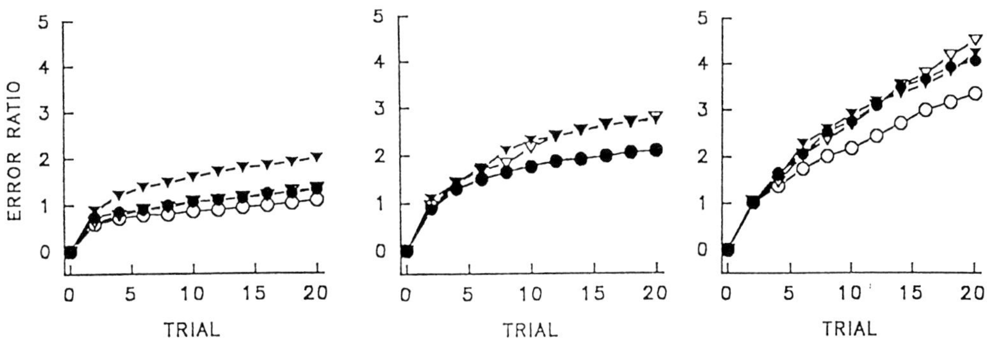

Figure 1. The effects of three doses of mecamylamine (5, 10, 20 mg) on error ratio (# errors/chain length) in young normals (left panel), elderly normals (center panel), and Alzheimer's disease patients (right panel). Modified from ref. 6.

Regarding psychomotor speed, mecamylamine produced dose-related slowing in a number of tasks that measured reaction time. These included increases in mean reaction time for the choice reaction time and manikin tasks. Although there were no significant dose by group interactions for the speed measures, older subjects tended to show proportionately greater increases in reaction time than the younger subjects did.

Physiological effects were consistent with peripheral ganglionic blockade. By contrast, there were minimal behavioral effects. Although there were small changes in the scores of some observer rating scales, these were clinically nonsignificant. No changes were seen in subject mood or physical side-effects ratings.

Nicotinic Agonist Studies in Alzheimer's Disease

Methods. This study examined the effects of intravenous nicotine on cognitive, behavioral, and physiologic functioning in both normal nonsmokers and patients with AD (10-12). We initially studied 12 moderately demented AD patients (seven female and five male) and 11 young normal nonsmokers (seven female and four male). Single blind infusions of saline placebo or nicotine bitartrate were given for 60 minutes at doses of 0.125, 0.25, and 0.5 μg/kg/min of nicotine base in a within-subjects design. Cognitive testing was performed at 0, 30, and 60 minutes, and at 4, 8, and 24 hours after the start of the infusion.

Results. Analysis of the cognitive effects in the AD group showed that there was no significant effect on immediate correct recall of a word list. However, there was a significant dose-related decrease in intrusion errors on this task, with a "U"-shaped dose response curve, i.e., the middle dose (0.25 μg) produced the biggest decrease in errors (Figure 2). The decline in errors was apparent for words presented at both 30 and 60 minutes after the beginning of the infusion. This decrease in intrusions was not simply due to suppression of responding as total word production was not different across doses. Analysis of long-term recall showed that words that were immediately recalled on the 0.25 μg dose were significantly more likely to be recalled 8 hours later than words immediately recalled under other doses or placebo.

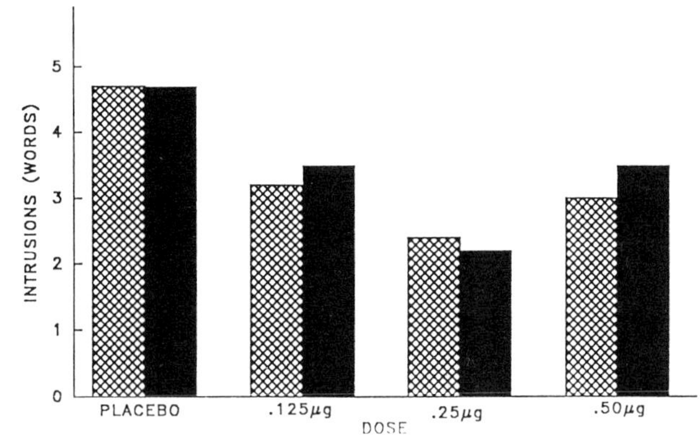

Figure 2. Intrusion errors in AD patients after intravenous nicotine at doses of 0, 0.125, 0.25, and 0.5 μg/kg/min. Hatched bars indicate testing at 30 minutes after start of infusion, filled bars indicate testing at 60 minutes after infusion start. Modified from ref. 20.

Behavioral measures showed significant dose-related changes. Depressive affect and anxiety self-ratings showed significant increases in both subject groups, particularly after the 0.5 μg dose. The AD group seemed more sensitive to adverse behavioral effects and showed significant increases in anxiety and depression at the 0.25 μg dose as well. These behavioral effects were closely linked to the drug infusion period and disappeared rapidly after the infusion was terminated.

Neuroendocrine measures tended to confirm that the doses used were active at CNS nicotinic receptors. ACTH and cortisol showed significant dose- and time-related increases in both groups of subjects. The ACTH increase was evident by 30 minutes, with the cortisol increase being delayed until 60 minutes suggesting no direct effect of nicotine on adrenal cortical cells. Reports of physical side-effects were generally minimal, although some subjects complained of a mild headache. There was no consistent report of nausea in either group.

Discussion

The studies of IV nicotine in normal non-smokers and AD patients demonstrated that acute treatment with nicotine that was sufficient to stimulate CNS nicotinic receptors could produce measurable changes in both short term recall errors and long-term recall consistency. Recent studies by Jones and colleagues (13) using subcutaneous administration of nicotine in AD patients, found that nicotine improved reaction time, sustained visual attention, and visual perception, but found no effects on memory. There appears to be general agreement that nicotine improves attention and may improve short term memory by facilitating attention to stimuli (14). However, there is evidence that nicotine may have a consolidative effect on memory as well, as evidenced by the increase in long term recall consistency in our study and studies by others which show that information learned prior to exposure to nicotine is better recalled at a later time (14-15).

The nicotinic antagonist mecamylamine produced cognitive impairment on several tasks. These effects included impairment of acquisition on the RAT, impaired recall on the Selective Reminding task, slowing of reaction time, and impairment of discrimination and liberalization of response bias on the Hi-Low Imagery task. Further, there was evidence that elderly normals were proportionately more sensitive to the effects of mecamylamine and AD patients were still more responsive suggesting a continuum of increasing sensitivity with increasing receptor loss. These results suggest that the loss of central nicotinic receptors seen in normal aging (16) and in AD and PD (3) have functional consequences. The deficits produced by mecamylamine resemble in several respects those seen in AD and to a lesser extent PD. Deficits in short- and long-term memory, impaired attention, liberal response bias, and decreases in reaction time are hallmarks of the dementing picture seen in these disorders. The age-related nature of some of the findings suggest that the decline in nicotinic receptors with age produces increased vulnerability to the effects of nicotinic blockade.

The possibility of using nicotine or other nicotinic agonists as a chronic therapeutic agent for cognitive enhancement for AD or PD is clearly worth further exploration. However, there are potential problems with nicotine itself. For example, nicotine provokes upregulation of its own receptor after chronic administration, perhaps through the mechanism of desensitization (17). Whether chronic nicotine would reproduce the positive effects seen after acute administration is unclear, especially in light of the known neurochemical effects of chronic administration. If nicotinic receptors are primarily involved in modulating the release of other neurotransmitters, chronic agonist-induced desensitization may not have the net effect of increasing signal traffic. On the other hand, if chronic agonist exposure increases stimulated release of neurotransmitter, than the overall gain of such a system may be increased. This may be beneficial to a patient with a deteriorating neurotransmitter system and may in part explain the improvement seen with nicotine on attentional tasks. Further problems with nicotine itself are its narrow therapeutic index, relatively high acute toxicity, and effects on mood and anxiety.

These results suggest that nicotinic modulation may be of benefit for the alleviation or improvement of cognitive impairments in various dementing disorders which show a loss of nicotinic receptors. Nicotine is unlikely to be an ideal candidate for this task, due to its low therapeutic index, but other, more selective agonists may be more useful. ABT-418 is a novel, highly selective nicotinic agonist which appears to activate preferentially the $\alpha4\beta2$ nicotinic receptor subtype in the CNS (18). Preclinical testing suggests that ABT-418 has cognitive-enhancing and anxiolytic properties with a large therapeutic index, i.e. its effective dose range is well below its toxic dose. A series of anabaseine derivatives have been developed that appear to be selective nicotinic agonists (19). Some of these compounds appear to have cognitive-enhancing properties as well as being cytoprotective.

It is theoretically possible then that a nicotinic drug could be developed for clinical use that could have both cognition-enhancing and neuroprotective properties, making it ideal for use in certain degenerative neurologic conditions such as AD or PD. It is also possible that other cognitive disorders that involve attentional impairment, such as Attention Deficit Disorder could benefit from nicotinic stimulation.

***Acknowledgements**: This work was supported by NIMH grant R29-46625 to P.N. and by GCRC M01-00109.*

References

1. Schwartz R. Autoradiographic distribution of high affinity muscarinic and nicotinic cholinergic receptors labeled with [^{3}H]acetylcholine in rat brain. Life Sciences 1986; 38: 2111-2119.
2. Whitehouse P, Martino A, Antuono P, Kellar K. Nicotinic acetylcholine binding sites in Alzheimer's disease. Brain Res 1986; 371: 146-151.
3. Aubert I, Araujo DM, Cécyre D, Robitaille Y, Gauthier S, Quirion R. Comparative alterations of nicotinic and muscarinic binding sites in Alzheimer's and Parkinsons's diseases. J Neurochem 1992; 58: 529-541.
4. Weingartner H, Cohen R, Sunderland T, Tariot P, Thompson K, Newhouse P. Diagnosis and assessment of cognitive dysfunctions in the elderly. in Melzer H (ed), Psychopharmacology, the Third Generation of Progress. New York, Raven Press, 1987: 909-919.
5. Martin BR, Onaivi ES, Martin TJ. What is the nature of mecamylamine's antagonism of the central effects of nicotine? Biochem Pharmacol 1989; 38: 3391-3397.
6. Newhouse P, Potter A, Corwin J, Lenox R. Age-related effects of the nicotinic antagonist mecamylamine on cognition and behavior. Neuropsychopharmacol 1994; 10: 93-107.
7. Thompson D. Repeated acquisition as a behavioral base line for studying drug effects. J Pharmacol Exp Ther 1973; 184: 506-514.
8. Corwin J, Peselow E, Fieve R, Rotrosen J. Memory in untreated depression: severity and task requirement effects. Presented at ACNP Annual Meeting, December, San Juan 1987.
9. Thorne D, Genser S, Sing H, Hegge F. The Walter Reed performance assessment battery. Neurobehav Toxicol Teratol 1985; 7: 415-418.
10. Newhouse P, Sunderland T, Tariot P, Blumhardt CL, Weingartner H, Mellow A, Murphy DL. Intravenous nicotine in Alzheimer's disease: a pilot study. Psychopharmacol 1988; 95: 171-175.
11. Newhouse P, Sunderland T, Narang P, Mellow AM, Fertig JB, Lawlor BA, Murphy DL. Neuroendocrine, physiologic, and behavioral responses following intravenous nicotine in nonsmoking healthy volunteers and patients with Alzheimer's disease. Psychoneuroendocrin 1990; 15: 471-484.
12. Newhouse P.A. Cholinergic drug studies in dementia and depression. Advan Exp Med Biol 1990; 282: 65-76.
13. Jones GMM, Sahakian BJ, Levy R, Warburton DM, Gray JA. Effects of acute subcutaneous nicotine on attention, information processing and short-term memory in Alzheimer's disease. Psychopharmacol 1992; 108: 485-494.
14. Warburton DM, Wesnes K, Shergold JM. Facilitation of learning and state dependency with nicotine. Psychopharmacol 1986; 89: 55-59.
15. Colrain IM, Mangan GL, Pellet OL, Bates TC. Effects of post-learning smoking on memory consolidation. Psychopharmacol 1992; 108: 448-451.
16. Court JA, Piggott MA, Perry EK, Barlow RB, Perry RH. Age associated decline in high-affinity nicotine binding in human brain frontal-cortex does not correlate with the changes in choline-acetyltransferase activity. Neurosci Res Comm 1992; 10: 125-133.
17. Marks MJ, Burch JB, Collins AC. Effects of chronic nicotine infusion on tolerance development and nicotine receptors. J Pharmacol Exp Ther 1983; 226: 817.
18. Arneric SP, Sullivan JP, Decker MW, Brioni JD, Buccafusco JJ, Briggs CA, Donnelly-Roberts D, Williams M. Cholinergic Channel Activators (ChCAs) as therapeutics for CNS disorders: ABT 418 as a prototype ChCA to treat Alzheimer's disease. Neuropsychopharm 1994; 10(3S): 395S.
19. Meyer EM. A nicotinic approach for neuroprotection and memory enhancement: studies with anabaseine derivatives. Neuropsychopharmacol 1994; 10(3S): 393S.
20. Newhouse PA, Potter A, Lenox RH. The effects of nicotinic agents on human cognition: Possible therapeutic applications in Alzheimer's and Parkinson's Diseases. Med Chem Res 1993, 2:628-642.

Awarded Posters

Effects of Nicotine on
Biological Systems II
Advances in Pharmacological Sciences

DISTRIBUTION OF NICOTINIC ACETYLCHOLINE RECEPTOR SUBUNIT IMMUNOREACTIVITIES ON THE SURFACE OF CHICK CILIARY GANGLION NEURONS

P.B. Sargent and H.L. Wilson

Departments of Stomatology and Physiology and the Neurosciences Graduate Program, University of California, San Francisco, CA 94143, USA

Summary: Immunocytochemical analysis of the expression of nicotinic acetylcholine receptor (AChR) subunits in the chick ciliary ganglion showed that virtually all cells in the ganglion express detectable levels of AChR-like immunoreactivity (LI) corresponding to the $\alpha3$ and/or $\alpha5$ subunit, the $\alpha7$ subunit, the $\beta2$ subunit, and the $\beta4$ subunit. Surface staining of subunit-specific monoclonal antibodies showed that a receptor containing $\alpha3$- and/or $\alpha5$-LI is located preferentially at synaptic sites, while a receptor containing $\alpha7$-LI was located extrasynaptically. The existence of two distinct patterns of cell surface staining with anti-AChR mAbs implies the existence of multiple classes of functional AChRs on ciliary ganglion neurons (1).

Introduction

Neurons in the embryonic chicken ciliary ganglion express mRNAs for 5 of the 8 putative neuronal AChR genes that have been identified thus far in chickens (expressing $\alpha3$, $\alpha5$, $\alpha7$, $\beta2$, and $\beta4$; (2)). This study was initiated to examine whether all neurons in the ganglion express protein products for these genes and whether the expression of individual subunit LIs at the neuronal surface reveals evidence for AChR heterogeneity (1).

Materials and Methods

Ciliary ganglia were dissected from euthanized chicks at 18-21 days of embryonic development. Frozen sections were prepared either from unfixed ganglia frozen by immersion in isopentane cooled to -170 °C or from ganglia fixed in 0.25-1.00% formaldehyde in 120 mM sodium phosphate, pH 7.2, for 1 hour at 22 °C and subsequently cryoprotected in 30% sucrose in physiological chick saline prior to gradual freezing in OCT embedding compound. Whole mounted ganglia were prepared by fixing ganglia in 0.25-1.00% formaldehyde for 1 hour at 22 °C and cutting them into halves or fourths.

AChR subunit LIs were detected using rat and mouse monoclonal antibodies (mAbs), kindly provided by Drs. Jon Lindstrom (University of Pennsylvania) and Darwin Berg, (University of California, San Diego) and cyanine 3.18- (Cy3) and 5.18- (Cy5) labeled secondary antibodies (Jackson Immunoresearch). Among the primary mAbs used were those specific, in Western blots, for the following AChR subunits: α2 (mAbs 321, 323), α3 (mAbs A3-1, A3-16), α5(α1) (mAbs 35, 210), α7 (mAbs 306, 318, 319), α8 (mAb 305), β2 (mAb 270), and β4 (mAbs B4-1, B4-2).

Synaptic sites were labeled with a monoclonal antibody (10h) to the synaptic vesicle-associated glycoprotein SV2 ((3); kindly provided by Dr. Steven Carlson, University of Washington). In a typical experiment in which both AChR-LI and synaptic boutons were visualized in a whole mount, ganglia are preincubated with phosphate-buffered saline containing 5% normal goat serum (hereafter, PBS-GS) for 1 hour, incubated with an anti-AChR mAb overnight at 4 °C, washed for 30 minutes, incubated with Cy3-goat anti-rat IgG (not cross-reactive with mouse IgG), permeabilized by incubation with 0.3% Triton X-100 in PBS, incubated with mouse mAb 10h for 2-3 hours, washed for 30 minutes, and incubated with Cy5-goat anti-mouse (not cross-reactive with rat IgG) for 2-3 hours. All solutions were made up in PBS-GS, and all steps were performed at room temperature, unless noted otherwise. After a final wash in PBS-GS, tissue was mounted in 90% glycerol/10% PBS containing an anti-oxidant to retard photobleaching. Control experiments included incubating tissue with mAbs recognizing AChR subunits whose mRNA is not detected in the ganglion (e.g., α8) and incubating tissue with each primary antibody and the inappropriate secondary antibody, to insure that there is no cross-talk between the first and second 'sandwiches'. Experiments on frozen sections were done similarly, except that the incubations and washes were shortened and the permeabilization step was omitted.

The relative location of subunit immunoreactivities and of synaptic sites was assessed using a Bio-Rad MRC-600 laser scanning confocal microscope having a novel filterset designed for dual detection of Cy3 and Cy5 (4). This system affords an opportunity to view two antigens simultaneously and in registration while optically sectioning through individual neurons.

Results and Discussion

The embryonic ciliary ganglion contains two classes of neurons: ciliary neurons, which innervate skeletal muscle in the iris and ciliary body, and choroid neurons, which innervate arterial wall smooth musculature. All neurons could be stained with mAbs having specificity for the following AChR subunits: α3 (mAb A3-1), α3/α5 (mAb 210), α7 (mAb 318), β2 (mAb 270), and β4 (B4-1). [MAb 210 is specific for the α5 subunit in blots (among those expressed by ciliary neurons) but may recognize α3 as well as α5 subunits when they are expressed on the cell surface, D.K. Berg, personnel communication.] Figure 1 shows panels of cells labeled with an anti-α8 mAb (305), which indicates the level of non-specific staining, and cells labeled with an anti-α3/α5 mAb (210).

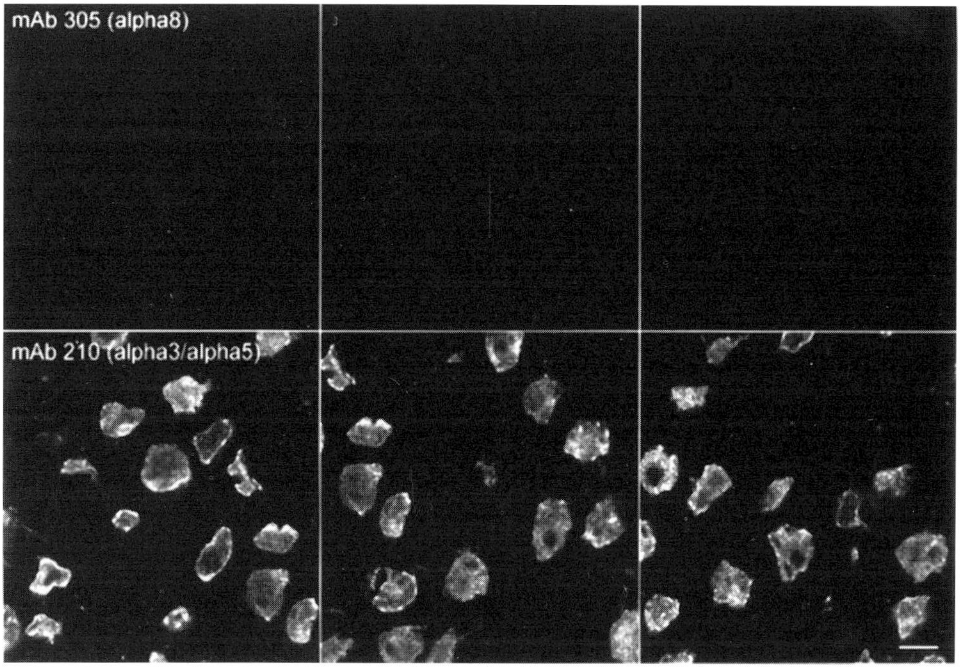

Figure 1. Ciliary ganglion neurons are stained with mAb 210 (α3/α5, bottom panels) but not with mAb 305 (α8, top panels). Scale bar, 15 mm.

Some of the mAbs that label neurons in frozen sections stained the surface intensely (e.g., Figure 1, bottom panels). We confirmed that this staining represents surface immunoreactivity by demonstrating that it can be obtained in unpermeabilized whole mounts, under conditions where mAbs presumably do not have access to the cell interior. Figure 2 shows an optical section taken through the interior of a ciliary ganglion neuron, demonstrating staining for mAb 210 ($\alpha3/\alpha5$) in the left image (Cy3 channel) and for mAb 10h (synaptic sites) in the right image (Cy5 channel). MAb 210 stains punctate areas of the cell surface, many or which correspond in position to synaptic sites. Since mAb 210 may recognize either $\alpha3$ or $\alpha5$ AChR subunits, this result suggests that $\alpha3$ and/or $\alpha5$ subunits are incorporated into AChRs located preferentially at synaptic sites, a conclusion also reached by Jacob et al. (5), who studied binding of mAb 35, which has a subunit specificity similar to mAb 210. Vernallis et al. (1) found by sequential immunoprecipitation that $\alpha3$ and $\alpha5$ subunits associate into common AChR oligomers, and so the question of whether mAb is recognizing $\alpha3$ or $\alpha5$ subunits on these surfaces is presumably moot.

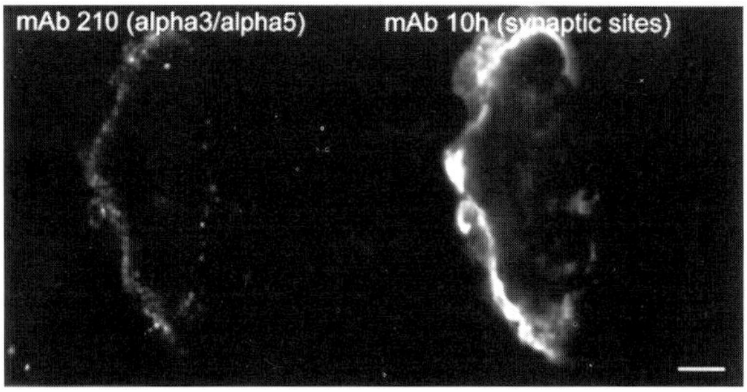

Figure 2. Optical section of a single ciliary ganglion neuron stained with mAb 210 ($\alpha3/\alpha5$, left panel) and with mAb 10h (synaptic sites, right panel). The two images represent Cy3 (left) and Cy5 (right) signals from the same field. Clusters of mAb 210 LI are found preferentially at synaptic sites (compare signals indicated by arrows in left and right images). Scale bar, 5 μm.

A second class of AChRs on ciliary ganglion neurons, in addition to the mAb 35-AChRs, are those recognized by α-bungarotoxin. To examine whether α-bungarotoxin-AChRs contain the $\alpha7$ subunit (6, 7), we stained frozen sections and whole mounts of the ciliary ganglion with both rhodamine-α-bungarotoxin (Molecular Probes) and with mAb 318 followed by Cy5-labeled goat anti-rat IgG. The results (Figure 3) indicate that both α-bungarotoxin and mAb 318 bind to similar regions of the neuronal surface, a result that is consistent with their recognizing a common AChR.

The quality of the surface staining obtained with rhodamine-α-bungarotoxin or anti-$\alpha7$ mAbs is different than that obtained with mAb 35 or 210. One difference is that mAb 35 or mAb 210 staining tends to produce a larger number of intensely-stained surface clusters

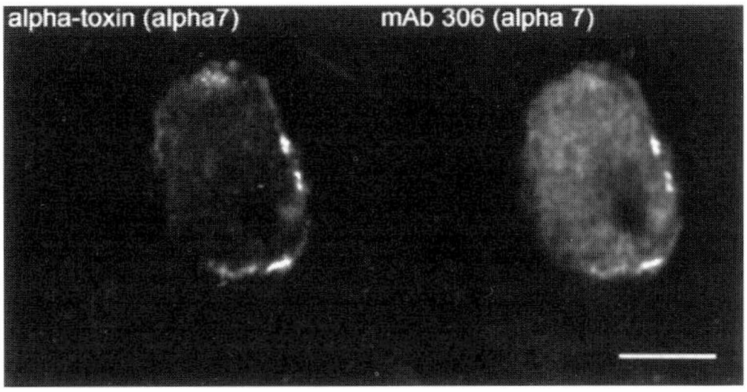

Figure 3. Optical sections of a single ciliary ganglion neuron stained with rhodamine-α-bungarotoxin (left, Cy3 channel) and mAb 318 (α7) followed by Cy5-goat anti rat IgG. The two reagents produce similar patterns of surface stain. Scale bar, 10 μm.

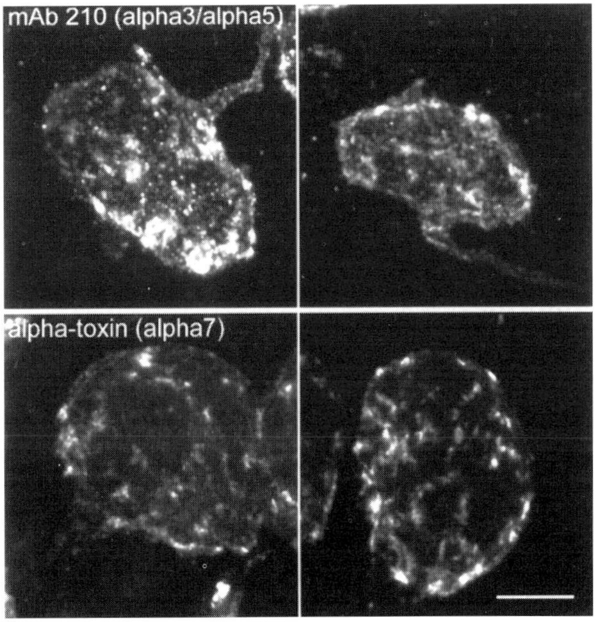

Figure 4. Differences in the appearance of staining obtained with mAb 210 (top panels) and with α-bungarotoxin (bottom panels). Unlike Figures 1-3, this figure contains reconstructions of all the optical sections through the cell body to produce a single image of all surface clusters. Clusters of α3/α5-AChRs are qualitatively different than those of α7-AChRs in having more numerous, punctate clusters. Scale bar, 10 μm.

than do reagents that reveal α7-LI. This is illustrated in Figure 4 for neurons that have been optically sectioned and reconstructed to produce 'through-focus' views of the entire cell.

Rhodamine-α-bungarotoxin or mAb 318 (α7) recognize areas of the surface that are largely extrasynaptic, as revealed in optical sections illustrated in Figure 5. Curiously, however, these reagents tend to stain patches of the cell surface that lie immediately adjacent to synaptic sites, as illustrated for optical sections through the middle of the cell (Figure 5, top panels) as well as for optical sections taken from the upper surface of the cell (Figure 5,

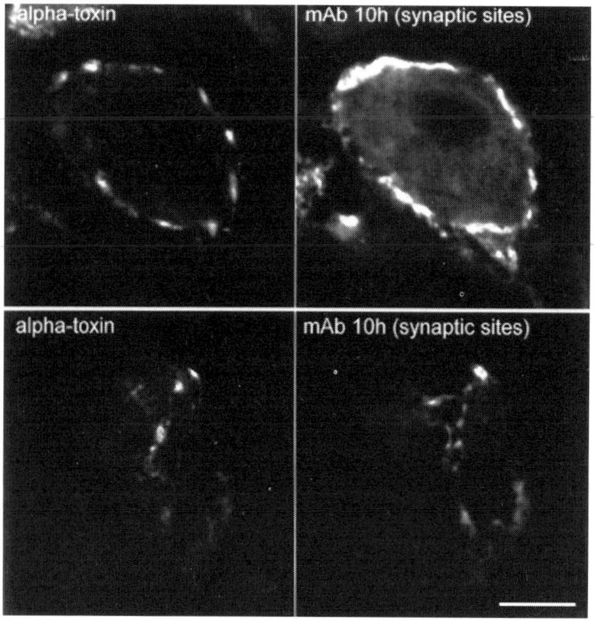

bottom panels). In some instances (Figure 5, bottom panel) α7-LI interdigitates with synaptic staining. α7-LI bears a special relationship to synaptic sites and can be described as perisynaptic (8).

Figure 5. Comparison of α-bungarotoxin binding sites (left images) with synaptic sites (right images). In optical sections taken through the interior of the cell (top images) and the cell surface (bottom images) α-bungarotoxin (α7) stains areas of the cell surface adjacent to synaptic sites. Scale bar, 10 μm.

Simultaneous imaging of α7-LI and α5-LI reveals areas of the cell surface staining for each subunit as well as areas staining for both (Figure 6). The presence of distinct patterns of staining support the results of Vernallis et al. (1), who found that α5 and α7 subunits are located within distinct populations of AChR oligomers in the ciliary ganglion.

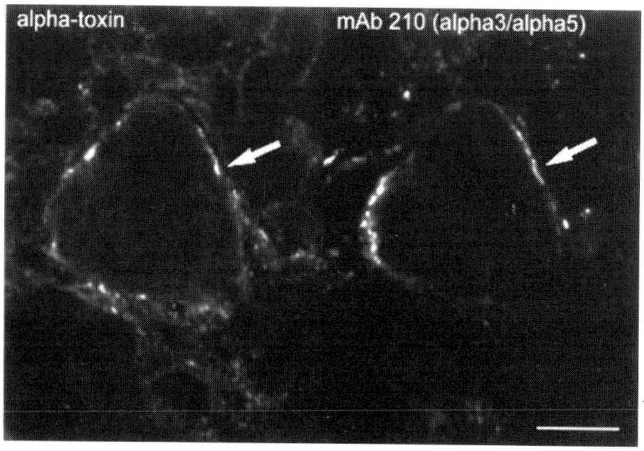

Figure 6. Comparison of α3/α5-LI with sites of α-bungarotoxin binding (α7). Areas of overlap (arrows) presumably lie extrasynaptically. Scale bar, 10 μm.

Conclusion

There appears to be little cellular diversity within the ciliary ganglion at the level of expression of AChR subunit immunoreactivities: virtually all neurons express detectable levels of immunoreactivity for the α3/α5, α7, β2, and β4 AChR subunits. The pattern of surface staining for α5-LI and α7-LI indicate that AChRs containing these subunits have distinct distributions. This suggests that distinct classes of AChRs exist on the surface of ciliary ganglion neurons, as indicated by immunoprecipitation studies on extracts (1).

Acknowledgments: *This work was supported by NIH NS24207. We thank Drs. Jon M. Lindstrom, Darwin K. Berg, and Steve Carlson for mAbs.*

References

1. Vernallis AB, Conroy WG, Berg DK. Neurons assemble acetylcholine receptors with as many as three kinds of subunits while maintaining subunit segregation among receptor subtypes. Neuron 1983; 10: 451-464.
2. Corriveau RA, Berg DK. Coexpression of multiple acetylcholine receptor genes in neurons: quantification of transcripts during development. J. Neurosci. 1993; 13: 2662-2671.
3. Buckley K, Kelly RB. Identification of a transmembrane glycoprotein specific for secretory vesicles of neural and endocrine cells. J. Cell Biol. 1985; 100: 1284-1294.
4. Sargent PB. Double label immunofluorescence with the laser scanning confocal microscope using cyanine dyes. NeuroImage 1994; in press.
5. Jacob MH, Berg DK, Lindstrom JM. Shared antigenic determinants between *Electrophorus* acetylcholine receptor and a synaptic component on chicken ciliary ganglion neurons. Proc. Natl. Acad. Sci. USA 1984; 81: 3223-3227.
6. Schoepfer R, Conroy WG, Whiting P, Gore M, Lindstrom J. Brain α-bungarotoxin binding protein cDNAs and MAbs reveal subtypes of this branch of the ligand-gated ion channel gene superfamily. Neuron 1990; 5: 35-48.
7. Couturier S, Bertrand D, Matter J-M, Hernandez M-C, Bertrand S, Millar N, Valera S, Barkas T, Ballivet M. A neuronal nicotinic acetylcholine receptor subunit (alpha7) is developmentally regulated and forms a homo-oligomeric channel blocked by alpha-BTX. Neuron 1990; 5: 847-856.
8. Jacob MH, Berg DK. The ultrastructural localization of α-bungarotoxin binding sites in relation to synapses on chick ciliary ganglion neurons. J. Neurosci. 1983; 3: 260-271.

Effects of Nicotine on
Biological Systems II
Advances in Pharmacological Sciences

AUTORADIOGRAPHIC DISTRIBUTION OF NICOTINIC RECEPTOR SITES LABELLED WITH [^{3}H]CYTISINE IN THE HUMAN BRAIN

Isabelle Aubert, Danielle Cécyre, Serge Gauthier and Rémi Quirion

Douglas Hospital Research Center, 6875 Lasalle Blvd, Verdun, Québec, Canada H4H 1R3 and Department of Neurology & Neurosurgery, McGill University (I.A., D.C., S.G., R.Q.) Department of Psychiatry, McGill University and Center for Studies in Aging, Montréal, Québec, Canada H3G 1A4 (S.G., R.Q)

Summary: The autoradiographic distribution of [^{3}H]cytisine, a putative α4β2 nicotinic receptor radioligand, was studied in the human brain. Highest levels of [^{3}H]cytisine sites were found in the lateral geniculate body and different thalamic nuclei. The middle layers of the frontal cortex, as well as the substantia nigra and striatum contained relatively high densities of [^{3}H]cytisine binding. Moderate amounts of [^{3}H]cytisine sites were observed in other laminae of the frontal cortex, in the temporal cortex, stratum lacunosum moleculare of the hippocampus, outer-molecular layer of the dentate gyrus, and molecular layer of the cerebellum. Additional laminae of the Ammon's horn, dentate gyrus and cerebellum have low levels of [^{3}H]cytisine binding sites.

Introduction

Recently, [^{3}H]cytisine was suggested to be a useful radioligand for the study of α-bungarotoxin insensitive nicotinic receptors, probably of the α4β2 subtype, in the mammalian brain (1-3). The pharmacological profile and regional distribution of [^{3}H]cytisine binding in membrane preparations of rat (1,3) and human (2) brains revealed that [^{3}H]cytisine likely binds to the same population of sites as those of [^{3}H]acetylcholine (ACh; under nicotinic conditions), [^{3}H]*N*-methylcarbamylcholine (MCC) and [^{3}H]nicotine. However, the use of [^{3}H]cytisine offers several advantages over other radioligands; among them, its stability and low levels of nonspecific labelling are most important (1-3).

In the human brain, the autoradiographic distribution of nicotinic sites was previously described using [^{3}H]nicotine, [^{3}H]ACh and [^{3}H]MCC (4-9). However, the sharpness of the autoradiographic images generated by these radioligands is not always optimal and their use requires high concentrations of competing drug (from 1 000 to 30 000 times the concentration of [^{3}H]ligands) in order to obtain significant amounts of specific binding. Therefore, the aim of the present study was to evaluate the usefulness of [^{3}H]cytisine in the localization and quantification of nicotinic receptor sites in various regions of the human brain.

Materials and Methods

Materials: [^{3}H]Cytisine (39.3 Ci/mmol) was obtained from New England Nuclear (Boston, MA, U.S.A.) while other chemicals were from usual sources (10).

Human Brain Tissues and Histopathological analyses: Human brain tissues were provided by the Brain Bank of the Douglas Hospital Research Center. Twenty normal controls, aged between 65 and 95 years old (76 ± 2 years), were used for this study. They had no neurological or psychiatric disorders based on a screening of hospital charts and/or phone interview with siblings when required. After autopsy, fresh brain weight was noted (1284 ± 40 g). One hemisphere was used for biochemical assays and the other for histological and pathological examinations (11).

[^{3}H]Cytisine Autoradiography: Sections (15 µm-thick) of frontal (Brodmann area 6) and temporal (Brodmann areas 20, 21 and 22) cortices, hippocampal formation, striatum (caudate and putamen nuclei), thalamus, substantia nigra and cerebellum were then cut at -18°C, thaw-mounted onto gelatin-coated slides, air-dried overnight and then stored at -80°C until used for receptor autoradiography.

Using assay conditions as described before (12), brain sections were incubated for one hour in Tris (50 mM)-HCl buffer (22°C, pH 7.4) containing 20 nM [^{3}H]cytisine in the presence or absence of unlabelled nicotine (2 µM) to determine the amount of specifically bound ligand. After 5 washes of 2 min each in Tris (50 mM)-HCl buffer (4°C, pH 7.4) and a dip in cold distilled water, slides were dried and juxtaposed to films, alongside with [^{3}H]-labelled standards, for 9 months. Films were then developed and radioactivity signals in various brain regions quantified using computerized image analysis.

Results

[^{3}H]Cytisine generated high quality autoradiograms in all regions of the human brain investigated (Figure 1 A - G). Under the described assay conditions, with a concentration of

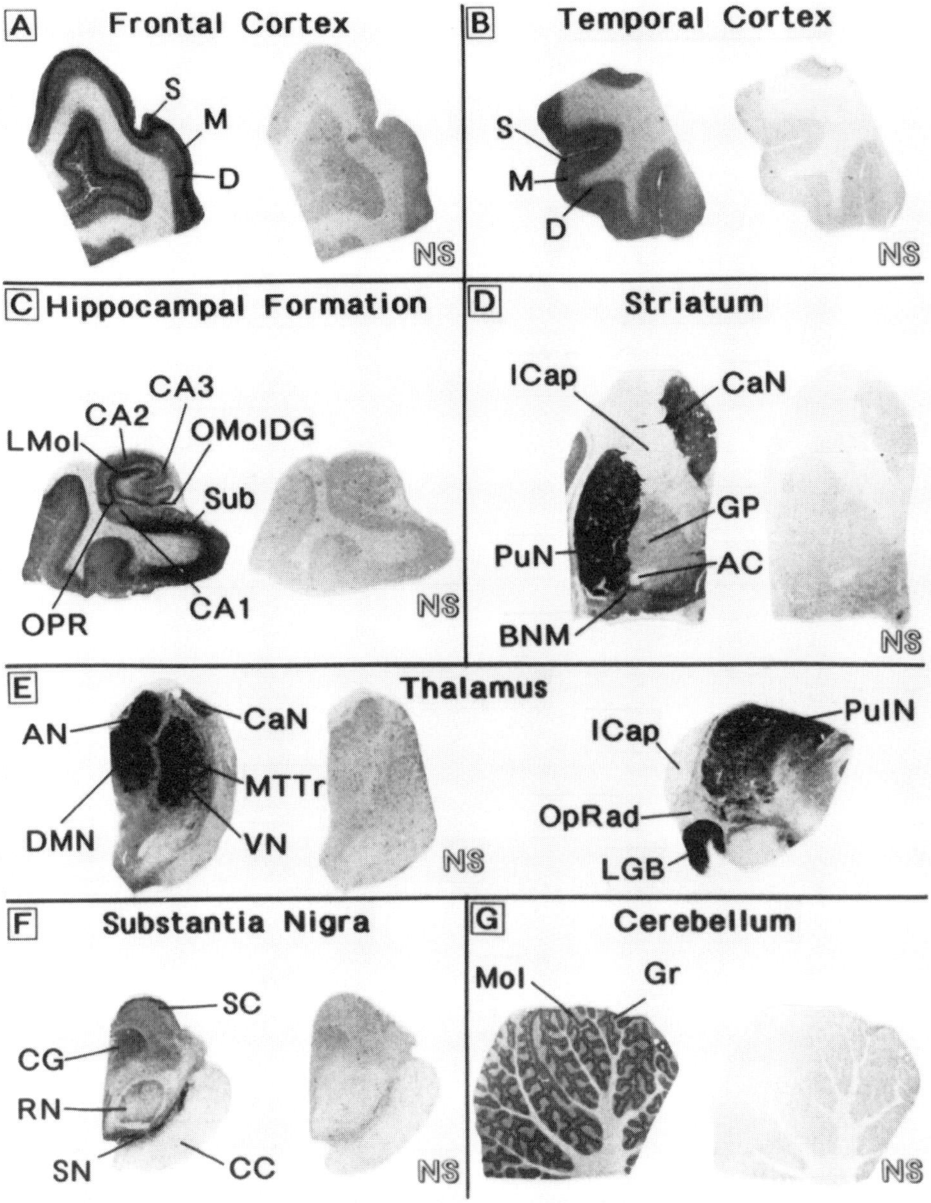

Figure 1. Representative photomicrographs of the autoradiographic distribution of [^{3}H]cytisine in various regions of the human brain. Abbreviations: AC, anterior commisure; AN, anterior nucleus; BNM, basal nucleus of Meynert; CA1-CA3, CA1-CA3 fields of the Ammon's horn; CaN, caudate nucleus; CC, crus cerebri; CG, central gray; DMN, dorsomedial nucleus; GP, globus pallidus; Gr, granular layer; ICap, internal capsule; LMol, stratum lacunosum moleculare; LGB, lateral geniculate body; Mol, molecular layer; MTTr, mammillothalamic tract; OMolDG, outer-molecular layer of the dentate gyrus; OPR, oriens, pyramidal and radiatum layers of the Ammon's horn; OpRad, optic radiations; PulN, pulvinar nucleus complex; PuN, putamen nucleus; RN, red nucleus; S, M and D, superficial, middle and deep laminae, respectively; SC, superior colliculus; SN, substantia nigra; Sub, subiculum; VN, ventral nucleus.

competing nicotine not exceeding 100 times that of [^{3}H]cytisine, the percent of specific binding ranged between 30 and 75% depending on the brain area (Fig. 2).

A distinct labelling pattern of different laminae of the frontal cortex was observed in 9 out of 11 samples of control patients (Fig. 1A), while in the temporal cortex, only one (not shown) out of 10 samples (Fig. 1B) revealed a clear laminar distribution. In the hippocampal formation, the clear labelling of the lacunosum moleculare and the outer-molecular layer of the dentate gyrus, as well as the middle layers of the subiculum, was apparent for all cases (Fig. 1C).

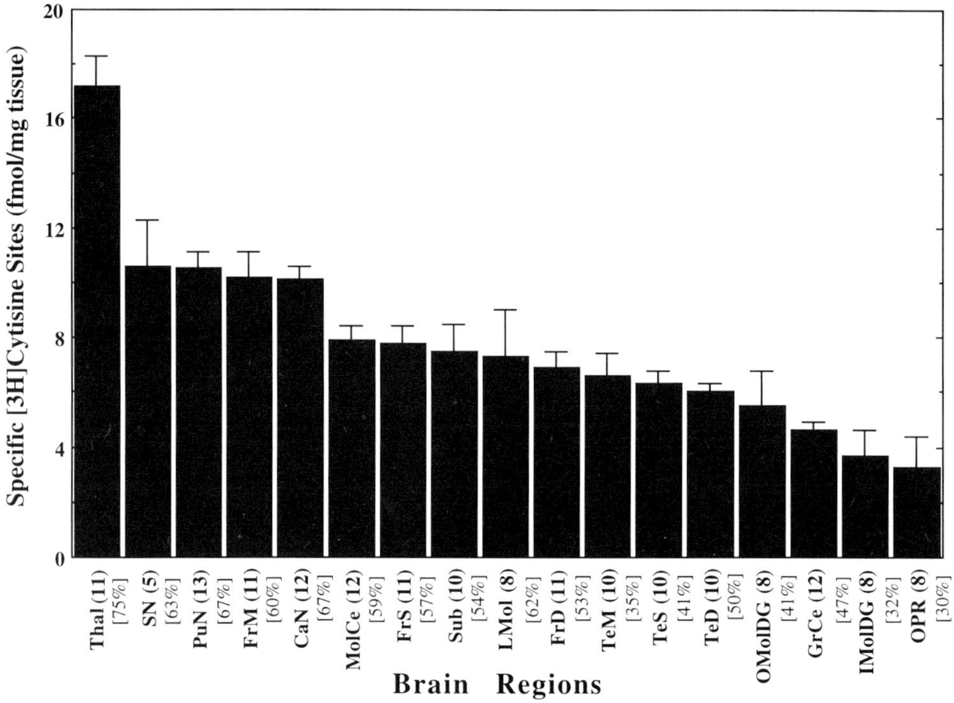

Figure 2. Quantitative autoradiographic distribution of [^{3}H]cytisine binding sites in selected areas of the human brain. The individual n and percentage of specific binding for each regions are given in parentheses (n) and brackets [%], respectively. Abbreviations: CaN, caudate nucleus; FrD, FrM and FrS, deep, middle and superficial layers of the frontal cortex, respectively; GrCe, granular layer of the cerebellum; IMolDG, inner-molecular layer of the dentate gyrus; LMol, stratum lacunosum moleculare; MolCe, molecular layer of the cerebellum; OMolDG, outer-molecular layer of the dentate gyrus; OPR, oriens, pyramidal and radiatum layers of the hippocampus; PuN, putamen nucleus; SN, substantia nigra; Sub, subiculum; TeD, TeM and TeS, deep, middle and superficial layers of the temporal cortex, respectively; Thal, thalamus.

Highest levels of [^{3}H]cytisine sites were found in the lateral geniculate body and various thalamic nuclei (Figs 1E, 2). In the frontal cortex (Figs 1A, 2) and subiculum (Fig. 1C), a distinct denser band was generally apparent (middle layers) and revealed high densities of [^{3}H]cytisine binding. The caudate and putamen nuclei (Figs 1D, 2) and the substantia nigra (Figs 1F, 2) also had high levels of [^{3}H]cytisine sites. Moderate amounts of [^{3}H]cytisine binding were found in the superficial and deep layers of the frontal cortex (Figs 1A, 2), throughout the temporal cortex (Figs 1B, 2), in the stratum lacunosum moleculare of the Ammon's horn (Figs 1C, 2), outer-molecular layer of the dentate gyrus (Figs 1C, 2), superficial and deep laminae of the subiculum (Fig. 1C) and in the molecular layer of the cerebellum (Figs 1G, 2). Low levels of [^{3}H]cytisine binding were observed in the oriens, pyramidal and radiatum layers of the Ammon's horn of the hippocampus (Figs 1C, 2), in the inner-molecular and granular layers of the dentate gyrus (Figs 1C, 2) and in the granular layer of the cerebellum (Figs 1G, 2).

Discussion

Various studies have previously investigated the quantitative distribution of nicotinic sites in human cortex (4-7,9). However, very few of them mentioned the Brodmann areas of the cortical regions analyzed (9). This is of considerable importance since the laminar distribution of these sites varies within Brodmann's areas (9). We report here on the localization of [^{3}H]cytisine sites in the frontal cortex (Brodmann area 6). In this area, the superficial and particularly middle layers are enriched in nicotinic sites. With previously used nicotinic radioligands, no data are available on this Brodmann area. However, a similar laminar pattern of [^{3}H]nicotinic sites is observed in the primary motor cortex (Brodmann area 4) and in the association areas of the frontal cortex (Brodmann area 8) (9). A clear laminar distribution in the temporal cortex (Brodmann areas 20, 21 and 22) was observed only for one case, in which the middle layers were more intensively labelled. This case could most likely reflect the distribution of nicotinic sites in Brodmann area 21 since, according to Perry et al. (9), the middle layers of the temporal cortex are markedly labelled with [^{3}H]nicotine in Brodmann area 21 but not in area 20 where nicotinic sites are more homogeneously distributed. To date, several studies have suggested that nicotinic receptor sites are enriched predominantly in layer IV of the neocortex (thalamocortical inputs) of the human brain (4,5,9) as it is observed in the rat (13). However, a systematic study on the localization of nicotinic sites on given cell types of different Brodmann areas still needs to be performed to test this hypothesis. The use of [^{3}H]cytisine, in parallel with histochemical procedures, should facilitate such detailed analysis.

The distribution and relative abundance of [^{3}H]cytisine sites in the hippocampal formation are similar to data previously obtained with other nicotinic radioligands (7,9). However, the sharpness of the autoradiograms generated by [^{3}H]cytisine allows for better visualization of the various laminae. In the present study, it is clear that the outer-molecular layer of the dentate gyrus is enriched with [^{3}H]cytisine sites. As recently suggested for the rat brain, nicotinic sites in this area might represent heteroreceptors of the perforant path (10).

The apparent densities of [^{3}H]cytisine sites in the caudate and putamen nuclei are similar. However, Adem et al. (7), using [^{3}H]nicotine, observed that the apparent densities of nicotinic sites were two times greater in the putamen than the caudate. These differences could be due to the particular level at which the caudate and putamen nuclei are analyzed. In the rat striatum (caudate and putamen nuclei), nicotinic receptors are located, in a significant proportion, on dopaminergic terminals of the nigrostriatal pathway (14). This may also be the case in the human brain as it was reported that [^{3}H]MCC binding sites are significantly decreased in the striatum of Parkinsonian patients (11).

In the thalamus, [^{3}H]cytisine binding had a similar autoradiographic distribution when compared to those of either [^{3}H]nicotine or [^{3}H]ACh (8). However, the level of specific binding obtained with [^{3}H]cytisine (75%) is greater that of previously used radioligands (50%). Adem et al. (7), have described the substantia nigra and the cerebellum as areas enriched with [^{3}H]nicotine sites. This is in accord with the present study using [^{3}H]cytisine. In addition, the use of [^{3}H]cytisine seems to provide a sharper definition of the autoradiograms in these brain regions (especially allowing the distinction of the granular and molecular layers of the cerebellum).

Conclusion

Human brain autoradiograms generated by the use of [^{3}H]cytisine were of sharper definition than those obtained with previous radioligands. Thus, [^{3}H]cytisine should be a useful ligand to map in detail, alongside with histochemical procedures, putative nicotinic receptor sites of the $\alpha4\beta2$ subtype in various subregions of the human brain. In addition, the quantification of [^{3}H]cytisine binding in precise subregions and laminae of brains from patients who had suffered from Alzheimer's and Parkinson's diseases should provide further understanding on the alterations of nicotinic receptors in these disorders.

Acknowledgments: *This work was supported by a Medical Research Council of Canada's grant to R.Q. who is a "Chercheur Boursier" of the "Fonds de la Recherche en Santé du Québec". I.A. received a studentship from the Faculty of Medicine, McGill University. The authors would also like to thank Dr. Hélène Le Jeune for her expert photographic skills and Dr. Anurag Tandon for critical reading of the manuscript.*

References

1. Pabreza LA, Dhawan S, Kellar KJ. [^{3}H]Cytisine binding to nicotinic cholinergic receptors in brain. Mol Pharmacol 1991; 39:9-12.
2. Hall M, Zerbe L, Leonard S, Freedman R. Characterization of [^{3}H]cytisine binding to human brain membrane preparations. Brain Res 1993; 600:127-133.
3. Anderson DJ, Arneric SP. Nicotinic receptor binding of [^{3}H]cytisine, [^{3}H]nicotine and [^{3}H]methylcarbamylcholine in rat brain. Eur J Pharmacol 1994; 253:261-267.
4. Whitehouse PJ, Martino AM, Wagster MV, et al. Reductions in [^{3}H]nicotinic acetylcholine binding in Alzheimer's disease and Parkinson's disease: an autoradiographic study. Neurology 1988; 38:720-723.
5. Kellar KJ, Whitehouse PJ, Martino-Barrows AM, Marcus K, Price DL. Muscarinic and nicotinic cholinergic binding sites in Alzheimer's disease cerebral cortex. Brain Res 1987; 436:62-68.
6. Araujo DM, Lapchak PA, Collier B, Quirion R. *N*-[^{3}H]methylcarbamylcholine binding sites in the rat and human brain: relationship to functional nicotinic autoreceptors and alterations in Alzheimer's disease. Prog Brain Res 1989; 79:345-352.
7. Adem A, Nordberg A, Jossan SS, Sara V, Gillberg P-G. Quantitative autoradiography of nicotinic receptors in large cryosections of human brain hemispheres. Neurosci Lett 1989; 101:247-252.
8. Adem A, Jossan SS, d'Argy R, Brandt I, Winblad B, Nordberg A. Distribution of nicotinic receptors in human thalamus as visualized by ^{3}H-nicotine and ^{3}H-acetylcholine receptor autoradiography. J Neural Transm 1988; 73:77-83.
9. Perry EK, Court JA, Johnson M, Piggott MA, Perry RH. Autoradiographic distribution of [^{3}H]nicotine binding in human cortex: relative abundance in subicular complex. J Chem Neuroanat 1992; 5:399-405.
10. Aubert I, Poirier J, Gauthier S, Quirion R. Multiple cholinergic markers are unexpectedly not altered in the rat dentate gyrus following entorhinal cortex lesions. J Neurosci 1994; 14:2476-2484.
11. Aubert I, Araujo DM, Cécyre D, Robitaille Y, Gauthier S, Quirion R. Comparative alterations of nicotinic and muscarinic binding sites in Alzheimer's and Parkinson's diseases. J Neurochem 1992; 58:529-541.
12. Aubert I, Rowe W, Meaney MJ, Gauthier S, Quirion R. Cholinergic markers in aged memory-impaired and unimpaired Long-Evans rats. Neuroscience (submitted).
13. Clarke PBS, Schwartz RD, Paul SM, Pert CB, Pert A. Nicotinic binding in rat brain: autoradiographic comparison of [^{3}H]acetylcholine, [^{3}H]nicotine, and [^{125}I]-α-bungarotoxin. J Neurosci 1985; 5:1307-1315.
14. Clarke PBS, Pert A. Autoradiographic evidence for nicotinic receptors on nigrostriatal and mesolimbic dopaminergic neurons. Brain Res 1985; 348:355-358.

Effects of Nicotine on
Biological Systems II
Advances in Pharmacological Sciences

EPIBATIDINE

Malgorzata Dukat[1], M.I. Damaj[2], Daniel Dumas[1], William Glassco[2], Everette L. May[1,2], Billy R. Martin[2], and Richard A. Glennon[1,2]

Departments of Medicinal Chemistry[1], and Department of Pharmacology[2]
Medical College of Virginia/Virginia Commonwealth University
Richmond, Virginia 23298 U.S.A.

Summary: The structure of (-)epibatidine was compared with that of (-)nicotine. Results of molecular modeling reveal that epibatidine can approximate the nicotine structure but that the internitrogen distance in epibatidine is longer than that calculated for nicotine. The higher [^{3}H]nicotine-labeled nicotine receptor affinity for (-)epibatidine (Ki = 0.055 nM) relative to (-)nicotine (Ki = 1.5 nM) suggests that the commonly accepted nicotine binding pharmacophore is in need of revision.

Introduction

Epibatidine [1] is an exciting new natural product that is attracting considerable scientific interest. It was first isolated from the South American frog *Epipedobates tricolor* by Daly and co-workers. An examination of alkaloids present in the frog resulted in the isolation of a crude fraction that potently produced a Straub-tail response when injected in mice; the active constituent was subsequently identified as epibatidine [1](1,2). Epibatidine was found to be 500 times more potent than morphine in producing the Straub-tail reaction, and 200 times more potent than morphine as an analgesic in the hot plate test (2). A more complete pharmacological profile of the natural (+)epibatidine and of its (-)enantiomer has been recently published by Badio and Daly (3).

The demonstration that epibatidine is a potent analgesic that seems to act via a non-opioid mechanism, coupled with a lack of natural material for additional evaluation, led to several total syntheses both of racemic epibatidine (4-6) and of its optical isomers (7,8).

Inspection of the epibatidine structure reveals a similarity with the structure of nicotine

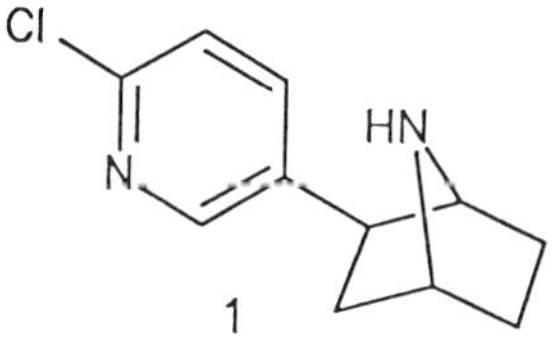
1

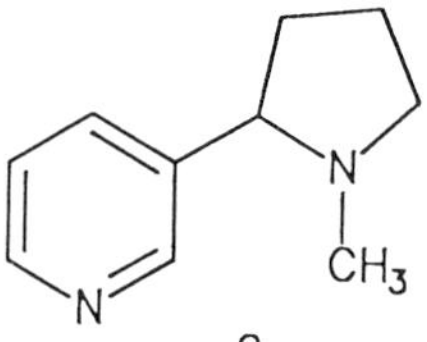
2

[2]. Indeed, spectral data for the nicotine analog anabasine were used for comparison in the original structure elucidation of epibatidine (2). This prompted us (i) to examine the binding of epibatidine at nicotine receptors, and (ii) to conduct molecular modeling studies in order to more closely examine the structural relationship between epibatidine and nicotine.

While our work was in progress, Badio and Daly (3) and Qian et al (9) published the results of their radioligand binding studies on epibatidine at [^{3}H]nicotine-labeled and [^{3}H]cytis-inelabeled nicotine receptors, respectively. Some of our results have also been published (10).

Materials and Methods

Molecular modeling: Details of the modeling studies have been published (10). In brief, structures of (-)nicotine and (-)epibatidine were built from fragments and standard bond lengths and angles using version 6.04 of SYBYL. The structures were minimized with a Tripos force field and full conformational searches were performed. A number of iow-energy conformations were identified with variation about the pyridine-pyrrolidine bond. In general, these conformers could be divided into two groups where the pyrrolidine nitrogenatom was slightly behind, or slightly forward, of the plane of the pyridine ring. Measurement were made using the lowest energy conformer of nicotine and epibatidine.

In the superimposition studies, (-)nicotine and (-)epibatidine were overlayed using the MULTIFIT command in SYBYL. Points selected for fitting were the pyridine nitrogen atoms, the pyrrolidine nitrogen atoms, and the pyridine centroids; basis for this selection are related to the previously published nicotinic pharmacophore (described below). (-)Nicotine was flexibly superimposed on the lowest energy conformer of (-)epibatidine. The final (i.e., superimposed) conformation of (-)nicotine was not significantly energetically different from the lowest energy conformer.

Results and Discussion

Epibatidine, like nicotine, possesses a substituted pyridine nucleus attached to a pyrrolidine ring. The methyl group of nicotine appears to contribute to binding and its removal, resulting in nornicotine, decreases receptor affinity by about 20- to 30-fold. Unlike nicotine, epibatidine lacks the pyrrolidine N-methyl group; this may bode a low receptor affinity for epibatidine. Also unlike nicotine, epibatidine possesses an aromatic chloro group and a two-carbon, rather than one-carbon, spacer between the pyridyl nucleus and the pyrrolidine ring. Thus, although the structure of epibatidine is reminiscent of nicotine, there are certain obvious differences. Can the structure of epibatidine mimic the structure of nicotine?

Results of superimposition studies: The final superimposed structures of (-)nicotine and (-)epibatidine are shown in Figure 1. In this superimposition, the two pyridine nitrogen atoms are within 0.25 Å, the two pyrrolidine nitrogen atoms are within 0.42 Å, and the two centroids are within 0.13 Å of one another. This represents a reasonable, if less than ideal, superimposition. That is, it would appear that the pyrrolidine nitrogen atom of (-)epibatidine is somewhat further removed from the pyridine centroid than is the corresponding pyrrolidine nitrogen in (-)nicotine. Nevertheless, the structure of (-)epibatidine is such that (even with the two-carbon spacer) its conformationally restricted azabicycloheptane ring locates important structural features in a manner that they roughly mimic those of nicotine. These results support a structural similarity between (-)epibatidine and (-)nicotine.

The nicotine receptor pharmacophore: What structural features are important for the binding of nicotinic ligands to nicotine receptors? Although its presence is consequential, the intact pyrrolidine ring of nicotine is not required for binding. That is, nicotine derivatives where a portion of the pyrrolidine ring has been excised retain affinity for nicotine receptors (11). For example, simple aminomethylpyridines derivatives, [3], bind in the nanomolar range, and affinity is highly dependent upon the nature of the R and R' substituents (11). These aminomethylpyridines, like nicotine, are conformationally flexible molecules; the exact conformation required for binding is unknown. In order to obtain some information as to what would constitute a more likely conformation for the binding of the aminomethylpyridines at nicotine receptors, we prepared and examined two conformationally-restricted aminomethylpyridine derivatives: the 2,7-naphthyridine derivative [4a], and the 1,6-naphthyridine derivative

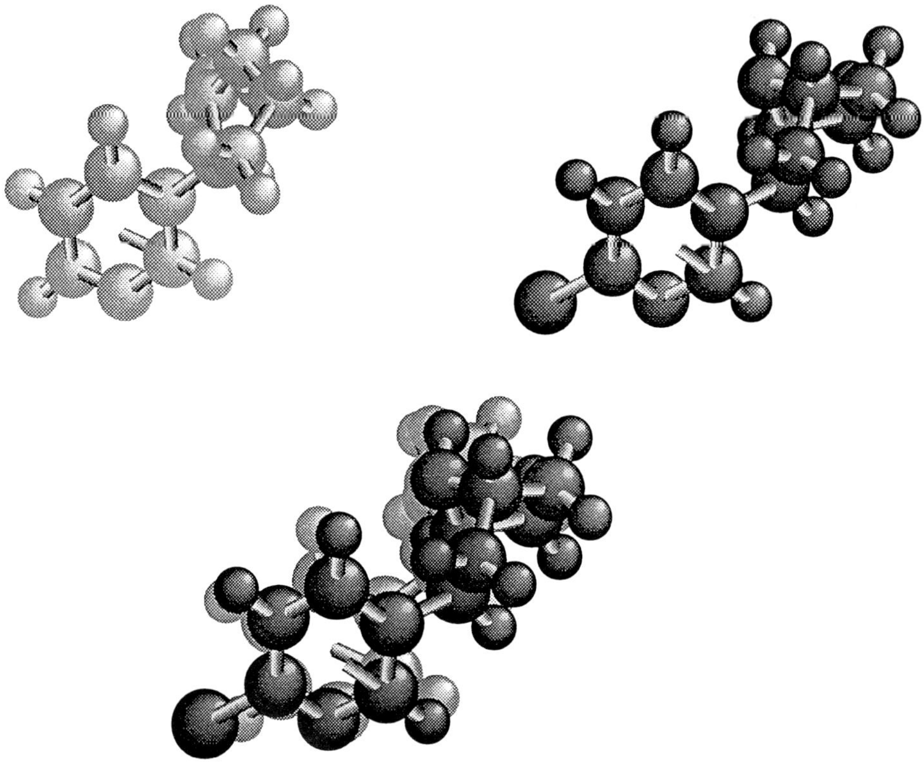

Figure 1. Structures of (-)nicotine (upper left), (-)epibatidine (upper right), and (-)nicotine flexibly superimposed on (-)epibatidine (center) showing the structural resemblance between the two molecules.

[4b]. These two compounds represent extreme conformations where the amino group is locked into an "up" and a "down" conformation, respectively. Compound [4a] (Ki = 18 nM) binds with about ten-fold higher affinity than [4b] (Ki = 165 nM) (11). Neither may represent the actual receptor-relevant conformation of the aminomethylpyridines, but the higher affinity of

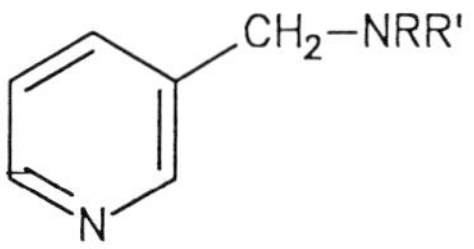

3

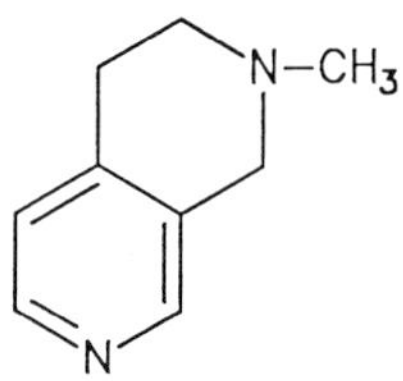

4a

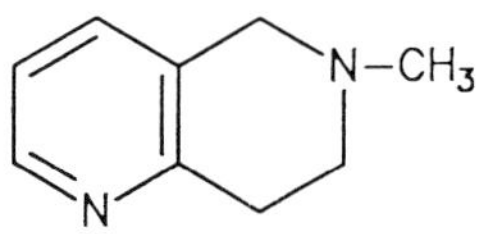

4b

the "up" conformationally-restricted analog [4a] suggests that this is a better approximation of the conformation required for binding.

Using a distance geometry approach, Sheridan and co-workers have formulated a model-nicotine pharmacophore. Due to the paucity of nicotine receptor ligands, the model was necessarily derived from data on a small number of agents. The model focusses on three structural components that are thought to be important for nicotinic activity — *A*: a cationic center (e.g. the pyrrolidine nitrogen atom of nicotine), *B:* an electronegative atom that likely participates in a hydrogen bond interaction with the receptor (e.g. the pyridine nitrogen atom of nicotine), and *C*: an atom that forms a dipole with the electronegative atom (e.g. a dummy atom representing the pyridine centroid of nicotine). Optimal distances for these features are shown in Figure 2.

Nicotine itself, one of the agents from which the pharmacophore was derived, fits the model quite well. The centroid/pyrrolidine-nitrogen or C-A distance (calculated above) of 3.77 Å falls within the range proposed by the model (i.e., 3.7 to 4.3 Å). Likewise, the internitrogen, or B-A, distance in nicotine (4.87 Å; Figure 2) is consistent with the model range (4.5 to 5.1 Å). Due to the presence of the pyridine ring, the C-B distance is fixed at 1.2 Å. It might be noted that the C-A distance and the B-A distance for the 2,7-naphthyridine [4a] (3.79 and 4.91 Å, respectively), are also within the range proffered by the model, and are essentially identical to those for (-)nicotine itself.

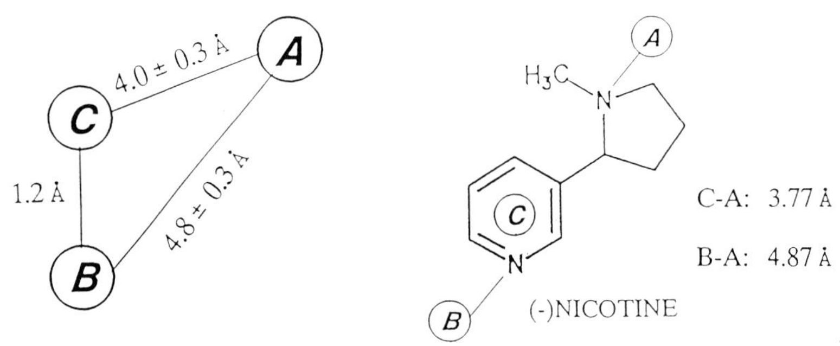

Figure 2. Nicotine pharmacophore as proposed by Sheridan et al (12). Features *A, B*, and *C* are, respectively: a cationic center, an electronegative H-bonding feature, and an atom (or pyridine centroid) that forms a dipole with the electronegative atom. Representative distances, as calculated in the present study, are shown for (-)nicotine.

We next examined these same distances for (-)epibatidine. The centroid/pyrrolidine-nitrogen distance in (-)epibatidine (4.28 Å) is slightly longer than that calculated for (-)nicotine, but is still within the range of the model. In contrast, the internitrogen distance (5.51 Å) is longer than that calculated for (-)nicotine, and is beyond the limits of the model. The conclusion, then, if this model is an accurate representation of the distances required for nicotine receptor binding, is that (-)epibatidine should be of significantly lower affinity than nicotine because the internitrogen distance significantly exceeds the optimal distance (and even the range) prescribed by the model. Radioligand binding data (Ki values) for nicotine and epibatidine are as follows: (-)nicotine, 1.5 ± 0.2 nM; (+)nicotine, 30.9 ± 3.4 nM; (-)epibatidine, 0.055 ± 0.013 nM, and (+)epibatidine, 0.055 ± 0.011 nM. Thus, epibatidine not only binds at nicotine receptors, it actually binds with about 25 times the affinity of nicotine. Consistent with these results, others have now obtained similar binding data with epibatidine (3,9).

How can the high affinity of epibatidine be rationalized? One explanation is that the pharmacophore model requires modification. Figure 1 shows that the structure of epibatidine can approximate the structure of nicotine; as discussed, the superimposition is reasonable if less than perfect. The major difference between the two structures is the longer internitrogen distance found in epibatidine relative to nicotine. Thus, we focussed next on this issue. Obviously, epibatidine was unavailable to those who initially formulated the nicotine binding pharmacophore. With the availability of binding data on epibatidine, this information *(assuming relatively common modes of binding for these and related compounds)* needs to be incorporated into the pharmacophore model.

Perhaps the pharmacophore model requires a simple extension of the limits proposed for the internitrogen distance. That is, the internitrogen (B-A) distance may represent an extreme limit. We prepared and examined isonicotine and isonornicotine ([5] and [6], respectively). These compounds may be viewed as epibatidine derivatives where a portion of the azabicycloheptane ring has been excised. Both of these agents ([5]; Ki = 12.5 nM; [6], Ki = 7.3 nM) bind with high affinity, and yet their calculated internitrogen distance (5.75 Å) far exceeds that of nicotine (4.87 Å). This distance also exceeds that of epibatidine (5.51 Å), and that proposed by the pharmacophore model (4.8 Å). Interestingly, the affinity of isonornicotine [6] is several-fold higher than that of nornicotine (Ki = 25 nM). Assuming a common mode of binding, it would appear that the internitrogen distance found in epibatidine is optimal for binding, that that for nicotine is on the shorter end of the tolerated range, and that for iso(nor)-nicotine is on the longer end of the range.

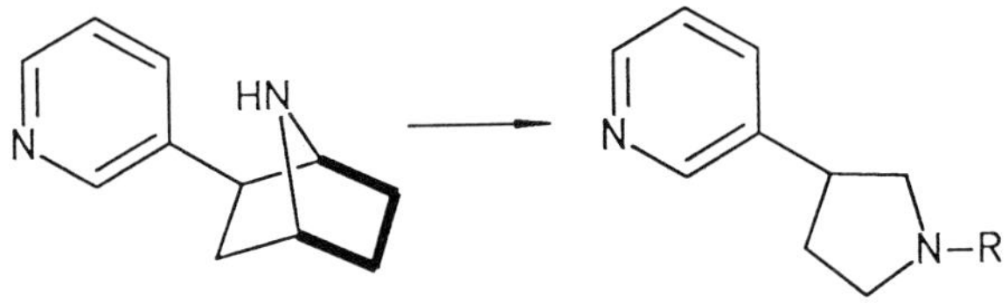

5 R = CH_3

6 R = H

One final question needs to be addressed. Is the enhanced affinity of epibatidine relative to nicotine related in any way to the presence of the chloro group? We prepared and examined 6-chloronicotine and found it to bind with higher affinity (Ki = 0.6 nM) than nicotine itself. However, its affinity is still 10-fold lower than that of (-)epibatidine. Recently, Badio and Daly have reported that deschloroepibatidine binds with an affinity comparable to that of epibatidine (3). The overall conclusion, then, is that the presence of the chloro group may contribute somewhat to binding, but that by itself it cannot account for the very high affinity of epibatidine.

In summary, we conclude that it is possible for the structure of epibatidine to approximate the structure of nicotine. This is supported by the results of radioligand binding studies that demonstrate a high affinity of epibatidine for nicotine receptors. However, the internitrogen distance calculated for epibatidine is somewhat longer than that calculated for (-)nicotine. It would seem that the high affinity of epibatidine is related, at least in part, to its conformationally restricted nature that holds the pyrrolidine nitrogen atom at a distance of about 5.5 Å from the pyridine nitrogen. The corresponding distance in the lower affinity nicotine and nornicotine is more than 0.5 Å shorter; the internitrogen distance in isonicotine and isonornicotine exceed that of nicotine and nornicotine by nearly a full Ångstrom. The presence of a chloro group, although it seems to slightly enhance the affinity of nicotine (vide supra), makes little contribution to the binding of epibatidine (3). Finally, the role of the N-methyl group should be mentioned. Demethylation of nicotine to nornicotine reduces affinity; the corresponding change in isonicotine [5] has little effect. It has been reported that N-methylation of epibatidine also has relatively little effect (13). So, even though all of these agents seem to bind at the same receptor, there is evidence that there may be some differences in the specific orientations with which they interact. Nevertheless, epibatidine offers us a new high-affinity tool with which to investigate nicotine receptors.

Acknowledgements: *This work was supported in part by the funds from the Technology Development Center, Virginia Center for Innovative Technology.*

References

1. Daly JW, Secunda SI, Garraffo HM, Spande TF, Wisnieski A, Nishihira C, Cover JF. Variability in alkaloid profiles in neotropical poison frogs (Dendrobatidae): Genetic versus environmental determinants. Toxicon 1992; 30:887-8.
2. Spande TF, Garraffo HM, Edwards MW, Yeh HJC, Pannell L, Daly JW. Epibatidine: A novel (chloro-pyridyl)azabicycloheptane with potent analgesic activity from an Ecuadoran poison frog. J Am Chem Soc 1992; 114:3475-8.
3. Badio B, Daly JW. Epibatidine: A potent analgetic and nicotine agonist. Mol Pharmacol 1994; 45:563-9.
4. Huang DF, Shen TY. A versatile total synthesis of epibatidine and analogs. Tetrahedron Lett 1993; 34:4477-80.
5. Broka CA. Total synthesis of epibatidine. Tetrahedron Lett 1993; 34:3251-4.
6. Clayton SC, Regan AC. A total synthesis of (±)-epibatidine. Tetrahedron Lett 1993; 34:7493-6.
7. Corey EJ, Loh T-P, AchynthaRao S, Daley DC, Sarshar S. Stereocontrolled total synthesis of (+) and (-)epibatidine. J Org Chem 1993; 58:5600-02.
8. Fletcher SR, Baker R, Chambers MS, Hobbs SC, Mitchell PJ. The synthesis of (+)- and (-)-epibatidine. J Chem Soc Chem Commun 1993; 15:1216-18.
9. Qian C, Li T, Libertine-Garahan L, Eckman J, Biftu T, Ip S. Epibatidine is a nicotinic analgesic. Eur J Pharmacol 1993; 250:R13-R14.
10. Dukat M, Damaj MI, Glassco W, Dumas D, May EL, Martin BR, Glennon RA. Epibatidine: A very high affinity nicotine receptor ligand. Med Chem Res 1994; 4:131-139.
11. Fiedler W, Dukat M, Damaj MI, Martin BR, Glennon RA. Nicotine: Structure-affinity studies; Development of novel agents. Abstract (P18), International Symposium on Nicotine: The Effects of Nicotine on Bioligical Systems II. Montreal, Canada, July 21-24, 1994.
12. Sheridan RP, Nilakantan R, Dixon JS, Venkataraghavan R. The ensemble approach to distance geometry: Applications to the nicotine pharmacophore. J Med Chem. 1986; 29:899-906.
13. Wypij DM, Shen TY. Epibatidine and related analogs compete with [^{3}]Hcytisine with high affinity and for binding to rat brain cortical membrane preparations. Abstract (P31), International Symposium on Nicotine: The Effects of Nicotine on Bioligical Systems II. Montreal, Canada, July 21-24, 1994.

Effects of Nicotine on
Biological Systems II
Advances in Pharmacological Sciences

MEASURING CHANGES IN MESOLIMBIC DOPAMINE AND RELATED SUBSTANCES IN RESPONSE TO SYSTEMIC NICOTINE USING IN VIVO ELECTROCHEMISTRY

Graham C. Parker and Paul B.S. Clarke

Dept. of Pharmacology & Therapeutics, McGill University, Montreal, Canada H3G 1Y6

Summary: Several psychopharmacological properties of tobacco smoking are believed to be mediated by central nicotinic receptors on mesolimbic dopamine (DA) neurons. Using in vivo electrochemical probes we monitored the extracellular levels of DA alone and also the levels of DOPAC and ascorbate, in the nucleus accumbens following systemic administration of nicotine (0.4 mg/kg SC). These studies used nicotine-naive, unanaesthetized freely-moving rats. Preliminary data suggest an initial (<20 min) decrease in accumbal DA following the first injection of nicotine. This decrease was not observed on the second injection day when the DA signal reached a higher maximum and had a longer duration of effect.

Introduction

One of the goals of neurobiology is to understand how nervous activity in the brain is related to mental functioning and behaviour. Great advances have been made in investigating such psychological constructs as reinforcement, learning and memory by studying the release of neurotransmitters putatively involved in the neurochemistry of such processes. Such behaviours have been demonstrated repeatedly to be affected by manipulations of the catecholamine dopamine (DA) and its release in the nucleus accumbens (NAcc).

Technological advances have allowed us to progress from measuring transmitter levels *post mortem* to techniques such as microdialysis whereby samples can be extracted for analysis *ex vivo* from freely moving animals (4). Now we can measure the levels of various neurochemicals *in vivo* using electrochemistry (5,6). Fortunately, certain monoamines, notably DA and serotonin are electroactive. This means that when a certain voltage is applied across two electrodes, these substances will oxidise, producing a current that can be measured with sensitive equipment. Electroactive substances differ in the voltage at which they will give up electrons, hence the identity of the substance can be determined by its oxidative potential, while its concentration local to the electrode surface can be estimated from the size of the resulting current.

Materials and Methods

Male Long-Evans rats (Charles River, Lachine, Quebec) weighing 300-350 g at time of surgery were housed individually with a 12 hr light/dark cycle (0700/1900 hrs) and ad libitum access to lab chow and water. Under sodium pentobarbital anaesthesia (60 mg / kg IP), working electrodes were implanted bilaterally into the NAcc (level skull, Bregma +1.2, Mid ± 0.6, Ctx -6.5). The working electrodes were constructed using Teflon-coated stainless steel (0.008" bare 0.011" coated, A-M Systems, Inc.). The Teflon was moved proud of the steel to form a well. The other end of the electrode was soldered to an amphenol pin (Electrosonic). The well was packed with USP grade graphite powder (Ultra Carbon Corp.) that had previously been mixed with 0.8 ml 0.7 mg / ml Nujol (Aldrich). Once the well was properly packed the electrode was impacted on a clean flat piece of glass to produce an even compression then expansion of the material producing a spherical electrode surface. The electrode was then dipped in Nafion (in a solution of ethanol/water, see Aldrich Cat no. 27,470-4 for details). The reference electrode comprised Teflon-coated silver wire (0.01" bare, 0.013" coated) exposed to a length of 1/2". The wire was then coiled flat. The other end of the silver wire was soldered to an amphenol connector. The electrode was then chlorided to produce an Ag/AgCl reference. It was necessary to carefully drill away the skull above the target coordinates so as to allow a window in the dura to be opened without causing bleeding on the cortex surface. This was to keep the working electrodes as clean as possible when lowered. The implantation of the reference electrode was based on an idea proposed to protect it from deleterious effects of exposure to brain tissue (7). An area of skull 0.01" in diameter was removed leaving the dura exposed but intact. A circle of gelfoam soaked in 0.9% NaCl was placed onto the dura to cover the area. The reference electrode was then lowered onto the gelfoam using a stereotaxic manipulator, another piece of gelfoam placed on top, then covered with dental cement. The auxiliary electrode was a length of stainless steel wrapped round a skull screw implanted so as to be in contact with the dura. By first sweeping through a range of voltages using a potentiostat (E-Chempro Instrument, GMA Computer Technology Ltd.)

we were able to calculate the voltage required to oxidise DA when we switched to chronoamperometry. Samples were taken every 20 sec without depleting the local extracellular concentration.

Discriminating between various electroactive substances which share a similar oxidative potential is a great challenge in electrochemistry. Ascorbic acid and DOPAC (a metabolite of DA), would swamp any changes in the level of DA observed at a 'plain' electrode because they are present in far greater concentration in the extracellular fluid. To provide the required chemical selectivity, the carbon paste matrix of the working electrode was coated with the perfluorosulfonated polymer Nafion. Since Nafion shares the same charge as ascorbate and DOPAC, it tends to screen these substances out (5), while the oppositely charged DA has access to the electrode surface.

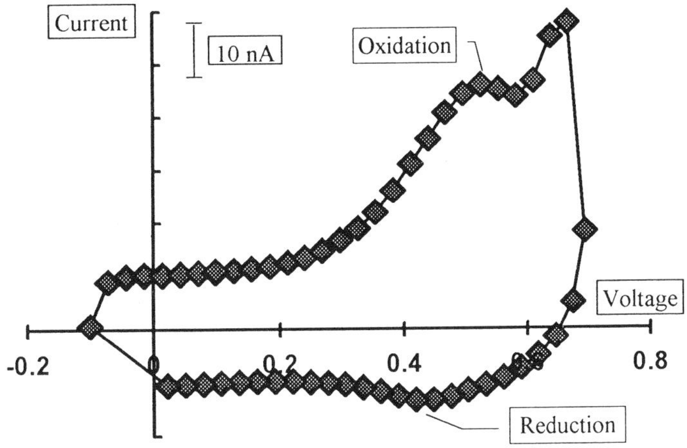

Figure 1. Cyclic voltammogram scanned at 130 mV/s from a working electrode in the NAcc.

The monoamine oxidase inhibitor, pargyline (100 mg/kg IP), blocks the degradation of DA into DOPAC. Hence it is an excellent test for the selectivity of our electrodes. As shown above, an unmodified electrode 'sees' the decrease in the DOPAC signal, while the Nafion-coated electrode only 'sees' the increase in DA. Although the unmodified electrode will also record changes in DA, the higher concentration of DOPAC swamps this signal. Note also the longer time course of effect on DOPAC than DA. Similarly, the potent D1 and D2 agonist pergolide (0.6 mg/kg IP), decreases the release of DA while increasing levels of ascorbic acid (8).

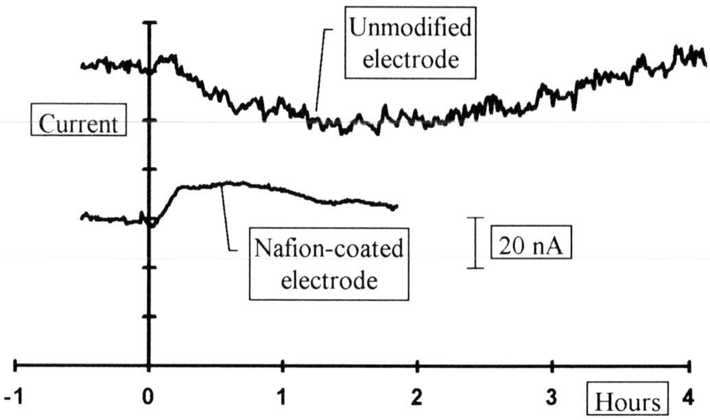

Figure 2. Oxidative current measured in the NAcc following pargyline (100 mg/kg IP).

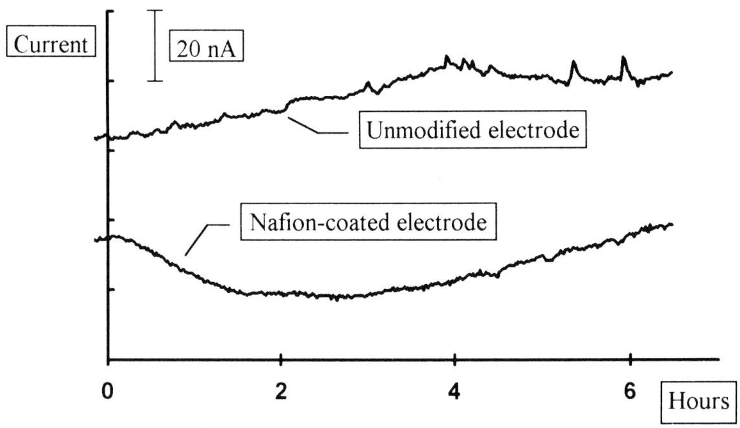

Figure 3. Oxidative current measured in the NAcc following pergolide (0.6 mg/kg IP).

Results

Figure 4 shows a steady increase in the oxidative current, reaching a maximum at 2 hr and not returning to baseline until after 4 hr measured using unmodified electrodes in the NAcc. The increase in current will be produced by increases in DA and/or ascorbic acid, but is probably mostly due to increases in DOPAC.

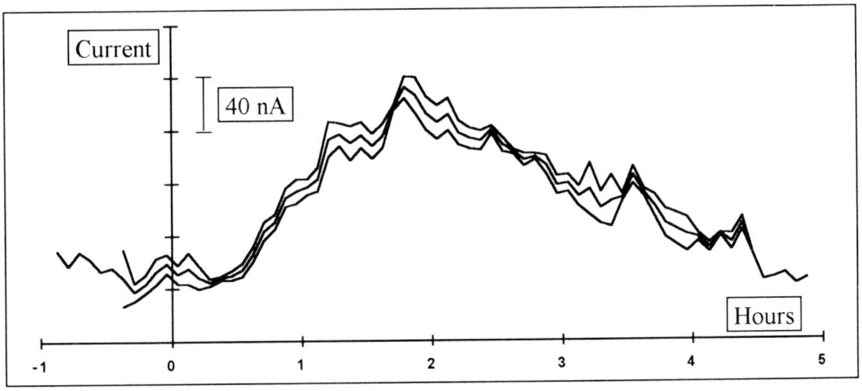

Figure 4. Oxidative current measured at accumbal unmodified electrodes in response to nicotine (0.4 mg/kg SC calculated as base, pH balanced, n=4 rats, mean ± SE).

In contrast, the electrodes modified to provide selectivity for DA showed a much shorter duration of effect in the NAcc (Figure 5). On days when a control (saline) injection was given, no response was seen following the injection. However, following the first injection of nicotine there was an initial *decrease* in the signal which recovered after 15 min, followed by

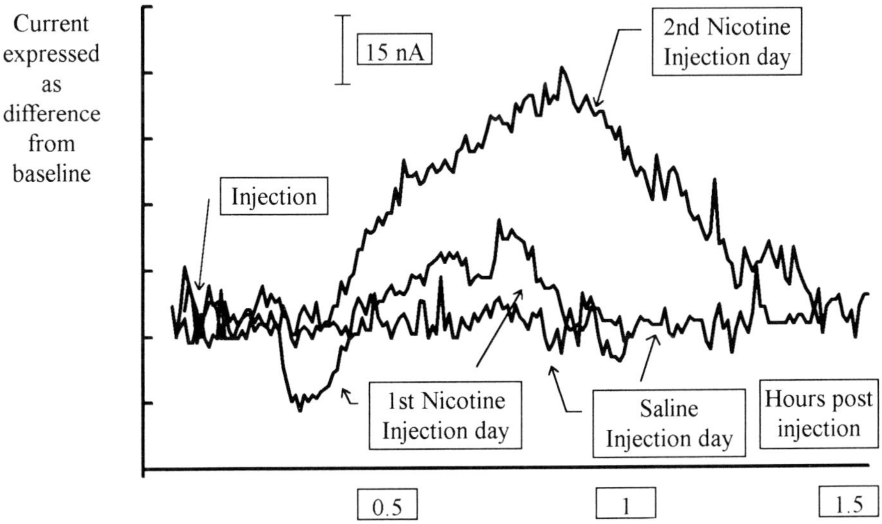

Figure 5. Response at Accumbal Nafion-modified electrodes to systemic injections of saline or nicotine (0.4 mg/kg SC calculated as base). Data shows response to saline and to injections of nicotine on three separate days (n=4 rats).

an increase lasting approximately 1 hr. On the second nicotine day, there was no decrease in signal and a much larger increase that was maximal at 45 min and returned to baseline 90 min after injection.

Discussion

The high temporal resolution afforded by electrochemistry has allowed us to capture a rapid and transient decrease in mesolimbic DA levels following systemic administration of nicotine to nicotine-naive rats. This decrease is an interesting finding of these preliminary data. That the response on the second injection day should be a more rapid increase to a greater maximum finds parallels with the behavioural responses to systemic nicotine and is consistent with observations made using microdialysis (4).

Electrochemistry clearly offers a useful tool for monitoring transmitter levels in drug and behavioural tests. However, there are certain caveats to its use. Obviously it can only measure levels of substances that are electroactive. Furthermore, it does not have the neurochemical discrimination of microdialysis. The careful verification of such a technique is vital before it can usefully be used to answer neurobiological questions.

***Acknowledgements**: This work was supported by a grant from the National Science and Engineering Research Council of Canada.*

References

1. Corrigall WA, Franklin KJB, Coen KM, Clarke PBS. The mesolimbic dopaminergic system is implicated in the reinforcing effects of nicotine. Psychopharm 1992; 107: 285-289.
2. Stolerman IP, Shoaib M. The neurobiology of tobacco addiction. TiPS 1991; 12: 467-473.
3. Clarke PBS. Dopaminergic mechanisms in the locomotor stimulant effects of nicotine. Biochem Pharm 1990; 40: 1427-1432.
4. Benwell MEM, Balfour DJK. The effects of acute and repeated nicotine treatment on nucleus accumbens dopamine and locomotor activity. Br J Pharmacol 1992; 105: 849-856.
5. Gerhardt GA, Oke AF, Nagy G, Moghaddam B, Adams RN. Nafion-coated electrodes with high selectivity for CNS electrochemistry. Brain Res 1984; 290: 390-395.
6. Lane RF, Blaha CD, Hari SP. Electrochemistry in vivo: monitoring dopamine release in the brain of the conscious, freely moving rat. Brain Res Bull 1987; 19: 19-27.
7. Kruk ZL J Neurosci Meth (submitted).
8. Zetterstrom T, Wheeler DB, Boutelle MG, Fillenz M. Striatal ascorbate and its relationship to dopamine receptor stimulation and motor activity. Eur J Neurosci 1991; 3: 940-946.

Effects of Nicotine on
Biological Systems II
Advances in Pharmacological Sciences

EFFECT OF NICOTINE AND COTININE ON NNK METABOLISM IN RATS

C. Kutzer, E. Richter, C. Oehlmann and S.E. Atawodi

Walther Straub-Institut für Pharmakologie und Toxikologie, Ludwig-Maximilians-Universität München, Nussbaumstr. 26, D-80336 München, Germany

Summary: [5-^{3}H]-4-(methylnitrosamino)-1-(3-pyridyl)-1-butanone (NNK) was infused subcutaneously at a daily dose of 7.2 nmol/kg with or without 26 μmol/kg nicotine or cotinine, respectively, over four weeks in male F344 rats. Coadministration of nicotine or cotinine had no effect on the excretion of NNK and its metabolites. However, at the end of the experiment binding of radioactivity to hemoglobin in nicotine- and cotinine-treated rats was reduced by 50% compared to NNK-only treated rats. The inhibition of hemoglobin adduct formation by tobacco alkaloids may explain the lower than expected differences in the adduct levels of tobacco-specific nitrosamines between smokers and nonsmokers as observed in several biomonitoring studies.

Introduction

The tobacco-specific nitrosamine NNK, a strong lung carcinogen in rodents, has been suggested to contribute to the increased risk of lung cancer in smokers (1). NNK and its reduction product NNAL are procarcinogens requiring metabolic activation by α-hydroxylation. *N*-oxidation of both NNK and NNAL, as well as glucuronidation of NNAL are supposed to be detoxification pathways (Fig. 1). Exposure of smokers and passive smokers to NNK has been verified by the detection of NNAL and its glucuronide in urine (2,3). The presence of hemoglobin adducts resulting from pyridyloxobutylation has also been taken as evidence of exposure to NNK and *N*-nitrosonornicotine (4,5). However, the differences in adduct levels between smokers and nonsmokers are much lower than would be expected on the basis of exposure to NNK and NNK. One possible reason for this could be an inhibition of metabolic NNK and/or *N*-nitrosonornicotine activation by nicotine. Coadministration of a 500-fold higher dose of nicotine has been shown to significantly inhibit the metabolic activation of NNK in rats given a single dose of NNK (6). However, smokers take up NNK together with more than 10,000-fold higher amounts of nicotine in multiple low doses. In order to better simulate the conditions in humans, the effect of a 3,600-fold higher dose of nicotine or cotinine on the metabolism of NNK was studied using continuous low dose subcutaneous administration in rats.

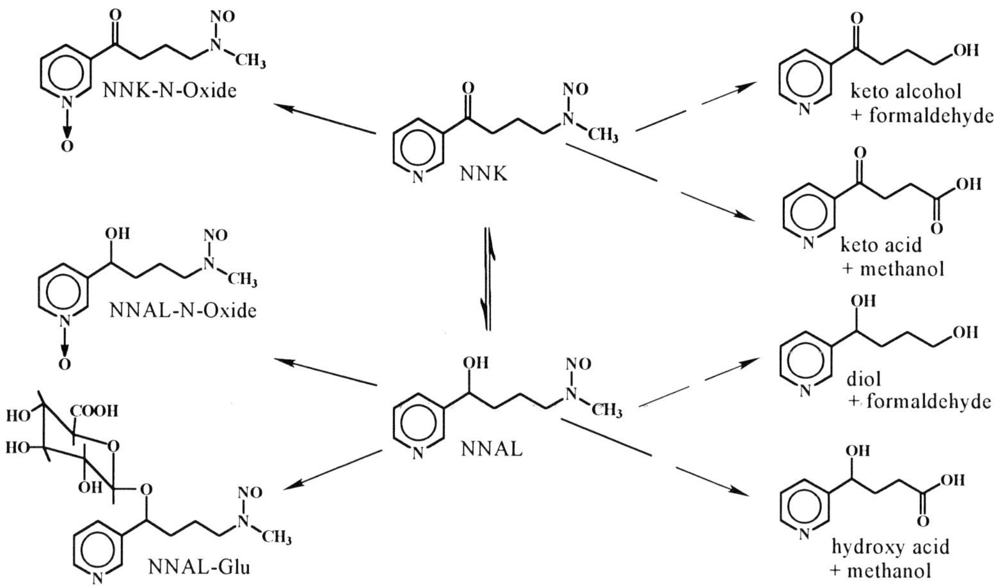

Figure 1. Metabolic scheme of NNK. Intermediate structures of α-hydroxylation are not shown.

Materials and Methods

Osmotic pumps (2ML4 ALZET) delivering a daily dose of 7.2 nmol/kg [5-^{3}H]NNK (>99% pure with a specific activity of 2.89 Ci/mmol) without or with 26 μmol/kg nicotine or cotinine over four weeks were implanted subcutaneously in male Fischer F344 rats (200 g): control (NNK only; n=8); nicotine + NNK (n=8) and cotinine + NNK (n=4). The animals were placed in metabolism cages and urine and faeces collected separately. After 28 days the rats were bled from the aorta abdominalis and plasma separated from blood cells. After lysis of the erythrocytes the hemoglobin solution was dialyzed against 3 changes of water for determination of hemoglobin binding.

Total radioactivity was determined in daily samples of urine and faeces and after 28 days in blood compartments. Urine samples were chromatographed on a 25 cm LiChrosorb RP18 SelectB column using a gradient of 20 mM phosphate buffer, pH 7.2, and acetonitrile (Fig. 3) at a flow rate of 0.7 ml/min. Metabolites were detected by HPLC solid-phase radioactivity monitoring and identified by cochromatography with unlabeled reference compounds using UV detection at 254 nm. Nicotine and cotinine were determined in plasma samples by radioimmunoassay.

Results

The pumps did not deliver the total calculated dose and delivery decreased over the experimental period. The plateau of total excretion reached only about 50% of the expected dose after one week and decreased slightly towards the end of the experiment (Fig. 2). Coadministration of nicotine or cotinine had no significant effect upon excretion of total radioactivity in urine or faeces (Table 1).

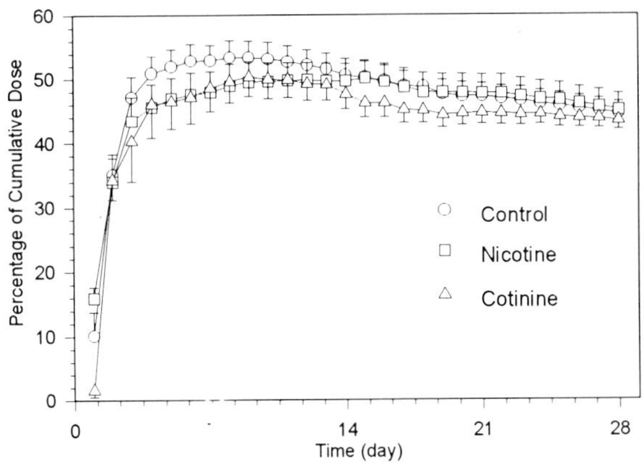

Figure 2. Excretion of radioactivitiy in urine and faeces of rats during continous infusion of a daily dose of 7.2 nmol/kg [5-^{3}H]NNK without or with 26 μmol/kg nicotine or cotinine. Mean ± SE of eight (control and nicotine) or four (cotinine) rats, respectively.

Keto acid was the major metabolite in rat urine (Fig. 2, Table 1). Hydroxy acid, NNK-*N*-Oxide, NNAL-*N*-Oxide and NNAL were also detected consistently, whereas NNAL-Glu and keto alcohol were at the limit of detection contributing <1% to the radioactivity in urine. The metabolic profile in urine did not change over time and was only marginally influenced by coadministration of nicotine or cotinine.

Table 1. Effect of nicotine and cotinine on the excretion of NNK metabolites in 24 hour urine after four weeks of s.c. infusion of a daily dose of 7.2 nmol/kg [5-^{3}H]NNK with or without 26 μmol/kg nicotine or cotinine to male F344 rats. Mean±SD; values labeled are statistically significantly different ($p<0.01$) from control (a) or nicotine-treated (b) rats, respectively.

	Control	Nicotine	Cotinine
Total radioactivity on day 28 (Percent of cumulative dose)	*(n = 8)*	*(n = 8)*	*(n = 4)*
Urine	38.1 ± 5.9	37.5 ± 5.9	36.9 ± 2.7
Faeces	6.4 ± 1.5	7.6 ± 2.2	6.5 ± 0.8
Urine + Faeces	44.5 ± 5.4	45.1 ± 7.4[a]	43.4 ± 2.8
Metabolites in 24 h urine on days 1-28 (Percent of total radioactivity)	*(n = 224)*	*(n = 224)*	*(n = 112)*
Hydroxy acid	19.3 ± 2.2	18.8 ± 3.4	19.0 ± 2.7
Keto acid	43.3 ± 4.7	45.2 ± 6.5	46.5 ± 5.9[a]
NNAL-Glu	0.2 ± 0.5	0.2 ± 0.6	0.5 ± 1.5[a,b]
NNAL-*N*-Oxide	14.9 ± 3.9	14.2 ± 3.9	13.2 ± 3.3[b]
NNK-*N*-Oxide	19.1 ± 2.7	19.0 ± 3.8	18.8 ± 3.8
Keto alcohol	0.4 ± 0.5	1.1 ± 1.0[a]	0.5 ± 0.7[b]
NNAL	2.7 ± 1.1	2.3 ± 1.2[a]	2.8 ± 2.4
Sum of Alpha hydroxylation	63.1 ± 10.5	65.2 ± 6.5[a]	66.0 ± 6.2[a]
Sum of *N*-Oxidation	34.0 ± 4.4	33.2 ± 6.0	32.0 ± 5.3[a]

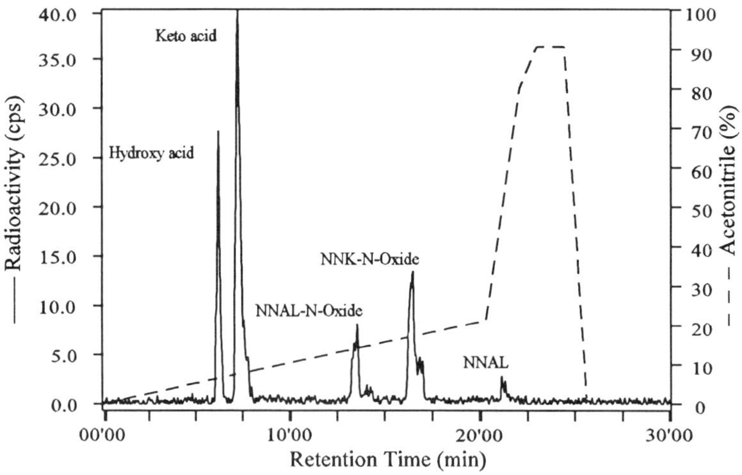

Figure 3. HPLC radiochromatogram of rat urine

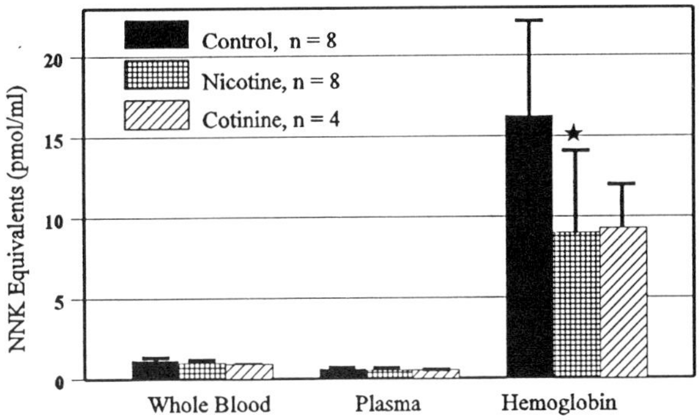

Figure 4. Distribution of radioactivity in blood after four weeks of s.c. infusion of a daily dose of 7.2 nmol/kg [5-^{3}H]NNK with or without 26 μmol/kg nicotine or cotinine to male F344 rats. Mean±SD; the asterix indicates a significant difference from control value, $p<0.05$.

After 28 days, a high proportion of the radioactivity in blood was bound to hemoglobin (Fig. 4). Nicotine and cotinine treatment reduced the hemoglobin binding about 50% and this difference was significant for nicotine ($p<0.05$) and at the borderline of significance for cotinine ($p=0.0602$). Since NNK was labeled in the 5-H position, only hemoglobin binding via pyridyloxobutylation is determined.

The plasma of nicotine and cotinine concentrations in nicotine-treated rats were 63±11 and 145±19 ng/ml, respectively, while only 843±93 ng/ml cotinine was detected in cotinine-treated rats. On a molar basis these concentrations are 740- and 1600-fold higher than the NNK equivalents measured as total radioactivity in plasma of nicotine-treated rats, and 9500-fold higher in cotinine-treated rats, respectively.

Discussion

Continuous subcutaneous coadministration of a 3600-fold higher dose of nicotine or cotinine with [5-^{3}H]NNK to rats did not change the overall excretion of radioactivity and the metabolite pattern in urine. This is in contrast to the result of an acute experiment where a significant ($p<0.001$) reduction of α-hydroxylation (72%) of NNK was observed (6). This may be due to the much lower concentrations of nicotine and cotinine administered by infusion in the present study as compared to the single s.c. injection in our previous experiment. Another explanation could be that a different rat strain (male Wistar rats of about 100 g) was used in the acute experiment (6).

The metabolite pattern in urine compares favourably well with results from experiments with rats and mice given single low doses of NNK (2,8) and is different from the pattern seen in patas monkeys where NNAL-Glu contributes a much higher percentage (20%) to the total urinary metabolites (9).

Hemoglobin binding via pyridyloxobutylation of [5-^{3}H]NNK-derived radioactivity was reduced by about 50% in rats receiving coadministration of nicotine or cotinine. This may explain the observed lower than expected hemoglobin adduct levels found in smokers as compared to nonsmokers [4,5]. However, additional studies need to be performed to determine the exact mechanism of this inhibitory effect of tobacco alkaloids.

***Acknowledgements**: Dr. S. E. Atawodi is an Alexander von Humboldt Research Fellow. We extend special thanks to Dr. D. Hoffmann (American Health Foundation, Valhalla, NY) for providing reference compounds, and Dr. G. Scherer (Analytisch-biologisches Forschungslabor Prof. F. Adlkofer, München) for nicotine and cotinine determination in plasma. This work was supported by VERUM.*

References

1. Hoffmann D, Brunnemann KD, Prokopczyk B, Djordjevic MV. Tobacco-specific *N*-nitrosamines and *Areca*-derived *N*-nitrosamines: chemistry, biochemistry, carcinogenicity, and relevance to humans. J Toxicol Environ Health 1994; 41:1-52.
2. Carmella SG, Akerkar S, Hecht SS. Metabolites of the tobacco-specific nitrosamine 4-(methylnitrosamino)-1-(3-pyridyl)-1-butanone in smokers' urine. Cancer Res 1993; 53:721-4.
3. Hecht SS, Carmella SG, Murphy SE, Akerkar S, Brunnemann KD, Hoffmann D. A tobacco-specific lung carcinogen in the urine of men exposed to cigarette smoke. N Engl J Med 1993; 329:1543-6.
4. Carmella SG, Kagan SS, Kagan M, Foiles PG, Palladino G, Quart AM, Quart E, Hecht SS. Mass spectrometric analysis of tobacco-specific nitrosamine hemoglobin adducts in snuff-dippers, smokers, and non-smokers. Cancer Res 1990; 50:5438-45.
5. Falter B, Kutzer C, Richter E. Biomonitoring of hemoglobin adducts: aromatic amines and tobacco-specific nitrosamines. Clin Investig 1994; 72:364-71.
6. Richter E, Tricker AR. Nicotine inhibits the metabolic activation of the tobacco-specific nitrosamine 4-(methylnitrosamino)-1-(3-pyridyl)-1-butanone in rats. Carcinogenesis 1994; 15:1061-4.
7. Langone JJ, Gijka HB, Van Vunakis H. Nicotine and its metabolites: radioimmunoassays for nicotine and cotinine. Biochemistry 1973; 12:5025-30.
8. Morse MA, Eklind KI, Toussaint M, Amin SG, Chung F-L. Characterization of a glucuronide metabolite of 4-(methylnitrosamino)-1-(3-pyridyl)-1-butanone (NNK) and its dose-dependent excretion in the urine of mice and rats. Carcinogenesis 1990; 11:1819-23.
9. Hecht SS, Trushin N, Reid-Quinn CA, Burak ES, Jones AB, Southers JL, Gombar CT, Carmella SG, Anderson LM, Rice JM. Metabolism of the tobacco-specific nitrosamine 4-(methylnitrosamino)-1-(3-pyridyl)-1-butanone in the patas monkey: pharmacokinetics and characterization of glucuronide metabolites. Carcinogenesis 1993; 14:229-36.

Effects of Nicotine on
Biological Systems II
Advances in Pharmacological Sciences

CONCLUDING REMARKS

THE FUTURE OF NICOTINE-CONTAINING PRODUCTS

M.J. Rand

Department of Medical Laboratory Science, Royal Melbourne Institute of Technology, Melbourne, Victoria, Australia

Summary: Nicotine has an established and continuing role as a valuable tool in physiological and pharmacological studies. Recent observations suggesting a beneficial effect in some disorders indicate that nicotine may have a place in therapeutics. The issue of smoking continues to be hotly debated, with a trend to a shift in emphasis to the alleged addictive action of nicotine. Nevertheless, the association between smoking and various severe diseases remains as a major matter requiring elucidation of mechanisms, which would allow various measure to be put in place to reduce health risks, possibly without depriving smokers of their enjoyment of the habit.

Scientific value of nicotine

It has been pointed out elsewhere (1) that studies with nicotine have played an vitally important part in elucidating physiological and pharmacological mechanisms. Briefly recapitulating, these were establishing the pathways of autonomic innervation of various organs, differentiating between major types of acetylcholine receptors (leading to the term 'nicotinic'), the establishment of chemical transmission in the central nervous system and thence the discovery of amino acid transmitters, the recognition of non-adrenergic non-cholinergic neuroeffector transmission, and the molecular biology of nicotinic cholinoceptors, about which we heard much in the Symposium.

The scientific importance of nicotine as a tool for investigating and elucidating physiological processes does not have an immediate bearing on the future use of nicotine-containing products by the populace, but it can be predicted with some certainty that nicotine will continue to be studied and the outcome of the studies is likely to be of value to mankind.

The role of nicotine in therapeutics

The lower prevalence of ulcerative colitis and Parkinson's disease in smokers suggests that a component of tobacco smoke has a prophylactic effect. The hypothesis has the scientific appeal of being testable by prospective studies, and the ethical dilemma posed by such studies in the present climate of medical opinion about smoking could be overcome if the effect were due to an identified component of tobacco smoke. In this Symposium, nicotine was postulated to be the effective agent and appeared to be of benefit in preliminary clinical studies on patients with ulcerative colitis (Thomas & Rhodes) and in animal models of Parkinson's disease (Janson et al.).

In Tourette's syndrome, the beneficial effect of nicotine in ameliorating the signs were rapid and dramatic (Sanberg & Silver). It would be of some interest to know whether smoking suppresses the manifestation of Tourette's syndrome.

The possibility of a beneficial effect of nicotine in Alzheimer's disease is suggested by the findings that cholinergic neurotransmission is selectively impaired and that nicotine has been shown to improve cognitive performance in animal and human studies. Again, it would be of interest to have epidemiological studies to determine whether there is an inverse relationship between a history of smoking and the development of Alzheimer's disease, as well as prospective studies to determine whether nicotine is in fact of clinical value in its treatment..

The possibility that smoking has prophylactic and/or beneficial effects in a number of disorders raises a question about the extent to which smokers are self-medicating, albeit unwittingly, to ward off the early signs and symptoms these disorders. If this should be the case, the reduction in tobacco-related diseases that is the rationale for restrictions on tobacco use may well be offset to some extent by increases in the incidence other diseases, which may be large in the elderly in whom Parkinson's disease and Alzheimer's disease are most prevalent.

The major present use of non-tobacco nicotine-containing products (gum, dermal patches, nasal sprays) is as aids to cessation of smoking. Their clinical efficacy is either unimpressive (gum, patch) or remains to be convincingly demonstrated (nasal spray). In some countries, nicotine gum and patches are supplied only on prescription, but in others they are available as over-the-counter preparations from pharmacies. Even when they are readily obtainable, they do not appear to be a satisfactory substitute for smoking.

The question of addiction to nicotine

The pros and cons of this question were discussed by several speakers at the Symposium. If it is a scientifically meaningful question, the meaning of the term addiction must be clearly stated in terms that allow for operational measurement of the state. Until such a definition exists, the debate will continue with regard to the addictiveness of nicotine or the lack thereof.

Those who have followed the work of the relevant WHO committees on the subjects of drug dependence were somewhat perplexed by the use of the term addictive in relation to nicotine in cigarette smoke by the US Surgeon General report in 1988 (2). The reason for this departure from what had previously been internationally accepted terminology is not clear. The report itself vacillates in places between the terms dependence and addiction, and there is the implication that the terms are identical in meaning. A possible reason for the US Surgeon General's preference for the term addiction is that it is more pejorative than dependence, and thus offers more vigorous support for campaigns against smoking. Another is that the term addiction may have some special legal meaning or significance in the USA that would allow the Food and Drug Administration to regulate cigarettes as nicotine-delivery systems.

The criteria for establishing that a drug is addicting are given in the 1988 report as follows:

> *"The central element among all forms of drug addiction is that the user's behavior is largely controlled by a psychoactive substance (i.e., a substance that produces transient alterations in mood that are primarily mediated by effects in the brain). There is often compulsive use of the drug despite damage to the individual or to society, and drug-seeking behavior can take precedence over other important priorities. The drug is "reinforcing" - that is, the pharmacologic activity of the drug is sufficiently rewarding to maintain self-administration. "Tolerance" is another aspect of drug addiction whereby a given dose of a drug produces less effect or increasing doses are required to achieve a specified intensity of response. Physical*

dependence on the drug can also occur, and is characterized by a withdrawal syndrome that usually accompanies drug abstinence. After cessation of drug use, there is a strong tendency to relapse."

The report goes on to state that "most smokers admit that they would like to quit but have been unable to do so", and adduces this as evidence for compulsive use. However, the 1989 report (3) states that "Nearly half of all living adults [in the USA] who have ever smoked have quit". It is difficult to reconcile these two statements, or to know which to choose in establishing whether or not nicotine is addictive. According to a WHO report in 1993 (4), the proportion of smokers in developed countries has dropped considerably in recent years, during a period in which there has been a dramatic increase in the use of cocaine. As for the other criteria: the reinforcing effect of nicotine can be demonstrated in animal experiments, but only under rigorously defined conditions that do not normally have a counterpart in human smoking; tolerance is well known as a pharmacological effect of nicotine, but in relation to smoking its development is largely to effects that are perceived as unpleasant by the novice smoker, such as nausea; escalation of the intake of nicotine is to a ceiling which is well short of intoxication, in sharp contrast to the case with opiates, cocaine, ethanol and many other drugs that are deemed to have a dependence liability by the WHO; evidence for physical dependence, as manifested by a withdrawal syndrome, is largely anecdotal and the withdrawal syndrome is not well characterized, in contrast to the syndrome following abstinence from prolonged use of large doses of opiates, ethanol, and many sedative-hypnotic drugs, and is only poorly assuaged by non-tobacco nicotine.

Another view about addiction is to place the emphasis not on an agent which is thought to sustain compulsive use, but on the user. Thus, the focus would be shifted from the drug; rather it would be on people who have a propensity for recourse to a drug. This has been studied in alcoholics and problem drinkers, but not to any significant extent in habitual drinkers who rarely if ever are severely intoxicated. For drugs such as opiates and cocaine, studies on the characteristics of habitual users have been for the most part on those incarcerated for criminal possession or use, and the validity of extrapolation of the findings to the vast majority of users who have not run foul of the law is problematic. There have been investigations of smokers to determine whether there are factors which characterize them and demarcates them from non-smokers (see 5) and it might be of some value to have more modern studies using the technologies that have emerged in the past 20 or so years.

If, by special pleading or by selective choice of data, it is deemed that nicotine is addictive, this raises the basic question: What should society do about addictions in general and nicotine addiction in particular? In practice, there are two main, diametrically opposed attitudes.

The first is prohibition. In general, prohibitions on substances (e.g., alcoholic beverages) or practices (e.g., gambling) that are deemed to be addictive have produced many more problems than they have solved by creating a new class of law-breakers served by criminal elements, resulting in flourishing of organized crime and corruption of enforcement agents. There can be little doubt that prohibition of tobacco would bring about the same disastrous consequences.

The second is to cater for the addiction of individuals in a regulated way and to take whatever benefits may accrue to society at large by various forms of taxation. This approach has been widely adopted with respect to gambling and to alcoholic beverages. Taxation or monopolistic control by the state of tobacco products is already a major source of revenue in most countries, and was long before any questions were raised about the adverse effects of smoking on health, let alone addiction.

If a laissez-faire attitude of controlled and taxed availability of tobacco products and their use is adopted, even though they may be deemed to be addictive, there still remains the question of whether habitual tobacco use is to be regarded as a disorder or disease. There are authoritarian views that it is. Thus, the tenth revision of the International Statistical Classification of Diseases and Related Health Problems (ICD-10) recognizes the following psychoactive drugs, or drug classes, the self-administration of which may produce mental and behavioural disorders, including dependence: Alcohol, Opioids, Cannabinoids, Sedatives or hypnotics, Cocaine, Other stimulants, including caffeine, Hallucinogens, Tobacco, Volatile solvents, Other psychoactive substances, and drugs from different classes used in combination. The fourth edition of the American Psychiatric Association's Diagnostic and Statistical Manual of Mental Disorders (DSM-IV) also included excessive use of nicotine and caffeine, as well as many other substances, as possible psychiatric disorders. With regard to tobacco or nicotine (and caffeine and caffeine-containing beverages), the criterion appears to be the existence of dependence (ICD-10) which is apparently regarded in itself as a behavioural disorder, or excessive use (DSM-IV) which requires a value judgement. An alternative view is that smoking, like many other activities, is considered as pleasurable, satisfying or rewarding by those who indulge in it, despite the fact that it entails a degree of risk to health and, increasingly, social disapproval in some quarters.

Harmful effects of tobacco smoking

These can be put into two main categories: physical health of the smoker and health effects of environmental tobacco smoke to others (passive smoking).

Health risks to smokers. Since the 1950s, many epidemiological studies have demonstrated an association between cigarette smoking and lung cancer, and since the 1970s similar studies demonstrated an association between cigarette smoking and coronary heart disease and various other disorders. The assumption derived from these data, that there is a proven causal relationship, has been critically examined by various experts and rejected (see Eysenck, 6). In fact, it can be asserted that the chains of events that would establish unequivocally a causal relationship in terms of the toxicological actions of substances in tobacco smoke to cancer, coronary heart disease and most other ill-effects associated with cigarette smoking has not been rigorously demonstrated. It is of the utmost importance to identify the components of cigarette smoke that do cause the adverse effects claimed by adherents to the perceived view.

At the twentieth meeting of the WHO Expert Committee on Drug Dependence in 1974 (7), the focus was on reducing or eliminating harm, and the value of this approach was endorsed at the twenty-eighth meeting (4). Some years ago, after the association between smoking and various serious diseases was established in epidemiological studies, there was a move that had as its goal the development of safer cigarettes (8). There is little doubt that exposure to identified hazardous components of tobacco smoke could be minimized by technology ranging from introducing genetic variants of tobacco plants to construction of cigarettes, and such developments could have a significant impact on the risk. But these ventures have not come to fruition largely because the vociferous anti-smoking lobby had as its goal the elimination of the use of tobacco, but also to some extent because this approach was not pursued vigorously from the outset by the tobacco industry.

The twenty-eighth WHO Committee (4) cited nicotine replacement for cigarette smoking as an example of harm minimization, but there is little evidence that either nicotine-containing chewing gum or patches are a satisfactory substitute for cigarettes. The pragmatic view is that the move *Towards a Safer Cigarette* should be resumed, which demands the identification of the factors responsible for any adverse effects, and the factors, presumably including nicotine, that are perceived as desirable by smokers.

The received view of the causal relationship resulted in so-called Health Warnings on cigarette packets. It can be seen as a form of harm reduction, although evidence that it has made any significant impact on smoking is lacking, and the overall effect may have been to reduce the effectiveness of various other forms of health warnings.

Environmental tobacco smoke. There is no doubt that many people, like King James, think that smoking is *a custom ... hateful to the Nose*. They clearly have a right to be free of this nuisance. This was provided in the past, in public transport for example, by having designated No Smoking areas. In the past decade there has been a change to prohibition of smoking in public transport and public buildings. This was largely due to the claim that all the adverse effects supposedly produced in smokers were also visited to some extent on passive smokers, despite the weakness of the scientific evidence to support this contention. A secondary consequence arising from prohibition on smoking and reinforcing it is that cleaning costs were substantially reduced and, regrettably, smokers have themselves to blame for thoughtless disposal of ash, cigarette butts and other smoking paraphernalia.

The ban on smoking in the workplace (leaving aside the sensible restrictions on places where there would be a fire hazard) and public buildings was introduced largely because of the fear of litigation by those claiming ill-health as a consequence of passive smoking. Another solution to this problem would be to install adequate control measures to prevent mixing of air between areas designated 'Smoking Permitted' and 'Smoking Prohibited', which is surely well within the capacity of current technology. Of course, efficient air control would entail a cost, but would be perceived as a benefit by a significant proportion of the population, and there would be a reduction in the economic loss that follows from people leaving the workplace to smoke.

Future of nicotine-containing products

The main issue is the continued use of tobacco products. It seems probable that there will remain a hard core of people wishing to smoke. The challenge is to reduce the risks to their health and annoyance to non-smokers. The ideal of those passionately devoted to a tobacco-free world might be achievable when all mankind is in harmony, when human dignity prevails, and the stresses of modern-day life are eliminated. If such a millennium comes about, it is possible that recourse to tobacco (and also to alcoholic and caffeine-containing beverages, to say nothing of more potent drugs) may cease to exist. Until then, the solace offered by nicotine-containing tobacco products will ensure their future use.

Acknowledgement: *The attendence of the author at the Symposium was supported by the Smoking & Health Research Foundation of Australia.*

References:

1. Rand MJ. Neuropharmacological effects of nicotine in relation to cholinergic mechanisms. In: Nicotinic Receptors in the CNS, Vol. 79, Progress in Brain Research. Fuxe K, Holmstedt B, Sundwall A, editors. Amsterdam: Elsevier, 1989: 3-11.
2. The Health Consequences of Smoking: Nicotine Addiction. A report of the Surgeon General. Washington D.C.: US Department of Health and Human Services, 1988.
3. Reducing the Health Consequences of Smoking: 25 Years of Progress. A report of the Surgeon General. Washington D.C.: US Department of Health and Human Services, 1989.
4. WHO Expert Committee on Drug Dependence. Twenty-eighth Report. WHO Technical Report Series, No. 836, Geneva: WHO, 1993.
5. Differences between smokers and non-smokers. In: Smoking and Health Now. London: The Royal College of Physicians, 1971: 111-112.
6. Eysenck HJ. Smoking and health. In: Smoking and Society. Tollison RD, editor. Massachusetts: Lexington Books, 1985: 17-88.
7. WHO Expert Committee on Drug Dependence. Twentieth Report. WHO Technical Report Series, No. 551. Geneva: WHO, 1974.
8. Wynder EL, Hoffmann D, editors. Towards a Less Harmful Cigarette. National Cancer Institute Monograph No. 28. Washington D.C.: US Department of Health, Education and Welfare, 1968.

Author Index

Subject Index

Respiratory Pharmacology and Pharmacotherapy

D. Raeburn, *Rhône-Poulenc Rorer Ltd, Dagenham, UK*
M.A. Giembycz, *Royal Brompton National Heart and Lung Institute, London, UK (Eds)*

Airways Smooth Muscle: Structure, Innervation and Neurotransmission

1994. 328 pages. Hardcover
ISBN 3-7643-5010-5

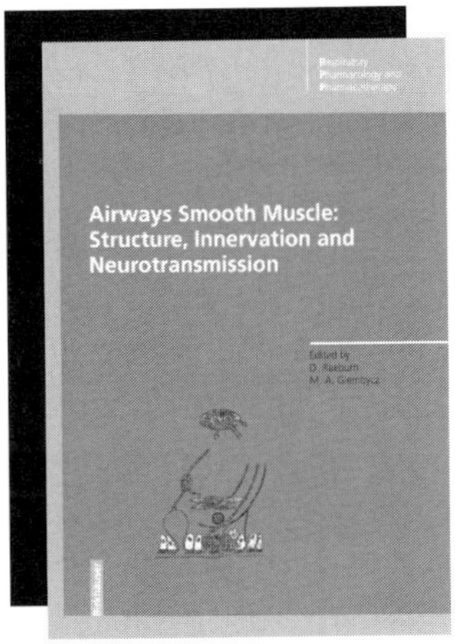

In the context of a growing incidence of respiratory disorders worldwide, particularly asthma and allergy, the importance of research into the respiratory system is increasingly recognized. The emphasis of this first volume in a new series of research monographs is on the anatomical aspects of airways smooth muscle, including its innervation and neurotransmission.

Scientists of international repute were invited to contribute chapters on anatomy, gross morphology and ultrastructure, sympathetic, parasympathetic and NANC innervation, vagal reflexes, prejunctional regulation of neurotransmission, and neural elements in airways smooth muscle.

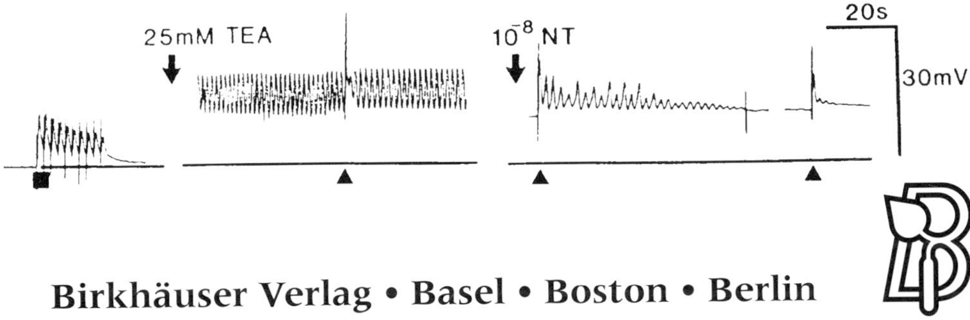

Birkhäuser Verlag • Basel • Boston • Berlin